BROADCAST–VIDEO ★ SOURCEBOOK I

1989–1990

Marilyn J. Matelski, Ph.D.
Boston College

David O. Thomas, Ph.D.
Ohio University

Focal Press
Boston London

Focal Press is an imprint of Butterworth Publishers.

Recognizing the importance of preserving what has been written, it is the policy of Butterworth Publishers to have the books it publishes printed on acid-free paper, and we exert our best efforts to that end.

Design and Production: LeGwin Associates
Cover Design: Hannus Design Associates
Cover Printer: New England Book Components
Text Printer: Sheridan Press, Inc.

ISBN 0-240-80067-2
ISSN 0959-1486

Butterworth Publishers
80 Montvale Avenue
Stoneham, MA 02180

10 9 8 7 6 5 4 3 2 1

Printed in the United States of America

Contents

Television *(continued)*

Preface and Acknowledgments

The *Variety* logo has long been recognized as a trademark for quality journalism, critical commentary, and insightful prediction. Begun as a vaudeville review in 1905, it has spanned the golden ages of film, radio, television, and theater, and continues to serve as an industry standard for professionals in all areas of entertainment.

The purpose of the *Variety Sourcebook* series is twofold: 1) to condense and critique some of the past year's most important mass communication trends; and 2) to expand *Variety*'s vast audience to include more students and teachers of higher education. Thus, we hope that this year's collection of data and accompanying editorial material will serve as a valuable resource for educators as well as industry professionals.

The *Variety Sourcebook* will be compiled and distributed annually, presenting broadcast-video information on even-numbered years and film-theater data on the years in between. The material used in all texts will be taken primarily from the *Variety* weekly trade magazine; it is copyrighted by them.

The authors would like to express their appreciation to the following individuals: Terry Pittman and Mike Reinemer of the Claritas Corporation, and to the Claritas Corporation for providing current data on the PRIZM Cluster System; Michael J. Weiss, author of *The Clustering of America*; Robin Page of Page & Associates; Paul Church, Research Director for the CBN Family Channel; Joel Makower of Tilden Press; Sharon Biggins, Research Director at WFAA-TV, Dallas/Fort Worth; Jerry Storseth of Mediacomp; Kathy Seaton, Jeanette Sullivan, Jim Colson, and Anne Braxton at Ohio University; Don Fishman, Gail McGrath, Dale Herbeck, and John Katsulas at Boston College; Nancy Street at Bridgewater State College; Peg Latham, Phil Sutherland, and especially to Karen Speerstra at Focal Press for initiating this project.

M.M.
D.T.

A Brief History of *Variety*

In 1905, the world began a century of massive change in geographic boundaries, political and sociological beliefs, and technological developments. In Asia, the Russo-Japanese war over territorial expansion ended in Japanese victory. *Bloody Sunday* came to be known as the beginning of an abortive revolution against the czar in Russia. In Vienna, Sigmund Freud published his *Three Essays on the Theory of Sexuality.* And in New York, the Hippodrome, the largest stage in the world for circuses, auto races, battle scenes, Wild West shows, and airplane dogfights, opened to the delight of millions of Americans.

Meanwhile, in a small office on 34th Street in New York City, showbusiness reviewer Sime Silverman decided to produce his own entertainment weekly as a reaction to other newspapers that he felt were slanted and overly influenced by advertisers. Sime called his new magazine-formatted newspaper *Variety,* and he pledged to keep its editorial content as fair and impartial as possible. This vow was most evident in the first issue (December 16, 1905), when Sime described his new publication:

> The first, foremost, and extraordinary feature of it will be FAIRNESS. Whatever there is to be printed of interest to the professional world WILL BE PRINTED WITHOUT REGARD TO WHOSE NAME IS MENTIONED OR THE ADVERTISING COLUMNS.
>
> "ALL THE NEWS ALL THE TIME" and "ABSOLUTELY FAIR" are the watchwords.
>
> The news part of this paper will be given over to such items as may be obtained, and nothing will be supressed which is considered of interest. WE PROMISE YOU THIS AND SHALL NOT DEVIATE.
>
> The reviews will be written conscientiously, and the truth only told. If it hurts it is at least said in fairness and impartiality.
>
> We aim to make this an artists' paper; a medium; a complete directory; a paper to which anyone connected with or interested in the theatrical world may read with the thorough knowledge and belief that what is printed is not dictated by any motive other than the policy above outlined.[1]

In addition to an editorial goal unique for its time, Sime and his wife, Hattie, also peppered their weekly newspaper with showbiz language, memorable headlines, and colorful entertainment reviews. In fact, many linguists have since developed research areas based on terms such as soap opera, whodunit, emcee and disk jockey—all of which first saw print in *Variety.*[2]

Stylistically, *Variety* also showed originality and aplomb with its format, logo, and presentational consistency. For example, Sime started a tradition of signing entertainment reviews with 3- to 5- letter code names in his first year. From that time on, only his wife could decipher who had *really* written the critiques. As for logo design, Hattie Silverman seemed to have developed the noteworthy " V— " for *Variety* as she was doodling on a nightclub tablecloth while watching a show.

In short, Sime and Hattie Silverman had begun a tradition with their new publication; and, as the entertainment business and media technology grew, so also did *Variety* and its commitment to reporting excellence. At each important step along the way, *Variety* has been on hand to present the newest, latest, and greatest moments in show business history worldwide.

When *Variety* was first published, vaudeville was still considered to be the most popular entertainment medium in America; however, by 1905, it had already begun to be challenged by the burgeoning film and music industries. In 1905, for example, Columbia introduced the two-sided recording disc; shortly before that time nickelodeons were sprouting up throughout the country. Of these two inventions, the nickelodeon was by far the most impressive.

The nickelodeons were so named because they were theaters where new motion pictures were shown for a five-cent admission fee. During this time, the average length of a movie was usually about twelve-and-a-half minutes long, and most nicks featured two to three films for each admission.[3] Needless to say, moviegoing soon became very popular with many Americans. In fact, by 1910, over 3,000 theaters could be found throughout the country.[4]

Since Sime Silverman prided himself on the ability to spot new trends, he decided to capitalize on this fast-growing phenomenon. As a result, *Variety* published its first film reviews on January 19, 1907, and included a comedy entitled *An Exciting Honeymoon* (7 minutes) as well as a melodrama called *The Life of a Cowboy* (13 minutes). Among his comments about the new medium, Silverman wrote:

> The fun is amusing; the picture well worked out and the audience remained in its seats until the close.[5]

Silverman was wise to recognize the motion picture trend early; for it wasn't long before movies captured the heart of America. . .and the world. In 1915, D.W. Griffith released his epic film, *Birth of a Nation,* and catapulted filmmaking to a new standard of excellence. By 1917, Charlie Chaplin, an English stage comedian, was making more than $1 million per year in his comedy feature. Through it all *Variety* noted the new talent as well as the new technology.

In 1918, a post World War I printer's union strike in New York brought most commercial printshops to a halt. This, of course, affected *Variety* profoundly. However, it also caused Sime to re-evaluate his decision about presenting show business news in a magazine format. Ironically, then, as a result of the strike, *Variety* moved to the newspaper format familiar to most readers today, which has proven to be very successful.

During the Roaring Twenties, the film industry continued to enjoy a huge amount of success. Jazz became a popular music form at neighborhood speakeasies, and radio broadcasting became the newest, most talked-about medium. Technologically speaking, in 1927, Warner Brothers released the first sound motion picture, *The Jazz Singer,* which *Variety* characterized accordingly:

> . . .it didn't do any more to the industry than turn it upside down, shake the entire bag of tricks from its pocket and advance Warner Brothers from last place to first in the league.[6]

In 1928, the Academy of Motion Picture Arts and Sciences staged its first award ceremony, and presented an Oscar for Best Picture to the film, *Wings.*

Likewise, in radio, entertainers were enjoying a great amount of good fortune. In 1928, for example, the first successful radio sitcom emerged, entitled "Sam and Henry;" it later became known as "The Amos 'n' Andy Show." "Amos 'n' Andy" helped to change our nation's working and leisure habits; in fact, some factories actually adjusted their hours so that employees could be home in time to hear the latest episode.

The jukebox was invented in 1927—an addition to the record industry that was both saleable to restaurants and nightclubs as well as promotional for home record purchases. As in radio and motion pictures, the music trade enjoyed great success with its new technological development. . .and continued to rise sharply until the Wall Street Crash of 1929.

The Great Depression affected every industry adversely, and the entertainment business was no exception. As *Variety* reported in its now famous article entitled, "Wall St. Lays an Egg,"

> Many people of Broadway are known to be wiped out. Reports of some in show business losing as much as $300,000 is not hearsay. One caustic comment to that was that the theatre is enough of a gamble without its people to venture into Wall street.[7]

In addition to the business woes of the entertainment industry during this time, *Variety* also suffered a severe blow when its founder, Sime Silverman, died in 1933—just one month after he had established *Daily Variety* in Hollywood. Sime's son, Sidne, became the weekly paper's new publisher, however, and both newspapers continued their success despite the tremendous loss.

During the 1930s and 1940s, moviegoing and radio listening reached all-time highs. The notion of genre in film and radio also became popular, as people flocked to see (or hear) their favorite Western, gangster flick, or soap opera. Big bands kept the music industry in high demand, and records continued to boom. However, no one truly recognized the power of the mass media until October 30, 1938, when Orson Welles incited national panic with his presentation of H.G. Wells' "War of the Worlds:"

> Near hysteria broke out all over the United States Sunday (30) night as a result of a fanciful invasion-from-Mars story by H.G. Wells which was broadcast over the Columbia Broadcasting System in the form of news flashes. One immediate effect of the strange behavior of the populace was to focus attention on various social and military implications.
>
> Persons close to the national defense branch of the Federal government expressed the view that besides revealing a jumpy state of nerves, brought on by the war clouds over Europe and Asia, the episode drove home how little prepared the nation is to cope with an abrupt emergency.[8]

Variety's approach to "War of the Worlds" was journalistic as well as entertainment-oriented, thus in keeping with Sime Silverman's original vow to print show-business news as impartially as possible. The editorial standard would continue to endure throughout the block booking disputes of the forties, the Communist witch hunt and the payola scandals of the fifties, the anti-war sentiment of the late sixties and the Watergate investigation of the seventies.

In 1950, Sidne Silverman (publisher of both *Variety* and *Daily Variety*) and Arthur Ungar (editor of *Daily Variety*) died within six months of each other. They had both seen the two newspapers grow through the golden ages of film and radio and had witnessed the introduction of the entertainment industry's newest medium—television.

In 1948, television became accessible to most Americans, and by 1950, both the radio and film industries felt a threat of extinction due to their formidible competitor. Moviemakers responded to the challenge by introducing such mass audience attention-grabbers as 3-D, Cinemascope, and smell-o-vision. Some of these techniques were successful; others were dismal failures. Radio, on the other hand, decided to narrowcast to small but loyal audiences by taking full advantage of offerings from the record industry. Both industries survived. . .and, along the way, *Variety* was witness to their triumphs and defeats.

In 1956, Sime Silverman's grandson, Syd, took over the publishing responsibilities for both *Variety* newspapers from Harold Erichs and Abel Green. Erichs and Green had managed the papers in the interim between Sidne's death and Syd's college education and military service. Syd embraced *Variety*'s editorial and stylistic policies as his father and grandfather had done, and carried on with the papers' unique point of view through developments of videocassettes, cable television, homevideo, satellite transmission, compact discs, and high-definition television. He also strengthened the papers' commitment to international news, which had begun before World War II and continues to be a very significant element today.

In October 1987, *Variety* was acquired by Cahners Publishing Company, and, while there have been several changes in both format and style to adjust to changing times, the new management has adhered faithfully to Sime Silverman's dream. As Sime said in his first editorial, "This paper is for variety and variety only in the broadest sense that term implies."[9] It is a goal for which *Variety* and *Daily Variety* continually strive.

References

1. *Variety* (December 16, 1905), p. 3.
2. Syd Silverman, "*Variety,* the Biz & Technology over the Years," *Variety* (January 11, 1989). p. 38.
3. Stan LeRoy Wilson, *Mass Media/Mass Culture* (New York: Random House, 1989), p. 149.
4. Linda J. Busby, *Mass Communication in a New Age: A Media Survey* (Glenview, IL: Scott, Foresman and Company, 1988) p. 207.
5. *Variety* (January 19, 1907), p. 9.
6. Wilson, p. 154.
7. "Wall St. Lays an Egg," *Variety* (October 30, 1929), p. 1.
8. "Radio Does U.S. a Favor," *Variety* (November 2, 1938), p. 1.
9. *Variety* (December 16, 1905), p. 1.

Bibliography

Busby, Linda J. *Mass Communication in a New Age: A Media Survey.* Glenview, IL: Scott, Forseman and Company, 1988.

Garatty, John A., and Peter Gay, eds. *The Columbia History of the World.* New York: Harper & Row Publishers, 1972

"Radio Does U.S. a Favor." *Variety,* 2 November 1938, pp. 1, 28.

Silverman, Syd. "*Variety,* the Biz & Technology over the Years." *Variety,* 11 January 1989, pp. 38, 42, 48.

Toll, Robert C. *The Entertainment Machine: American Show Business in the Twentieth Century.* New York: Oxford University Press, 1982.

Variety, 16 December 1905, p. 1.

Variety, 19 January 1907, p. 9.

"Wall St. Lays an Egg." *Variety,* 30 October 1929, pp. 1, 64.

Wilson, Stan LeRoy. *Mass Media/Mass Culture.* New York: Random House, 1989.

How to Read *Variety*

On December 16, 1905, Sime Silverman published his first weekly edition of *Variety* magazine. It was known originally as an entertainment review journal (*Chicot and Sime's Reviews of the Week),* with relatively few news items ("Fynes Quits Proctor—The First 'Real' Story").[1] However, within its first year of publication, Sime discovered that *Variety* could service the fledgling entertainment industry in a much grander way than he'd first envisioned. As a result, he developed his now famous method for disseminating show business news to professional colleagues as well as to educators and the interested public at large.

Variety has four basic characteristics which set it apart from most entertainment publications:

1. It covers the entire spectrum of show business. This includes theatre, film, and the broadcast and music industries, as well as international events, governmental influences, technological innovations and overall economic concerns. By providing so many areas of entertainment news in one journal, the reader is able to formulate a more complete picture of how each facet of show business is interrelated to the others. Thus, if one entertainment form is affected by economic adversity, for example, reading *Variety* might help to predict trends in other areas of show business.

2. It utilizes a unique vocabulary which is both colorful as well as descriptive. As the glossary at the end of this book indicates, the words and phrases used often invite active participation from the reader because they are so different from those written in other newspapers and magazines. Also, it should be noted that *Variety*'s novel use of language has created many new terms which have since been standardized to become accepted show business jargon.

3. It enjoys an ongoing reputation for identifying and highlighting significant events in the entertainment industry long before many other newspapers have begun to take notice. For example, block booking by motion picture companies was first mentioned by *Variety* in 1914. The issue was not resolved until 1948; however, *Variety* readers were made very aware of all sides in the controversy for at least three decades before the ultimate Supreme Court decision.

4. It focuses on *all* current events with regard to their potential influence on the entertainment industry. *Variety* prides itself on its ability to pinpoint the essential mass-communication issues that must be considered in all news events. In 1929, for example, the Wall Street crash caused trauma in every aspect of American life. *Variety* reported this tragedy as objectively as possible (with its now famous article, "Wall St. Lays an Egg"), by taking note of the general economic disaster as well as its potential impact on each aspect of the show business industry. Thus, during the Great Depression. . .and in other noteworthy periods of history. . .*Variety*'s analyses have proven to be detailed, insightful, and useful to entertainment professionals and observers.

To best illustrate *Variety*'s unique contributions to show business reporting, it seems most appropriate to provide a chronology of twentieth century events. . .along with some sample *Variety* headlines to describe them. Although this chronology is by no means exhaustive, it serves to illuminate *Variety*'s diversity in the entertainment industry, its colorful language, its ability to predict future trends, and, finally, its ability to adapt general news headlines to specific show business concerns. To most professionals, *Variety*, indeed, has earned its reputation as the "entertainment bible."

Partial Chronology

1908 **Thomas Edison seeks a trust for his sound and film inventions.**
"A Square Deal For All is Edison's Promise: In His First Statement to An Amusement Journal 'The Wizard' Tells *Variety* He Will Take Personal Charge of His Moving Picture Interest. Gives Opinion on Subjects." (June 20, 1908, p. 12)

1909 **The Motion Picture Patents Company—an alliance of nine film companies—is founded. Article announcing the decision in 1908.**
"Moving Picture War Over." (December 26, 1908, p. 8)

1912 **The sinking of the *Titanic* affects communication technology as well as the entertainment industry.**
"Paralyzing Titanic Terror Cast Pall Over Theatres: Amusements Suffering from Effects of

Sea Horror. Normal Conditions Not Looked For Inside of Another Week. Many Cancelling Reservations for Sailing." (April 20, 1912, p. 4)

The music publishing industry begins a price-slashing war, which causes near-bankruptcy for many sheet music producers.
"Cut-Rate Music Battle Brings Price Down to 3¢: Sheet Music Strikes its Lowest Level. 'Ten Cent' Stores Talking of Becoming Publishers for Protection in the Future. Song Writers Worried Over Royalty." (April 27, 1912, p. 6)

The first commercially marketable talking-picture machine is developed by Chrono-Kinetograph Co.
"Six Million Dollars Back of Chrono-Kinetograph Co.: John Cort and Moneyed Associates Incorporate Under New Jersey Laws to Manufacture and Operate New Talking Picture Device Invented by Dr. Kitsee." (November 15, 1912, p. 1)

1914 **Motion pictures become a part of dinner and nightclub entertainment.**
"Moving Pictures in Cabarets Added Restaurant Attraction: Broadway Places Considering Installing Films for Diners. Churchill's Putting in Kinemacolor. Shanley's May Also. New Broadway Gardens Announce Film, Food and Dancing. May Save and Make Money for House." (May 1, 1914, p. 19)

Mary Pickford receives $2,000 a week as film's highest-paid star.
"$104,000 Salary for Film Star. Mary Pickford's Contract: Noted Player in Feature Films Plays and Cast. Was Offered $200,000 for Same Period." (November 28, 1914, p. 3)

1915 **D.W. Griffith releases *The Birth of a Nation*, crowding box offices throughout America.**
"Griffith's $2 Feature Film Sensation of Picture Trade: *Birth of a Nation* at Liberty, New York, Will Do $14,000 at Box Office This Week. Theatre Seats 1,200. Capacity Crowds Thronging House Twice Daily. Griffith Co. Renting Theatre. First Week's Advertising Bill $12,000." (March 12, 1915, p. 3)

1922 **The motion picture industry begins self-regulation by establishing the Hays Office.**
"Probing the Hays Mystery: Reformers See Victory in Acceptance. Impression Zukor Dominated Move for Hays Appointment Now Prevails—Secret Sessions and Motives Questioned—What Do Statements of Pres. Harding and Former Cabinet Official Mean?" (January 20, 1922, pp. 1, 38, 39)

1927 **The Federal Radio Act sets up a presidentially-appointed commission to regulate broadcasting.**
"Radio Control Bill Drafted as a Political Compromise: Conference Measure Approved in House in 48 Hrs; Prompt Passage in Senate Forecast—Provides for New Kind of Commission, on Trial a Year." (February 2, 1927, p. 50)

***The Jazz Singer*, a talking picture produced by Warner Brothers, ushers in the technology of sound to movies. Article predicting future trends related to the specific event.**
"Actual 'Talking Pictures' Loom Up as Screen Possibility Shortly." (November 23, 1927, p. 9)

1927–28 ***Show Boat*, a landmark musical, opens on Broadway after much controversy. Articles describing the various stages of the show's development.**
"Miss Miller in *Show Boat* Next Season? Ziggy Ill—In Bed for Two Weeks—Reported Ferber Adaptation Put Over." (February 9, 1927, p. 35)

"Ziggy Dodging Lyric for His *Show Boat?*: Due December 21—Report Figuring on Ziegfield-Ettanger House for *Rosalie*." (December 14, 1927, p. 48)

Show Boat Music Wholly Restricted: Ziegfield and Music Publishers Trying to Prevent Over-popularity of Tunes." (January 4, 1928, p. 3)

1933–34 **Broadway shows experience their highest popularity in contemporary American theatre history.**
"25 Shows in the Money: '33-'34 Average Best in Years: 25 of 75 B'way Legits So Far Look Safe—Last Season's Average One Out of Five—Nine Hits and 15 Moderate Successes Indicated—Film Coin Helps." (January 30, 1934, p. 53)

1934 **The Federal Radio Act is revised and updated to include television. Articles describing the various stages of developing the Act.**
"Tightening Radio Grip: Communications as New Dept. Present Federal Control May Be Merged and Absorbed in Huge Bureaucracy with Dominion Over Telephone, Telegraph, Broadcasting—White House Toying with Idea." (November 14, 1933, p. 33)

"U.S. to Regulate Air Adv.: Tighter Gov't Radio Control. Congress Ordered to Survey Radio—Likelihood Present Commission Will Be Scrapped—President Roosevelt Plans New Seven-Man Board." (February 13, 1934, pp. 1, 42)

The Catholic Legion of Decency is established.
"Church's Film Offensive: Chase 'Dirt' Via 13,000 Theatres. Concerted Effort by National Catholic Welfare Conference to Clean Up Pix—Force Campaign Through Exhibitors." (April 10, 1934, pp. 1, 20)

1936 **Great Britain is the first country to broadcast regular TV programming. Article in reference to the potential track records of British television versus American TV.**
" 'Fatigue' of Television Handicaps; Predictions All '5 Years or More' " (December 31, 1938, pp. 1, 47)

1938 **Orson Welles' radio production, "War of the Worlds" brings America into an abrupt panic. Article noting the mass communication event as it relates to a broader political spectrum.**
"Radio Does U.S. a Favor: Preparedness Vs. Panic Issue. Strategists Taking Cognizance of That Sunday Night Broadcast—Dramatizes Vividly the Lack of Common Sense Should a Real Air Attack Ever Occur—Anti-Radio Press Overplayed it, for Its Own Reasons." (November 2, 1938, pp. 1, 28)

Father Coughlin stirs up controversy with his religious broadcasts.
"Coughlin's Hot Potato: Radio Priest Anti-Semitic? WMCA 'Footnotes' Coughlin's Views With Its Own Rebuttal Interpretations—Apart From the Religious Bias, It's a Tough Spot for Radio, on Heels of the Rutherford Stuff." (November 23, 1938, p.25)

CBS enters the record business by buying the American Record Company.
"CBS Takeover of Phono Co. Thurs. (15)." (December 14, 1938, p. 37)

"Consummate American Record Deal: Frank Walker, Asked Where He Stands, Denies Joining CBS Subsidiary." (December 21, 1938, p. 24)

1939 **Television is introduced to the United States at the New York World's Fair. Articles describing the various stages of developing broadcast television in the U.S.**
"Sight and Sound Over Phone Wires Onto Screen Claimed by General Electric: Television Amplified—Deal with FBO Made by RCA and Combo to Employ Studios on Coast—Edison's Picture and Speech Reproduced from Orange N.J. to Schenectady, N.Y.—Joe Kennedy Paid $480,000 for RCA Interest." (January 11, 1928, p. 5)

"Television Broadcasts in Two Cities, But General Release Not Yet Set." (June 27, 1928, p. 1)

"Televish's U.S. Touchoff: N.Y. Fair Will Signalize Bally. However, It's All Still Experimental, Conversational and Promotional." (April 12, 1939, pp. 7, 47)

***Gone With the Wind* premieres in Atlanta, Georgia.**
"Atlanta Normal Again After *Wind*'s Tornado Premier, State Holiday." (December 20, 1939, p. 8)

1943 ***Oklahoma* opens on Broadway.**
"Operettas' Popularity Gets New Impetus By *Oklahoma* Smash." (April 7, 1943, p. 41)

1944 **Frank Sinatra appears live at New York's Paramount Theater, causing pandemonium at Times Square.**
"Sinatra to Sparkplug War Bond Drive to Bobby Soxers; Glamor Pitch." (November 8, 1944, pp. 1, 31)

1947 **The first World Series is broadcast on television.**
"1st Televised World's Series Game Nips Broadway Theatre B.O. by 50%." (October 1, 1947, pp. 1, 52)

CBS and NBC experiment with color television.
"Jolliffe Sees Color Tele Supplanting Black & White; No Sudden Blackouts." (October 29, 1947, p. 31)

The Communist scare leads to Hollywood blacklisting.
" 'Red Herring' Just Another Fish Story, Major H'Wood Feels of Baiting. The House Committee on Un-American Activities has a Number of Homeoffice Execs of Major Film Companies on its List, It's Reliably Reported. J. Parnell Thomas, Chairman of the Committee, It's Understood Plans to Call These Pix Officials for Quizzing at Continued Hearings Either In Washington or New York. Roster of Probes Con-

sist of Film Bigwigs Who Are or Have Been Members of Such Liberal Groups as The Progressive Citizens of America and its Predecessor, the Independent Citizens Committee of the Arts, Sciences, and Professions." (May 21, 1947, p. 2)

"As the Witch Hunt Spreads, Radio Wonders If It, Too, Will Get Burned." (October 29, 1947, p. 25)

1948 **TV becomes accessible to consumers. Articles following the growing trend of television.**
"The Poor Like Their Tele Sets Also." (November 14, 1947, p. 29)

"Tele's Big Boom Worrying Radio." (January 7, 1948, p. 1, 46)

"U.S. Video Sets Now 484,350." (August 25, 1948, p. 30)

"U.S. Tele Sets Hit 612,000 Mark." (October 27, 1948, p. 23)

The Supreme Court ends vertical integration and block booking in the film industry. Articles describing the various stages of the vertical integration/block booking controversy.
"Attempted Feature Control May Change Picture Field: Manufacturers Watching Big Exhibitors' Efforts to Tie Up Feature Fim: Apt to Force Manufacturers to Become Exhibitors for Self-Protection. Film Makers Want Open Field." (May 22, 1914, p. 19)

"Block Booking Fight Starts." (February 28, 1927, pp. 4, 48)

"Hays Must Not Railroad: Party in Power Told by Exhibs. Reported Hays Organization Engineering Trade Conference—Expected Disorganized Exhibitors Unable to Submit Substitute Distribution Plan for Block Booking—Federal Trade Commisions See 'Out' for Themselves." (October 5, 1927, pp. 5, 15)

"Block Booking's 'Out?' " (October 14, 1927, pp. 25, 46)

"Blockbooking Not Showmanship." (December 14, 1938, p. 3)

"U.S. Verdict Shocks Film Biz: See Forced Sale Via Court Edict." (May 5, 1948, pp. 1, 18)

"Divorcement Won't Panic Biz: Theatre Loss NSB—Skouras. (May 19, 1948, pp. 5, 18)

The FCC initiates a freeze on TV station licenses to stabilize the industry. Articles describing the various developments leading up to the FCC decision.
"FCC Bids Now Hit Landslide." (November 26, 1947, pp. 27, 38)

"Video Bids Keep Pouring Into D.C." (December 10, 1947, pp. 31, 46)

"Flock of New Tele Bids Brings Total Up to 64." (December 17, 1947, pp. 29, 38)

"90 Video Stations Now Have Grants; 4 Hearings Set." January 21, 1948, pp. 27, 34)

"If FCC Was 'Too Fast' on the Tele Grant Trigger, It Wasn't For Long." (January 28, 1948, p. 31)

"TV Permits Near 100 Mark as Three More Get FCC Nod." (May 5, 1948, pp. 40, 42)

"FCC's Tele Freeze Won't Shove Current Stations Upstairs; Move Seen as Welcome TV 'Breather' ." (October 6, 1948, p. 29)

"Freeze Gives Time to Solve Tele Sticklers." (October 6, 1948, pp. 29, 34)

"TV Stations Still on Horns of Deep Freeze Dilemma." (November 17, 1948, pp. 27–38)

1951 **Microwave technology provides nationwide TV broadcasting.**
"CBS-TV Has First Crack at N.Y.-L.A. Micro Sept. 29." (September 19, 1951, p. 31)

1952 **The FCC lifts its freeze on television station licensing.**
"TV Freeze Lift Expected on Jan 1, But New Stations Unlikely in '52." (October 31, 1951, pp. 22, 36)

1955 **Film companies start releasing old films to TV networks and stations.**
"New Millions for Old Pix: O'Neill's Road to TV Riches." (March 2, 1955, pp. 1, 42)

1956 **Elvis Presley tops the record charts with his unique brand of rock 'n' roll.**

"If Elvis Ain't Sayin' Nothin', How Come He Sells Records: Hampton." (December 19, 1956, p. 52)

Videotape is introduced to TV producers amidst controversy. (Articles and commentary.)
"Live Celluloid Vs. Dead Electronics (A Drama in Three Acts with a Shavian Preface)—Commentary." (January 9, 1957, p. 100)

"Who's Gonna Control Tape? Lotsa Intrigue on Jurisdiction." (January 16, 1957, pp. 29, 48)

"Tape: Planned Parenthood. Everyone Wants to Cut Up Baby." (July 8, 1959, p. 33)

1959 **The payola scandal damages the entertainment industry. (Articles describing the various developments leading up to the payola scandals.)**
"TV Payola Rearing its Head: Prods., Names Play Angles." (July 13, 1955, pp. 41, 48)

"Payola Indictment of Randy Dixon, Philly Disk Jock." (July 17, 1957, p. 43)

"Too Many DJ's Payola Rises." (July 24, 1957, p.1)

"Teleblurb Industry Decries Payola, Vegas Cuffoed (Including Losses)." (July 31, 1957, p. 27)

"Indict Freedman in Quiz Scandal." (November 12, 1958, p. 23, 42)

"Quiz Baffler: TV Controls? Penalties Sure to Be Inflicted." (October 14, 1959, pp. 23, 35)

"FBI Now Eyes Quiz 'Fixes' " (November 11, 1959, pp. 1, 54)

"Double-Pronged Probe Into Payola Slows Charge of the 'Loot Brigade' " (November 11, 1959, pp. 1, 57)

"Howard Miller has 'Proof' of Deejay Payola." (November 11, 1959, p. 1)

" 'Don't Condemn a Whole Industry for Sins of Few:' Oren Harris." (December 30, 1959, pp. 1, 53)

Stereo sound is introduced to the public.
"Stereo's Big Sound at Fairs: See $500,000,000 Equipment Sales." (October 7, 1959, p. 61)

1962 **Telstar I launches a new age of satellite programming.**
"World TV: A Click Premier: But Plenty of Hurdles Remain." (July 18, 1962, pp. 27, 38)

"Telstar Seen Throwing Talent Unions into Global Tizzy: Who Shells Out Table for Performers?" (July 18, 1962, p. 27)

"D.C. Reflects on Telstar Opening a Pandora's Box." (July 18, 1962, pp. 27, 37)

1964 **The Beatles introduce a new era of rock 'n' roll.**
"Beatlemania Doesn't Stop With Disks, Mops Up with Fancy $2 Fan Club Too." (May 27, 1964, p. 49)

"Redcoats Still Rocking U.S.: But is Trend Nearing End?" (June 10, 1964, p. 49)

"Beatlemania's Second Wind: UA Pic, U.S. Tour Boom Disk Sales." (July 22, 1964, p. 73)

1967 **Public Television is begun by recommendation of the Carnegie Commission.**
"PTV: Is Everybody Happy?: Striking a Blow for Status Quo." (February 1, 1967, pp. 35, 55)

"Public Television's Obstacle Course." (February 1, 1967, pp. 35)

"Program Holes in Carnegie Board." (February 1, 1967, pp. 35, 56)

1968 **The MPAA develops a rating system for films.**
" 'Ratings' An All-Things Thing: Smoother-Upper of Sharp Angles." (September 18, 1968)

1969 **Television broadcasts the first American's steps on the moon.**
"Greatest Show Off Earth: TV-AM Webs Bet $13-Mil on Moon." (July 16, 1969, pp. 1, 54)

"CBS Winning Moondust Twins: Sky-High Stars: Wally & Walter." (July 23, 1969, pp. 33, 45)

1970 **New developments in video technology flood the broadcast marketplace.**
"Standardization Plus $100,000,000 In Programs Key to Vidcassettes Boom." (January 6, 1971, pp. 3, 68)

"See It Big; Will It Be Good?: CMA Exec Says New Era Is Here." (January 6, 1971, pp. 63, 68)

1971 **Cigarette advertising is banned from television and radio.**
"Cigs Ready to Blow Off TV-Radio by Sept. '70 to Kayo Ad 'Warnings' " (July 23, 1969, pp. 33, 43)

"Cig-Makers Tell FCC They Won't Use Smokescreen to Get Around Ban." (January 13, 1971, p. 40)

1976 **Thomson-CSF Laboratories introduces a new, eight-pound color television camera.**
"Newest Mini is a Microcam." (January 21, 1976, p. 77)

1981 **The FCC decides to deregulate radio.**
"House Report Swipes FCC's Dereg Express." (November 4, 1981, pp. 47, 60)

1988 **Film animation advances with special effects technology in Steven Spielberg's hit *Who Framed Roger Rabbit?***
"Buena Vista's *Roger* On Top For the Year." (January 11, 1989, p. 16)

1989 **The TV industry continues to expand its technology with advanced compatible television.**
"Advanced widescreen tube unveiled; NBC Sarnoff, Thomson show ACTV." (August 2, 1989, pp. 40, 47)

References

1. *Variety* (December 16, 1905), p. 1.
2. *Variety* (October 30, 1929), pp. 1, 64.

Traditional Research Methodologies in Mass Communication

A significant portion of this book includes statistics and/or informational data about the electronic media industry. Since information of this type is often collected by several different means, or methodologies, a brief discussion of traditional mass communication research seems appropriate as an introduction to the rest of this text.

Unlike other commercial endeavors, the electronic media came into the public eye without any visible means of reflecting programming success or failure. Department stores or mail order houses, for example, could always use sales slips or purchase records to determine their growth. Radio, however, lacked any available feedback mechanism. This occured despite the fact that broadcasters, like all other businesspeople, needed informational data for their future planning.

Recognizing the need to measure this intangible audience, early radio advertisers decided to form their own research organization—the Cooperative Analysis of Broadcasting (CAB)—in 1930. At that time, the CAB used telephone surveys as a means of assessing the size of radio audiences and the popularity of specific radio programming. Since those early days, however, other research companies have come forth, developing more diverse and sophisticated methodologies for gathering data on audience demographics, behavior, programming preferences, attitudes, psychographics, consumer patterns, and even lifestyle patterns.

Audience Research

Audience research, simply put, is "a systematic means for collecting and reporting objective, reliable information about the audience."[1] Research results are used by almost every sector of the mass communication industry in conducting program evaluations, format reviews, pricing and sales decisions, and promotion strategies. They are also studied by advertisers and media-buying companies as well as by program producers and packagers.

Most media concerns by industry executives involve two areas: the audience and program cost-effectiveness. Data collections in these areas rely upon two basic types of research: primary and secondary. Primary research involves *new* data-gathering, using at least one of three methodologies: 1) survey-based research; 2) focus group research; and 3) experimental research. Secondary research uses *existing* data from resources and databases such as the census, purchase records, and production budget figures.

Primary Research

Survey-based Research

Survey-based research is a data-gathering technique that presents a specific group of questions to a selected group of respondents. Extreme care is taken in the selection of specific questions as well as in the selection of a representative sample of respondents. The survey may be conducted by mail, by telephone, or in face-to-face situations.

Some of the most sophisticated examples of survey-based research include the diary services such as the peoplemeter and overnight audimeter readings used by Nielsen and Arbitron. These devices record each minute that the television is turned on and which channel has been selected at any given time.

Another fast-growing sector of survey-based research is music testing for radio stations. If two or three radio stations in the same market play the same format, the competition for listeners is considerable. Thus, research companies have become quite adept at addressing the issue of how to make each song count. Currently, two types of survey-based music testing are utilized: 1) the weekly callout; and 2) the Auditorium Music Test.

The weekly callout is a telephone survey of approximately 100 people per week in *every* market. Its purpose is to track the life cycle of a song and to assess what broadcasters call burn—whether the song has peaked, is now dying (burned out) or is still on the rise. To conduct weekly callouts, interviewers phone at random and ask individuals to listen to short pieces of music (normally nine to twelve seconds). Then they ask them a series of questions which can be tabulated to assess where a song is positioned that week in terms of its life cycle. If the weekly callout indicates possible burn-out for a song, research companies will usually recommend that stations discontinue playing the song. Songs which the weekly callout suggests are still on the rise are recommended for additional play.

The Auditorium Music Test is perhaps the most rigorous survey-based examination of audience music preferences. The test is administered to a carefully selected audience sample of 100-110 people, and is format-specific as well as market-specific. The sample audience is preselected according to format preference, and normally only one demographic group is tested at a time. Once the preselection has been completed, each sample audience assembles in an auditorium and fills out a survey with the usual demographic information (age, gender, etc.). The audience then listens to nine- to twelve-second clips of approximately 300 songs. After each song, the listeners indicate their preference ratings on the survey form. The test usually takes about ninety minutes, with breaks for the audience after each 100 songs.

The Auditorium Music Test has two important advantages. First, results can be tabulated and acted upon before the end of the ratings period and before the diaries are returned. Second, because the test is format-specific as well as market-specific, it can provide researchers with valuable data for further testing (like whether a song is rated higher if it follows a slower-paced song).

Focus Group Research

Focus group research is used primarily for three purposes: 1) to study the affect or subjective feel of a particular station, program, or format; 2) to identify attitudes and opinions that exist in the audience market; and 3) to identify new questions or a series of hypotheses that can be verified later through subsequent surveys. Each focus group is comprised of 10 to 12 people carefully chosen by the focus group director. Usually, the focus group meets in a room with a one-way mirror located on one wall. While the focus group director works with the selected individuals to identify attitudes, feelings and opinions about a particular station or program, a member of the station's management team watches from behind the one-way mirror. Results of a focus group are written up as a set of hypotheses to be tested later, generally through some type of survey-based telephone interviews.

Researchers have found that focus groups reveal hidden feelings that are not immediately apparent in quantitative (statistical) research. For example, in markets where the technologies of competing television stations are equal, differences in local news ratings often rest upon difficult-to-define feelings about the talent of the news team and the overall tone of the broadcast. In these cases, focus group methodologies are often used to identify the affective characteristics, such as warmth, friendliness and professional acumen, that separate the more popular stations from those which are less successful.

Experimental Research

Experimental research utilizes an empirical method to study specific correlations or relationships. The *control group* (one that is seen as the norm from which all other groups are tested) is the key to this type of methodology. If, for example, an advertising agency wants to test the effectiveness of a TV commercial for an upscale car with an audience of upscale consumers, that agency would also want to compare the results of this group with control group composed of television viewers from all economic bases. If the comparison with the control group shows no difference, the advertiser can then surmise that the commercial is probably not reaching its target audience. Other examples of experimental research include the effect of increased TV violence on audience behavior and/or preference; the importance of a laughtrack to the success of a comedy series; and the impact of television on children.

Secondary Research

Secondary research utilizes existing primary data to ask new questions, seek correlations, and provide new insights. Usually secondary research will combine two or more sets of data comparing, for instance, census information (regarding age, sex, and occupation) with purchase records. Such research is called *multifactor* or *multivariate* research because several variables are compared simultaneously.

Because audiences are so multidimensional, most media researchers combine several primary and secondary methodologies so that they may provide the most detailed, current information on today's consumer media habits. And, as computer access increases, the availability of large databases will continue to provide the mass communication industries with linkages that not only verify existing research but can be used to generate more knowledge about audiences in the future.

Statistical Differences between *Variety* and Other Sources

From time to time, readers may note that certain statistical representations in *Variety* differ from those found in other publications. These discrepancies can occur for several reasons, among them:

1) the data has been drawn from different sources;

2) certain factors have been analyzed in one study but not in another, despite the fact that the researchers utilized the same data base;

3) the collection of data was taken on significantly different dates; and/or

4) the statistical methodologies utilized by *Variety* differ from those used in other publications.

Therefore, as a service to our readers, statistical sources, methodological explanantions, and the dates of data collection are included whenever possible.

Reference

1. James Webster, *Audience Research* (Washington, D.C.: National Association of Broadcasters, 1983), p. 1.

New Trends in Audience Research

The increasing complexity and fragmentation of the broadcast audience has never been more apparent. Twenty-five years ago, the average television viewer had approximately 2.6 viewing options. Now, with cable reaching 55 to 57% of all households, that same television viewer has an average of 24 options and many households have over 60 options. Correspondingly, network share has been steadily decreasing in the face of increased viewing options. In addition, the remote control on television sets has replaced linear viewing: viewers can switch from one channel to the next, from network to local to cable, at the push of a button. With home videocassette units now being used by almost 70% of households, this fragmentation is exacerbated: VCR owners rent an average of 3 tapes a month and utilize their VCR's for the time-shifting of programs. As a result, many broadcasters, advertising agencies and media buyers have found that traditional research methodologies do not adequately respond to this increased audience fragmentation. A new look at the broadcast audience is required: one that provides an assessment not only of the demographics of a given dominant market area (DMA) but of the specific viewing and consumer patterns of sub-groups within any DMA.

Overview

Research firms such as Nielsen and Arbitron provide broadcasters with considerable demographic, viewing time, and program preference (share and rating) data for each DMA on a regular basis. For example, Nielsen reports indicate that the average hours of household (HH) television usage has increased from 6 hours and 11 minutes each day in 1975–1976 to 7 hours and 5 minutes in 1987 for a total weekly average in November, 1987 of 29 hours and 40 minutes.[1] Table 1 shows how weekly TV viewing breaks down at the national level by age and sex.

Accordingly, with the exception of male teens, women watch more television on average than men. Utilizing Nielsen's average minute audience, the national audience composition by selected program type can be assessed as shown in Table 2. Women also watch more informational prime-time programming than do men: 7,800,000 women (18 and over) watch news and informational programming while 5,970,000 men watch such programs.[2] A similar viewing pattern holds for morning television such as "Good Morning America" and the "Today Show."

Good Morning America (7:30–8:00am)	
Women (18+)	2,960,000
Men (18+)	1,810,000
The Today Show **(7:30–8:00am)**	
Women (18+)	2,870,000
Men (18+)	1,940,000

(Source: Nielsen Media Research, November 1987)

Table 1. Weekly TV Viewing by Age

	hrs/min viewed per wk		
Age	*Children*	*Female*	*Male*
6–11 years old	22 hrs, 46 min		
2–5 years old	25 hrs, 26 min		
teens		23 hrs, 13 min	24 hrs, 16 min
18–34 years old		28 hrs, 43 min	26 hrs, 15 min
35–54 years old		32 hrs, 14 min	26 hrs, 07 min
55 and over		41 hrs, 17 min	37 hrs, 07 min
Total persons	29 hrs, 40 min		

(Source: Nielsen Television Index, NAD Report, 1987)

Table 2. Prime-time Audience Composition by Selected Program Type*

Group	*General drama*	*Suspense/ mystery*	*Situation comedy*	*Adventure*	*Feature films*	*All regular network 7–11 pm programs*
Women 18+	10,170,000	9,000,000	11,870,000	7,870,000	10,630,000	10,080,000
Men 18+	6,380,000	6,830,000	7,310,000	6,390,000	7,720,000	7,400,000
Teens 12–17	1,160,000	1,200,000	2,470,000	1,180,000	1,570,000	1,530,000
Children	1,420,000	1,470,000	3,700,000	1,830,000	2,100,000	2,040,000
Total Persons	19,130,000	18,500,000	25,350,000	17,270,000	22,020,000	21,050,000

*All figures are estimates for period of November, 1987 and represent number of viewers.
(Source: Nielsen Media Research, 1988 Neilsen Report on Television)

Average television viewing time and audience composition is also measured in terms of daypart: the times that demographic groups watch television. Table 3 indicates the selected times that each demographic group watches television.

More Americans watch prime-time television (8:00pm to 11:00pm EST except Sunday which is 7:00pm to 11:00pm) on Sundays than on any other day, and Fridays and Saturdays draw the least viewers.

Monday	101.1
Tuesday	94.5
Wednesday	93.9
Thursday	97.0
Friday	89.4
Saturday	87.1
Sunday	105.8
Total Avg.	95.5

In millions of people. Average minute audiences November, 1987; excluding unusual days.
(Source: Nielsen Media Research, 1988 Nielsen Report on Television)

Interestingly, weekly rental activity of videocassettes mirrors prime time viewing. While Friday and Saturday prime-time viewing is the lowest in the week, video cassette rentals for Fridays and Saturdays account for 56% of all weekly rental activity:

Monday	9%
Tuesday	8%
Wednesday	8%
Thursday	8%
Friday	22%
Saturday	34%
Sunday	11%
Total	100%

(Source: 1989 Video Store Retailer Survey, based on survey data from 378 specialty video outlets; in *Video Store Magazine*, August 1989, p. 12)

Nielsen, Arbitron and similar firms provide broadcasters, advertisers and media buyers with specific data for each market that can be utilized to identify distinct audience trends, assess differences between the national averages and a given local DMA, and to identify specific viewing patterns of each DMA.

Geo-demographic Methodology

However, while research firms such as Nielsen and Arbitron provide reliable demographic and program preference data (share and rating) of the viewing audience for any given time and program in any DMA, market research firms are seeking ways to increase the discrimination of this research to target a given audience or consumer base with greater specificity. Psychographic methodologies introduced in the 1970s, while addressing audience values and attitudes, are often untargetable geographically. While attitudinal research has become increasingly sophisticated and is currently utilized at the network level where geographic definitions are not of primary importance, there is a clear need for an integrated geo-demographic methodology to define and target local broadcast audiences and consumers.

Integrated geo-demographic research methodologies attempt to address three primary concerns: 1) identifying the primary target audience/consumer, 2) identifying where they live and shop, and 3) identifying what media can best reach this target. Broadcasters and advertisers utilizing broadcast media recognize that the increased fragmentation of the broadcast audience is coupled with increased fragmentation of consumer markets and shopping patterns. It has been known for some time that the audience with the most buying power tends to watch the least television yet have the most viewing options. And conversely, the audience with the least buying power watches the most television yet has few viewing options. Accordingly, several research companies are developing and utilizing statistically precise, measurable and predictable research methodologies that can define and identify target audiences at a micro-geographic level (city blocks and rural routes). These methodologies can integrate with other known aspects of consumer behavior lifestyles as well as with media research data from companies such as Nielsen and Arbitron.

PRIZM Cluster Analysis

Computer scientist Jonathan Robbin pioneered the development of geo-demographic research methodology in 1974. Robbin designed a computer program that would sort America's 36,000 zip codes into forty "lifestyle clusters." The system became the backbone for Claritas Corporation and is called PRIZM for Potential Rating Index for Zip Markets. PRIZM has evolved into one of the most powerful integrated geo-demographic market research tools available.[3] While a number of research corporations are developing and utilizing similar geo-demographic/lifestyle stratification systems, this section will focus on the PRIZM system.

The Claritas PRIZM Cluster System is based on the assumption that people of similar cultural backgrounds, income levels, and perspectives naturally gravitate toward each other. They choose to live with their peers in neighborhoods that offer affordable advantages and compatible lifestyles. The PRIZM system assumes that once settled in, people naturally emulate their neighbors, exhibiting shared patterns of social values, tastes, expectations, and consumer behavior.

The Claritas PRIZM system goes beyond analysis of age, sex, income, median home value and other normal demographic data to describe the lifestyles of Americans. PRIZM attempts to describe what we buy, how we vote, what are our values, what are our habits, and how our neighborhoods (ZIPS) affect our lifestyles. Building on 1980 census data, PRIZM Cluster Analysis integrates a large number of varied databases from consumer market surveys, purchase records, broadcast audiences, and cable subscribers, compiled and direct-response mailing lists, data from hundreds of interviewers, buyer behavior patterns, product usage patterns, media patterns, and lifestyle habits. The result is what amounts to a factor analysis of forty lifestyle clusters that has been consistently tested and optimised against over sixty national consumer files. Each cluster has its own definable characteristics and can be assessed in terms of social rank, household composition, ethnicity, mobility urbanization, housing, media usage, purchasing patterns, product preferences, and percentage of population in any given area of dominant influence (ADI), Zip code area, or radio footprint as well as national population figures.[4]

Because the PRIZM System is designed for data integration, it cluster-codes and integrates resources including syndicated media-audience databases from firms such as A.C. Nielsen, Arbitron, Birch Radio, Simmons and Mediamark Research. These resources are then integrated with market facts data resources, on-line retrieval and planning models, and major compiled lists and other databases including product user profiles. Together they deliver a powerful marketing tool that not only segments and targets consumers but provides the information necessary to select appropriate media to reach that target.[5]

PRIZM research, on a micro level, evaluates not only ZIP codes but the four-digit block codes in major cities as well as route clusters on 7-digit postal carrier routes. Thus, not only can PRIZM help to identify lifestyle patterns for each cluster but can be used with considerable precision to define each block or route within each cluster. For most applications, ZIP clusters suit targeting needs.

The Forty Clusters

The forty clusters identified by Claritas Corporation are organized within twelve broad Social Groups: four suburban groupings, three urban groupings, three town groupings, and two rural groupings. Descriptions of each social group are provided in Table 4 with brief descriptions and demographic data of all clusters within each group.

Table 3. Average TV Viewing Times (hours and minutes per week)

Group	*Mon.-Fri. 10am-4:30pm*	*Mon.-Fri. 4:30pm-7:30pm*	*Mon.-Sun. 8-11pm*	*Sat. 7am-1pm*	*Mon.-Fri. 11:30-1am*
women					
18+	6:05	4:59	9:56	:36	1:19
18–24	4:49	3:16	6:58	:37	1:06
55+	7:48	7:07	12:06	:32	1:24
12–17	2:30	4:12	7:16	1:00	:38
Men					
18+	3:17	3:52	9:02	:35	1:19
18–24	3:29	2:51	6:26	:35	1:15
55+	5:06	6:09	11:19	:34	1:17
12–17	2:22	3:54	7:40	:59	:41
Children					
2–5	5:26	4:19	5:05	1:43	:19
6–11	2:22	4:25	6:07	1:43	:23

(Source: Nielsen Media Research, November 1987)

Table 4. The Forty PRIZM Clusters

Social rank	*Cluster nickname*	*Thumbnail description*	*%US HH*	*Median income*	*% College grads*	*Primary age/ 2nd age*	*Family type*
The S1 Group:	This group includes educated, affluent executives and professionals in elite metro suburbs. The three clusters in this group are characterized by top socio-economic status, college-plus educations, executive and professional occupations, expensive owner-occupied housing, and conspicuous consumption for many products, goods and services. Representing 5.22% percent of all households, this group contains 32% of the nation's $75,000-plus household incomes and an estimated third of America's net worth.						
1	Blue Blood Estates	America's wealthiest neighborhoods, upperclass professionals and executives, heirs to old money, accustomed to privilege and luxury. One in ten millionaires found here.	1.1	70,307	50.7	45–54/ 35-44	fam
2	Money & Brains	Sophisticated, upscale singles and childless couples living in swank townhouses, apartments and condos	0.9	45,798	45.5	45–54/ 55–64	f/s
3	Furs & Station Wagons	New money, well-educated, mobile professionals and managers with nation's highest incidence of teenage children.	3.2	50,086	38.1	35–44/ 45–54	fam
The S2 Group:	This group includes pre- and post-child families in upscale, white-collar suburbs. While significantly below S1 in socio-economic levels, this group displays all of the characteristics of success including high education levels, income, and home values.						
5	Pools & Patios	Empty nesters with good educations, high white-collar employment; double incomes insure the good life.	3.4	35,895	28.2	55–64/ 45–54	cpl
6	Two More Rungs	High concentration of multi-ethnic, older professionals living in multiple-unit housing. Conservative spending patterns.	0.7	31,263	28.3	65+/ 55–64	f/s
7	Young Influentials	Yuppie metro sophisticates with exceptional high-tech, white-collar employment levels. Mainly singles with childless couples, unrelated adults in condos, apartments or homes with open lifestyles.	2.9	30,398	36.0	18–24/ 25–34	f/s
The S3 Group:	Upper-middle class child-rearing families in outlying, owner-occupied suburbs. This group represents the traditional family: Mom, Dad and school-age children, double incomes, two or more cars, and single-unit housing. S3 is the essence of the traditional American Dream.						
8	Young Suburbia	Large, young families with school-age children; high white-collar employment and strong consumers of family products.	5.3	38,583	23.8	35–44/ 24–34	fam
10	Blue-Chip Blues	High-school graduates with blue-collar occupations and families. Double incomes and high employment make this the wealthiest blue-collar neighborhood.	6.0	32,218	13.1	35–44/ 25–34	fam

continued

Table 4. *continued*

Social rank	*Cluster nickname*	*Thumbnail description*	*% US HH*	*Median income*	*% College grads*	*Primary age/ 2nd age*	*Family type*
The U1 Group: Educated, white-collar singles and couples in upscale, urban areas. Upscale socio-economic status, cosmopolitan lifestyles, high degree of divorced and separated singles in big-city America.							
4	Urban Gold Coast	Upscale, usually New York white-collar single professionals living in high rise areas.	0.5	36,838	50.5	25–34/ 65+	sgl
11	Bohemian Mix	Urban integrated bohemian areas of singles; mix of students, writers, actors, artists, aging hippies.	1.1	21,916	38.8	25–34/ 18–24	sgl
14	Black Enterprise	The most family oriented of the U1 cluster group; about 70% black with median income well above average.	0.8	33,149	16	35–44/ 45–54	f/s
15	New Beginnings	Young, often recently divorced pre-child white-collar and clerical workers.	4.3	24,847	19.3	25–34/ 18–24	f/s
The T1 Group: Educated, young mobile families in exurban satellites and boom towns far from the major metropolitan areas. These are highly mobile, white-collar adults living in single-unit housing. This group has seen considerable growth due to urban exodus.							
9	God's Country	Highest socio-economic white-collar neighborhoods located outside major metropolitan areas. Well educated, highly mobile families who opted to live in America's most beautiful mountains and coastal areas.	2.7	36,728	25.8	35–44/ 25–34	fam
17	New Homesteaders	Like God's Country but nine rungs down on the socio-economic scale. High concentration of military personnel; highly mobile and growing family cluster.	4.2	25,909	15.9	18–24/ 25–34	fam
19	Towns & Gowns	These college towns usually consist of 75% locals and 25% students. Very high educational, professional, technical levels; modest incomes with taste for prestige items.	1.2	17,862	27.5	18–24/ 25–34	sgl
The S4 Group: Middle-class, post-child rearing families in aging suburbs and retirement areas. Each cluster in S4 represents a continuing U.S. trend towards post-child communities. This group contains many aging couples, widows, and retirees on pensions and social security incomes.							
12	Levittown, U.S.A.	Empty nest couples living in aging tract housing in middle-class suburbs. High white-collar employment and double incomes make for comfortable living.	3.1	28,742	15.7	55–64/ 65+	cpl

continued

Social rank	*Cluster nickname*	*Thumbnail description*	*%US HH*	*Median income*	*% College grads*	*Primary age/ 2nd age*	*Family type*
13	Gray Power	Primarily sunbelt communities of senior citizens who choose to uproot and retire among peers. These are the nation's most affluent elderly, retired, and widowed neighborhoods with predominantly multi-unit housing.	2.9	25,259	18.3	65+/ 55–64	cpl
20	Rank & File	The blue-collar version of Levittown, U.S.A. High concentration of duplex rows and multi-unit "railroad" flats. Leads nation in durable manufacturing.	1.4	26,283	9.2	55–64/ 65+	cpl
The T2 Group: The three clusters in the T2 Group are America's blue-collar, child-rearing familes. These neighborhoods have a high incidence of large families, average household incomes, and are primarily in factory towns and remote suburbs of industrial cities.							
16	Blue-Collar Nursery	Leads the nation in craftsmen and also top in married couples and households of three or more. Usually satellite towns and suburbs of small industrial cities.	2.2	30,077	10.2	35–44/ 25–34	fam
21	Middle America	Mid-sized, middle-class satellite suburbs and towns at the middle of the socio-economic scale.	3.2	24,431	10.7	55–64/ 45–54	fam
23	Coalburg & Corntown	Small midwestern towns and cities surrounded by farmland and populated by blue-collar families.	2.0	23,994	10.4	65+/ 35–44	fam
The U2 Group: Mid-scale families, singles and elders in dense, deteriorating urban row and high-rise areas. These middle-class neighborhoods show high concentrations of foreign born, working women, clerical and service occupations, and an increasing minority population. The differences between the clusters in the U2 Group are as important as the similarities.							
18	New Melting Pot	New immigrant neighborhoods of Hispanic, Asian, and Middle-Eastern populations mixed with older European stock. Mainly situated in port cities on East and West coast.	0.9	22,142	19.1	65+/ 55–64	f/s
22	Old Yankee Rows	Similar in age, housing and family composition to New Melting Pot but has high concentration of high-school educated Catholics of European origin and few minorities. Mainly in older industrial cities of the Northeast.	1.6	24,808	11	65+/ 55–64	f/s
27	Emergent Minorities	80% black neighborhoods with children, half of which in single-parent homes. Below average education and blue-collar and service occupations.	1.7	22,029	10.7	18–24/ 25–34	s/s

continued

Table 4. *continued*

Social rank	*Cluster nickname*	*Thumbnail description*	*%US HH*	*Median income*	*% College grads*	*Primary age/ 2nd age*	*Family type*
28	Single City Blues	High ethnic mix and range of classes; few children; night occupations; a poor man's Bohemia in multi-unit housing.	3.3	17,926	18.6	18–24/ 25–34	sgl
The R1 Group: Rural towns and villages amidst farms and ranches in Mid-America. These sparsely populated communities have lower-middle to downscale socio-economic levels, high concentrations of German and Scandinavian ancestries, large families, low education, and maximum stability.							
24	Shotguns & Pickups	Outlying townships and crossroads villages in the nation's breadbasket. Large families with school-age children headed by blue-collar craftsmen, equipment operators, and transport workers.	1.9	24,291	9.1	35–44/ 45–54	fam
26	Agri-Business	Usually prosperous ranching, farming, lumbering and mining areas marred by rural poverty. Geo-centered in the Great Plains and mountain states.	2.1	21,363	11.5	65+/ 55–64	fam
33	Grain Belt	High concentration of working farm owners and less affluent tenant farms. The nation's most sparsely populated rural communities with one in five of America's farmers.	1.3	21,698	8.5	65+/ 55–64	fam
The T3 Group: Mixed gentry and blue-collar workers in rustic mill and factory towns. These are America's smaller industrial cities, factory, mining and mill towns and includes rustic coastal villages. The T3 Group have lower-middle incomes, limited educations, and live in single-unit and mobile home housing.							
25	Golden Ponds	Small, rustic towns and villages in coastal resort, mountain, lake and valley areas where senior citizens choose to retire. While not as affluent as Gray Power, this cluster has high concentration of retirees.	5.2	20,140	12.8	65+/ 55–64	cpl
29	Mines & Mills	Appalachian mining and mill towns where light and heavy industry makes this the top ranking total manufacturing cluster. Blue-collar occupations; limited education.	2.8	21,537	8.7	55–64/ 45–54	fam
31	Norma-Rae-Ville	Concentrated in the South, these neighborhoods include hundreds of industrial suburbs and mill towns. These are country folk with limited educations. Cluster has a high population of black Americans, and leads the nation in non-durable manufacturing.	2.3	18,559	9.6	18–24/ 45–54	fam

continued

Table 4. *continued*

Social rank	*Cluster nickname*	*Thumbnail description*	*% US HH*	*Median income*	*% College grads*	*Primary age/ 2nd age*	*Family type*
32	Smalltown Downtown	Older industrial cities and factory towns that once boomed with heavy industry. High population density; while the primary age group is 65+, the secondary age group is 18–24.	2.5	17,208	10	65+/ 18–24	f/s
The R2 Group:	Landowners and migrants in poor rural towns, agrarian villages and hamlets. This group has low population densities, low socio-economic rankings, minimal educations, large, highly stable households with widowed elders, and peak concentrations of mobile homes. Employment is blue-collar and farm labor.						
30	Back Country Folks	Very remote, rural towns in the Ozark and Appalachian uplands. Strongly blue-collar with some farmers. Leads all clusters in concentration of mobile homes and trailers.	3.4	19,843	8.1	65+/ 35–44	fam
35	Share Croppers	Traditionally, areas devoted to tenant farming, chicken breeding, paper milling, but the sunbelt migration has attracted light industry and some growth.	4.0	16,854	7.1	65+/ 55–64	fam
38	Tobacco Roads	Found mainly in the South, these largely black farm communities are dependent upon agriculture. While above average in number of children of all ages, one-third live in single-parent households.	1.2	13,227	7.3	65+/ 55–64	f/s
39	Hard Scrabble	America's poorest rural areas. Leads all other clusters in concentration of adults with less than eight years education. Trails all clusters in concentration of working women.	1.5	12,874	6.5	65+/ 55–64	fam
The U3 Group:	Mixed, unskilled workers in aging, urban row housing and high-rise areas. This group has high concentrations of minorities, equipment operators, service workers, and laborers with low education levels, large families headed by single parents living in multi-unit housing. Chronic unemployment is a major element in the U3 Group as are high concentrations of singles and broken homes.						
34	Heavy Industry	While similar to Rank & File, this cluster is nine rungs down on the socio-economic scale and is hard hit by unemployment. Mainly concentrated in the industrial Northeast; highly Catholic; few children. High divorce rate.	2.8	18,325	6.5	65+/ 55–64	f/s
36	Downtown Dixie Style	Predominantly black neighborhoods in southern cities with duplex and multi-unit housing. High unemployment and solo parenting.	3.4	15,204	10.7	18–24/ 65+	s/s

continued

Table 4. *continued*

Social rank	*Cluster nickname*	*Thumbnail description*	*% US HH*	*Median income*	*% College grads*	*Primary age/ 2nd age*	*Family type*
37	Hispanic Mix	America's Hispanic barrios. Large families, young children often raised by solo parents. Cluster ranks first in short-term immigration and second in percent foreign born.	1.9	16,270	6.8	18–24/ 24–34	f/s
40	Public Assistance	The nation's poorest neighborhoods; 70% black with twice the average unemployment. Large solo-parent families in public high-rise buildings or tenements.	3.1	10,804	6.3	18–24/ 65+	s/s

Key: Social rank is in decreasing order from 1 to 40; % US HH are 1987 figures of percent of American Households in each cluster; Median income based on 1987 figures; Primary age is followed by the second largest age group in each cluster, family type means predominant family type where *fam* means married couples with children; *cpl* means married couples; *f/s* means a mix of families and singles; *sgl* means singles; and *s/s* means solo-parent families and singles.

(Source: Claritas Corporation, 1987, 1988)

Table 5. Cluster Group Television and Radio Preferences by Type and Format

Code	*%HH*	*Clusters included:*	*TV preferences by type*	*Radio format preferences*
S1	5.22	Blue Blood Estates Money & Brains Furs & Station Wagons	News; Latenight	Classical Easy Listening Progressive
S2	7.00	Pools & Patios Two More Rungs Young Influentials	News, Entertainment News; Latenight	Classical Easy Listening
S3	11.33	Young Suburbia Blue Chip Blues	Sitcoms	Progressive Easy Listening Classical
U1	6.67	Urban Gold Coast Bohemian Mix Black Enterprise New Beginnings	Latenight	Classical Black Radio Progressive
T1	8.02	God's Country New Homesteaders Towns & Gowns	News; Latenight	Classical Progressive Country
S4	7.37	Levittown, USA Gray Power Rank & File	Sports; News; Drama	Easy Listening Progessive Classical
T2	7.37	Blue Collar Nursery Middle America Coalburg & Coaltown	Drama; Sitcom; Game	Country Easy Listening Progresive
U2	7.58	New Melting Pot Old Yankee Rows Emergent Minorities Single City Blues	Music TV; Drama	Black Radio Classical Progressive

continued

Table 5. *continued*

Code	*%HH*	*Clusters included:*	*TV preferences by type*	*Radio format preferences*
R1	5.27	Shotguns & Pickups Agri-Business Grain Belt	Drama; News	Country Easy Listening
T3	12.85	Golden Ponds Mines & Mills Norma-Rae-Ville Smalltown Downtown	Drama; Game Shows	Country
R2	10.15	Back Country Folks Share Croppers Tobacco Roads Hard Scrabble	Drama; Game Shows	Country Black Radio
U3	11.13	Heavy Industry Downtown Dixie Style Hispanic Mix Public Assistance	Music TV; Soaps; Drama	Black Radio

* %HH means total percent of households in this cluster group. Only program types and radio format types with an index above 100 (100 = national average) are listed for each group.

Table 6. Heavy Viewing Clusters

Nickname	*Index*	*Fam.*	*Primary age*	*Employment*	*Group code*	*TV preference*
Downtown Dixie Style	146	s/s	18–24	blue collar	U3	soaps
Public Assistance	139	s/s	18–24	blue collar	U3	soaps; music
Tobacco Roads	132	f/s	65+	blue/farm	R2	music; drama
Mines & Mills	130	fam	55–64	blue collar	T3	soaps
Emergent Minorities	130	s/s	18–24	blue/service	U2	music; sports

An index of 100 equals the national average.
* Indicates the most predominant generalized types of television programs watched by each cluster.
(Source: Claritas Corp, 1987, 1988; Michael Weiss, p. 150)

Table 7. Light Viewing Clusters

Nickname	*Index*	*Fam.*	*Primary age*	*Employment*	*Group code*	*TV preference*
Young Suburbia	76	fam	35–44	white collar	S3	sitcom
Two More Rungs	76	f/s	65+	white collar	S2	news; late
Shotguns & Pickups	75	fam	35–44	blue/farm	R1	soaps
Furs & Station Wagons	74	fam	35–44	white collar	S1	news; late
Towns & Gowns	71	sgl	18–24	white collar	T1	late night

An index of 100 equals the national average.
* Indicates the most predominant generalized types of television programs watched by each cluster.
(Source: Claritas Corp 1987, 1988; Michael Weiss p. 150)

Table 8. Top Five Clusters Watching Television between 10:00am and 1:00pm

Cluster nickname	*Index*	*Primary age*	*Secondary age*	*Cluster Group*
Black Enterprise	205	35–44	45–54	U1
Shotguns & Pickups	199	35–44	45–54	R1
Grain Belt	182	65+	55–64	R1
Emergent Minorities	175	18–24	25–34	U2
Back Country Folks	166	65+	35–44	R2

(Source: Claritas Corporation, 1989. Measured against 1988 national average for US households)

Table 9. Bottom Five Clusters Watching Television between 10:00am and 1:00pm

Cluster nickname	*Index*	*Primary age*	*Secondary age*	*Cluster Group*
Urban Gold Coast	8	25–34	65+	U1
Blue Blood Estates	37	45–54	35–44	S1
Furs & Station Wagons	53	35–44	45–54	S1
Single City Blues	56	18–24	25–34	U2
Towns & Gowns	56	18–24	25–34	T1

(Source: Claritas Corporation, 1989. Measured against 1988 national average for US 1988 households)

The Claritas Corporation cautions that the cluster system is dynamic. As populations shift and age and the economy expands and contracts, cities and neighborhoods will expand and contract, resulting in cluster changes.

Television Viewing Patterns of Clusters and Cluster Groups

Each cluster and cluster group has a distinctive set of television viewing preferences and patterns. The affluent cluster of Blue Blood Estates enjoys late night talk shows and news programs while shunning soaps. The Mines & Mills cluster watches a large amount of soap operas but does not watch news oriented programs such as "The Today Show" or "Nightline." Older Americans have very diverse viewing preferences. Those living in wealthy clusters like Gray Power enjoy "Good Morning America," "Nightline," and "NBC Sports World," while those living in poorer clusters such as Golden Ponds watch "Another World," "Super Password," and "Days of Our Lives."[6]

Television viewing preferences of cluster groups can be related to radio format preferences. Table 5 indicates a generalization of television preferences by type and corresponding radio format preferences in descending order.

However, television viewing preferences need to be seen in relation to overall television use. According to Michael Weiss, the heaviest commercial television viewers have their sets on for an average of 35 hours per week and live in predominantly black neighborhoods.[7] The heaviest commercial television viewers are found in the five clusters listed in Table 6. This contrasts significantly with the more affluent clusters who tend to watch the least amount of commercial television, as shown in Table 7.

The increased specificity and discrimination of PRIZM Cluster Analysis allows broadcasters and researchers to get a more precise picture of specific viewing audiences. For example, the top five clusters watching daytime television between 10:00am and 1:00pm Mondays through Fridays are shown in Table 8.

These contrast significantly with the five clusters watching television the least during the 10:00am to 1:00pm period, as shown in Table 9.

Accordingly, the socio-economic nature of the 10:00am–1:00pm television viewing time slot is quite apparent.

Applications of Cluster Analysis

In the words of Robin Page of Page & Associates in Atlanta, the key to the future of audience research is the word "local."[8] Because of the ability of cluster analysis to assess a wide variety of local markets throughout the country, a number of companies and broadcasters are using cluster analysis and other forms of geo-demographic methodologies to assess their markets for a wide variety of purposes.

Cable companies utilize cluster research for three purposes: 1) to identify their target audience for programming purposes, 2) to identify potential subscribers, and 3) to plan expansions of cable services so that capital funding is used to bring cable service to those areas where there will be the highest potential for new subscribers.

The CBN Family Channel utilized cluster research to identify its primary target audience and with this research, changed the name of the cable channel from The Christian Broadcasting Network to The Family Channel. Paul Church, Director of Research for the CBN Family Channel, states that cluster analysis helped identify the 36% target audience for the CBN Family Channel as primarily young, middle-class blue- and white-collar families with children living often just outside suburbs and in the ex-urban fringe. Utilizing Claritas groupings, this primary target audience consists of cluster groups T1, T2, and S3.[9]

Cluster analysis revealed that to a lesser extent, four other groupings watch the CBN Family Channel: Rural America, Southern America, Upscale Suburban America who also watch the Arts and Entertainment Channel, and Inner-city Downtown America. The CBN Family Channel is utilizing cluster research to help in programming the channel to maximize impact on their target audience and to expand the more infrequent viewing of the four other groupings.[10]

Cable companies are also utilizing cluster research to identify potential new subscribers and plan expansions of cable systems in major cities. A cable company with a given amount of capital funds to lay new cable can maximize financial return by using cluster analysis to assess which blocks in a city would have the most potential for new cable subscribers. This is achieved by utilizing the nine-digit zip code which includes block codes. Each block is assessed as if it were a cluster and those block-clusters with a high probability for cable subscription can be readily identified.

Site analysis based on cluster research is also currently being used by a wide variety of major corporations to select new sites for future expansion. A major motion picture chain can use cluster analysis to select not only the sites but size of future theaters. Supermarket, fast-food, and retail corporations utilize PRIZM to select sites for new stores that will have the best probability of success. Again, the nine-digit block code is normally utilized in site research.

Product manufacturers and advertising firms are increasingly utilizing cluster analysis to assist in developing the product profile for their product and the media strategy to successfully advertise the product. The lifestyle research provided by geodemographic cluster analysis assists manufacturers in identifying potential buyers by cluster. If, for example, a wine company wishes to introduce a new quality wine, they might target the "Urban Gold Coast" neighborhoods such as the Upper East and Upper West side of Manhattan which have the highest utilization of specialty wines and the highest number of purchases of classical records and tapes. These neighborhoods also have the highest readership of *New York Magazine*, *The New York Times*, and *Metropolitan Home*, and watch "Nightline" and "Late Night with David Letterman" so an advertising strategy targeting this cluster can also be developed utilizing cluster research.[11]

Local television stations are increasingly using cluster research to better understand and target their DMA. Usually, local broadcasters combine existing cluster groups in their market to create new groupings of six or seven identifiable groups that represent their market. Dallas/Fort Worth station WFAA-TV, for example, has identified six PRIZM Target Cluster Groups for their DMA: *Affluentials*, which comprise 17.6% of Dallas/Fort Worth DMA households; *Greenbelt Families*, comprising 21.4% of DMA households, *Singles & Couples*, comprising 15.8% DMA households, *Blue Collar Country*, comprising 14% DMA households; *Rural Rustics*, comprising 17.9% DMA households; and *Urban Melting Pot*, comprising 13.4% DMA households.[12]

Stations like WFAA-TV can utilize PRIZM to assist in sales promotion as well as programming. By using cluster analysis, the sales department can meet with potential advertisers, provide them with a graphic cluster-group map and supporting materials on their DMA. This material and research then helps to make an effective argument that their station can best target the potential consumers that the advertiser wishes to reach.

Radio stations, unlike local television, have a more precise indication of their audience because they already have a clearly identifiable and assigned footprint. However, radio stations can also utilize PRIZM Cluster Analysis to better understand their audience, as an advertising sales tool, and as a means to fine-tune or change programming. Radio stations utilizing PRIZM Cluster Analysis can provide a service to local firms and advertisers by assisting them to identify and understand the clusters and cluster groups most directly affected by their product or services.

Perhaps the most sophisticated broadcast application of the PRIZM Cluster Analysis is now being carried out by Mediacomp in Houston. This media buying firm was conceived in the mid-seventies by Sylvan L. Brown and has evolved into the leading media buying firm utilizing PRIZM. Mediacomp provides a fully integrated system that promises at least a 25% increase in advertising revenue by reaching more of an advertiser's target audience per media dollar.[13]

The Mediacomp System first establishes a target profile for each product based on relative marketing value of each of 16 age and sex categories. Then, Mediacomp adds to this a qualitative psychographic description of the target consumer. The system then identifies appropriate segments of the PRIZM 40 Cluster System for the product, target consumer, and desired multivariate target audience of every television program within each selected market in such a way as to optimize reach and frequency within the overall budget. Mediacomp's planning module also determines the optimal media-mix of television, radio and newspaper and provides clients with the percent of target audience obtainable at any cost level.[14]

Essentially, the Mediacomp System operates at four successive levels: 1) the planning level, 2) the buying level, 3) a traffic and affidavit level, and 4) a post analysis level. Once the media buy plan is approved, the buy level executes all buy contracts and can send them via Fax to all appropriate media. The traffic and affidavit module handles all accounting and paperwork associated with the buy. The post-analysis module provides the client with an assessment of the effectiveness of the buy.

According to Jerry Storseth of Mediacomp, the promised 25% increase in effectiveness of a Mediacomp buy is actually very conservative. He states that averages range above 33% in increased effectiveness per media dollar spent.[15]

Broadcast Audiences in the 1990s

Technological, economic, social and demographic factors will change broadcast audiences and preferences in the 1990s. During the baby boom following World War II, almost one-third of the current U.S. population was born. The effects of the aging of this generation and its lower birthrate will impact all sectors of American life. The number of Americans over fifty will increase, the number of women living alone will increase, and the young adult population will decline significantly. Changes of this magnitude will affect the lifestyles and program preferences of broadcast audiences across America. Enhanced communications technologies including advances in computer and facsimile (FAX) technologies will allow more individuals to work away from urban centers.

Technological advances in broadcasting will not only include High Definition Television (HDTV) but, perhaps as importantly, pay-per-view. According to a recently study by Business Communications Company (BCC), Inc. in Connecticut, pay-per-view is expected to boost cable revenues from $11.2 billion to $28.billion in 1995. Pay-per-view, according to BCC, will be the fastest growing sector of the cable industry, growing from less than 1% of cable revenues in 1987 to 13.3% of cable revenues in 1995. BCC reports that this growth will come at the expense of home videocassette rentals and sales, where total share is expected to drop 12 points by 1995.[16] These changes will primarily affect young, affluent, urban, suburban and town clusters and have little impact on most rural clusters.

Regardless of the changes ahead, it seems certain that the utilization of cluster analysis to pin-point audiences and consumers will not only increase but achieve greater sophistication. Television stations and cable companies that once had perhaps one person doing research will need to increase staffing to remain competitive. Firms like Claritas Corporation will find substantial competition from other geo-demographic research companies such as Donnelley Marketing Information Services which is owned by the Dun & Bradstreet Corporation. In short, competition among retaillers, manufacturers, publishers, service companies, and advertisers, as well as broadcasters, will usher in a new era of market research where successive generations of integrated geo-demographic cluster analysis, for good or ill, will help shape as well as reflect future trends.

References

1. Nielsen Media Research, Nielsen Report on Television, November, 1987.
2. Nielsen Media Research, November, 1987.
3. For an excellent discussion of the Claritas PRIZM System, see Michael J. Weiss, *The Clustering of America.* New York: Tilden Press/Harper & Row, 1988.
4. Claritas Corporation, *How to Use PRIZM.* Alexandria, VA: Claritas Corporation, 1986. pp. 2–3.
5. *Ibid.*
6. Weiss, pp. 121, 308, 345.
7. Weiss, p. 150.
8. Statement by Robin Page, President, Page & Associates, in a telephone interview, April 4, 1989.
9. Statement by Paul Church, Director of Research, CBN Family Channel, in a telephone interview, July 25, 1989.
10. *Ibid.*
11. Claritas Corporation, *How to Use PRIZM*, p. 6 and Weiss, pp 279-281.
12. WFAA-TV, "The Dallas/Fort Worth Lifestyle Locater: Advertisers' Guide to Market Segmentation" (WFAA-TV, 1984).
13. Based on personal correspondence and telephone interviews with Jerry Storseth of Mediacomp, April through August, 1989.
14. *Ibid.*
15. *Ibid.*
16. Paul Sweeting (ed.) "Stiff Competition: PPV to Boost Cable at Expense of PRCS" in *Video Software Dealer,* July, 1989 pp.36–37.

Bibliography

Claritas Corporation, *How to Use PRIZM.* Alexandria, VA: Claritas Corporation, 1986.

Claritas Corporation, various reports and materials, 1985, 1986, 1987, 1989.

Church, Paul, Director of Research, CBN Family Channel; telephone interview, July 25, 1989.

Nielsen Media Research, Nielsen Report on Television, November 1987; Neilsen "Average Minute Audiences," November, 1987; and Nielsen NAD Report, 1987.

Page, Robin, President, Page & Associates; telephone interview, April 4, 1989.

Storseth, Jerry, Mediacomp; personal correspondence and telephone interviews, April through August, 1989.

Sweeting, Paul (ed.) "Stiff Competition: PPV to Boost Cable at Expense of PRCS" in *Video Software Dealer,* July, 1989. pp. 36–37.

Video Store Retailer Survey, in *Video Store Magazine.* August, 1989.

Weiss, Michael J. *The Clustering of America.* New York, Tilden Press/Harper & Row, 1988.

Television

The 1988–1989 TV season gave birth to the term "trash television," and Geraldo Rivera and Morton Downey, Jr. emerged as the two most identifiable personalities of "shock" talk shows. However, "talk trash" was not alone: other syndicated programming was equally affected. Shows like "A Current Affair" and "Inside Edition" became popular by adding the dimension of tabloid reporting to the traditional TV magazine format.

In general, broadcasters continued to compete heavily with cable programming, but the year's prime-time season seemed somewhat lacking in hit shows. "Roseanne" was seen as the most successful new program, and "Unsolved Mysteries" and "Midnight Caller" were also credited with tremendous promise. Meanwhile, advertising rates on both cable and broadcast TV climbed even higher than before, with daytime programming reflecting the largest capital gains.

Network promotion took on a new dimension in 1989, with combined, corporate campaigns involving discount department stores and restaurants. However, the success of these multichain efforts was yet to be determined by the premiere date of the 1989–1990 fall season.

Finally, new studies were released on televised violence and minority representation on the screen—both studies confirmed that TV needed much improvement in both areas.

More Mags Will Fly In Fall; Too Much of a Bad Thing?

VARIETY 4/12/89

By JOHN DEMPSEY

New York A soaring "Current Affair" and a revitalized "Entertainment Tonight" are the big stories amid the glut of tv-magazine series elbowing one other for time periods in firstrun syndication.

"A Current Affair," from 20th Fox TV Syndication, "is head and shoulders above the other tabloid magazine shows," says Mike Levinton, v.p. of programming for the Blair Television rep firm. "It was there first, and the other ones are just carbon copies."

John von Soosten, v.p. of programming for Katz Communications, says " 'A Current Affair' has written the book on the reality-based magazine format. It's the one against which all the others are measured."

Paramount's veteran "Entertainment Tonight" strip "is stronger than it was a year or two ago," says Levinton. "It's doing more features, and its pieces are more gossipy, more eyebrow raising than before. The beauty part for Paramount is that it's been able to gradually shift emphasis year to year as mores change, and this season the changes are right on target."

According to von Soosten, "Paramount has pumped new life into 'Entertainment Tonight.' The atmosphere of the show is brighter, snappier, breezier. It's now oriented more to the common man than to the entertainment-business insider."

America's 'Affair'

"A Current Affair" scored big numbers in the February Nielsen sweeps, harvesting solid (in some cases double-digit) percentage increases in rating and share and in the key adult demos compared with the last three sweep books (February, May and November 1988) in prime access. The access increases were even greater when Nielsen compared "A Current Affair" with the rating and share and demos of the lead-in strips.

One rep source familiar with the figures, who requested anonymity, says if the seven Fox stations that strip "A Current Affair" ponied up license fees commensurate with the rating the show delivers, Fox would rake in $500,000 a week. Fox also pockets the additional $350,000 a week that "Affair" generates from the two national 30-second spots in each half-hour, giving the company a total gross of $850,000 a week. That's a staggering figure when balanced against the production cost of the show, which one insider says comes to about $260,000 a week.

But that $260,000-a-week figure will balloon because Fox is negotiating with Maury Povich, the host, whose contract expires in the spring. Syndication sources say Povich makes about $450,000 a year, a number that should easily soar over the $1-million-a-year mark, particularly since Paramount has sounded him out about jumping to host its "Tabloid" magazine strip. That show makes its debut in September.

The production budget of the newer shows in that genre is probably in the $200,000-a-week range. Even though they're not doing nearly as well in the ratings as "Affair," GTG, which produces and distributes "USA Today On TV," and King World, the producer/distributor of "Inside Edition," have announced that their shows will be back for 1989-90.

Michael King, president-CEO of King World, says the way renewals are going so far, station license fees alone will cover the $200,000-a-week production nut on "Inside Edition." That would allow the company to make its profit on the two 30-second national-barter spots in each half-hour.

And "we're looking to break even next season" on "USA Today On TV," says Bud Rukeyser, senior v.p. of GTG Entertainment. "Year three is when we'd start the process of making a great deal of money."

Severest revenue problems for "USA Today" and "Inside Edition" are that they are shut out of good time periods in New York, Los Angeles and Chicago.

"USA Today" has never had a good time period in New York (it now runs at 5:30 a.m. on WNBC-TV), and it has lost access time period on WMAQ-TV Chicago to "Family Feud." Rukeyser gets infuriated when he thinks about Los Angeles, where KNBC-TV, which is getting respectable numbers with it at 7:30 p.m., will move it to a weaker time period in September to make way for Paramount's new "Tabloid" strip. Paramount has yet to sign either a host or an executive producer for "Tabloid." But according to Rukeyser, Paramount was able to package the guaranteed 7:30 p.m. access time period for the unknown quantity

Update on the tv mags currently airing nationally

Mary Hart

ENTERTAINMENT TONIGHT (Paramount TV)
The grandaddy of all of the magazine strips ... now in its eighth year and certain of renewal for a ninth ... revitalized this season with a brighter look, a faster pace and more provocative stories ... Feb. '89 numbers are up over the previous year's in all of the key demographic categories.

Maury Povich

A CURRENT AFFAIR (20th Century Fox TV)
Considered the slickest, best edited and wittiest of the tabloid-magazine shows ... host Maury Povich has become a major celebrity ... its February-sweep ratings were stronger than those of any other syndicated strip.

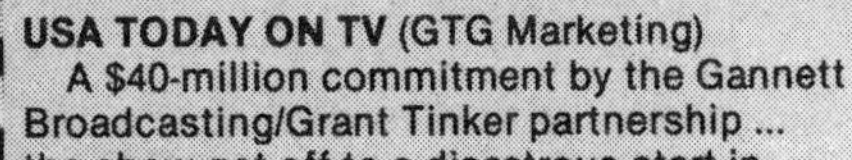
Bill Macatee

USA TODAY ON TV (GTG Marketing)
A $40-million commitment by the Gannett Broadcasting/Grant Tinker partnership ... the show got off to a disastrous start in September, in both ratings and content ... new exec producer Tom Kirby (the third, after Steve Friedman and Jim Bellows) still fine-tuning it.

Bill O'Reilly

INSIDE EDITION (King World)
Tried to buck the odds by premiering in January ... David Frost, the name host, fired after four weeks ... with frequent sex-and-violence-saturated pieces, the show is perceived as too much of a clone of "A Current Affair."

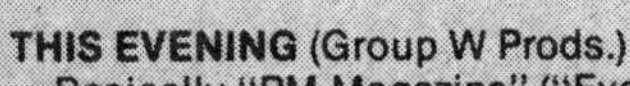
Nancy Glass

THIS EVENING (Group W Prods.)
Basically "PM Magazine" ("Evening" on the five Group W-owned stations) repackaged with a national host (Nancy Glass) ... got off to a bad start in the ratings at 7 p.m. on WCBS-TV New York ... cleared on only a handful of stations.

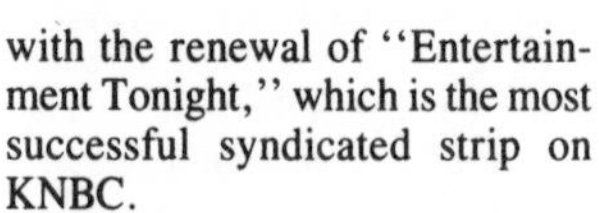

with the renewal of "Entertainment Tonight," which is the most successful syndicated strip on KNBC.

John Rohrbeck, v.p.-g.m. of KNBC-TV, says he made the commitment to "Tabloid" for 7:30 p.m. as part of a deal concluded before he placed "USA Today" in the time period. He adds that he's renewed "USA Today" for latenight slotting beginning September and that if "Tabloid" doesn't work in the ratings, he'll move "USA Today" back to 7:30.

WNBC-TV New York and WMAQ-TV Chicago have decided to cancel "Inside Edition" as of September. And King World will lose its 7:30 p.m. time period on KCBS-TV come September because KCBS has bought the rights to King World's powerhouse gameshows "Wheel Of Fortune" and "Jeopardy" for double-access slotting.

King World and GTG officials acknowledge they may have to pay compensation to tv stations in one or more of the top three markets to get "Inside Edition" and "USA Today" placed in decent time periods. Sources say it could cost King World or GTG as much as $60,000 a week out of pocket to get even an independent station in New York to run "Inside" or "USA" in access or at, say, 11 p.m.

Group W Prods. is not making many deals for "This Evening" because "the station community is very tentative right now," says Derk Zimmerman, president of Group W Prods. "There's so much turmoil in the marketplace, and stations have made so many changes since the start of the season that a lot of them are saying, 'Let's wait for another rating book.'

"I think there's an opportunity in the marketplace for a user-friendly magazine show, but we've got a lot of work to do."

Group W sources acknowledge "This Evening" is being hurt by the perception that "Inside Edition" and "USA Today" look as though they will not be folding their tents. The hope of Group W's salesmen when they started sounding out stations three months ago for "This Evening" was that the February-sweep books from Nielsen and Arbitron would put either "Inside Edition" or "USA Today" or both out of business and open us some good time periods. But as Blair's Levinton puts it, "The syndication marketplace just can't support so many magazine series. I don't hold out much hope for 'This Evening.' "

Another anchor?

Katz' von Soosten says Group W's producers might keep "This Evening" around if "they drop the single-anchor concept (Nancy Glass is the sole anchor) and hire a second person so the two can play off each other. The single anchor makes it too similar to 'Current Affair' and 'Inside Edition.' "

The strip that could be damaged most by the glut of magazine shows in syndication is "Inside Story," the WSVN-TV Miami-produced series to which MCA TV has picked up the rights. "The time periods just aren't there," says Levinton.

But Shelly Schwab, president of MCA TV, points to the strong ratings of "Inside Story" on WSVN. "Nobody is breaking down doors to renew 'USA Today' or 'Inside Edition,' " Schwab adds.

The consensus emerging among syndication experts is that in one form or another, all seven of the magazine strips will be on the air in September. But when Nielsen and Arbitron have compiled the numbers for the November 1989 sweeps, only "Entertainment Tonight" and "A Current Affair" are expected to be getting the kinds of ratings and demos that guarantee healthy profits. It is expected that a third show (identity still to be determined) will do well enough to soldier on for the rest of the '89-90 season.

If Talk is Cheap, Big Apple Must Be Bargain Heaven

VARIETY 4/19/89

By RALPH TYLER

New York More than 400 guests in an average week have their say on television talkshows or hybrid talk-and-news shows emanating from New York, turning the tv antenna into a Tower Of Babble.

There are 19 Monday-Friday tv talkshows produced in New York, not counting those on cable. Is that a case of overkill? Tribune Entertainment, which is adding Joan Rivers to the cacophony in the fall, obviously doesn't think so.

"Whatever you do, there's already too many," says Sheldon Cooper, the company's prez. "But the marketplace always needs a hit." The syndicated daytime show will be on WCBS-TV, taped at that station's studio.

New Yorkers can crawl out of bed to any of four wakeup shows 7-9 a.m. Three are network offerings: NBC's veteran "Today" vying for frontrunner with ABC's "Good Morning America," and "CBS This Morning" playing catchup in an attempt to make it a genuine 3-way race.

Nationally, "Today" placed 5.0, "Good Morning America" 4.5 and "This Morning" 2.3 in February ratings. In Gotham, "Today" and "Good Morning America" both had 3.5 ratings, while CBS scored 2.3.

Across from them is "Good Day New York" on WNYW, as downhome as its name implies, featuring frequent live remotes from around town and tapes from the previous night's events. It had a 1.5 local rating in February.

Early bird's delight

For the eager executive, there's an even earlier talkshow, CBS' "This Morning's Business," which quizzes figures in business and finance starting at 6 a.m. Monday-Friday. WPIX has two 6 a.m. once-a-week talkshows for early-risers, "Open Mind" on Wednesdays and "Weekend" on Saturdays.

Marty Ryan, exec producer of "Today," credits coanchors Bryant Gumbel and Jane Pauley, rather than the show's longevity, for its success. "They obviously do the best interviews, so publicists get us the best guests," he says.

The show has been on for 37 years — Pauley with it for 10, Gumbel for eight and Ryan for nine. The biggest change over the years, he says, is the impact of satellite technology. "Not everyone has to come to New York. They can be in London or Rome and still be on the program."

He also says there's been an increased emphasis on science and medicine, particularly as they apply to the consumer. " 'Today' has always done that, but it's become our staple, our bread and butter."

Doing the program live, he remarks, is "a walk on the wild side — a guest oversleeps, the car doesn't show up, he gets stuck in traffic. You have to be prepared to move things around. You don't have inventory, because you can't double-book."

The program is geared to an audience that is "up and moving around," per Ryan, meaning "Today" is as much listened to as watched. The producers also are aware that women viewers predominate in the second hour, and schedule their guests accordingly.

Greg Kasparian, director of network research at CBS-TV, has stats on the female-male ratio of the three web wakeups. The audience for the NBC and CBS programs is 54% women and 36% men, he says, with 10% non-adult. ABC's show is more heavily female, 61% to 33% male.

Skewing older

Women watching the CBS effort are "slightly skewed toward the older age segment," Kasparian notes. Females over 50 make up 55% of the "CBS This Morning," female audience, and males past the half-century mark make up an even higher proportion of 65%. For "Today," the ratio of over 50s among females is 53% and males 51%. Of the audience for "Good Morning America," 49% of the women and 56% of the men have reached their sixth decade.

Kasparian also notes the NBC and ABC wakeups have a slightly more upscale audience than CBS'.

To John Goodman, a senior producer at "CBS This Morning," which bowed Thanksgiving 1987, what counts is, "This program is an equal partner with the other two. For a lot of years, the CBS program was not on a par. It is now — not ratings-wise but certainly programming-wise. The competition for guests is keener than ever. We're not a threat in ratings, but we are when it comes to bookings."

He says the goal of David Corvo, exec producer of the show, is "one, make us competitive, and two, distinguish ourselves from the other shows." To the latter end, there is a segment every Friday in which the homes of leading business and entertainment figures are visited. Another way the show seeks to set itself apart is having director Peter Bogdanovich report every Thursday on movies, particularly classics, available on videocassette.

Head to head

Jack Riley, exec producer of "Good Morning America," points out that the 7-9 a.m. slot is the only time the three networks go head to head with such similar shows. "Traditionally, there's been room for all three of us," he says. "There are enough numbers, although there's a smaller audience available than at other dayparts. Some 15-20% of tv homes watch morning tv.

"The biggest problem facing us is trying to keep our share of the audience. The early audience tends to be people getting ready for work. They want weather, the top stories, and then they're on their way. People will say, for instance, 'I tune in at 7:20 every day.' People watch it conditioned by their own needs.

"The audience in the first hour is generally information-oriented and younger," Riley continues. "The change is not slam-bang sudden, but people at home watching from 8 to 9 skew a little older toward the latter part of the show, and men decline."

Segments are kept short, he says, because of commercial interruptions and also because the attention span of people getting ready for work is shorter. "They don't have the luxury of sitting around."

The program averages about 1½ celebrity interviews a show, done live or on tape. "That's something everyone else does," Riley says, "so we try to do something different. We go out behind the scenes months before a movie is finished and they're out flacking it."

He says the importance of celebrity interviews has declined because "everyone's doing it. Where 10 years ago, if we got an interview with James Garner — he didn't do many tv interviews — it was an exclusive. Now an actor may be on 'Entertainment Tonight,' 'Donahue,' one of the morning shows or 'Live At Five.' His options nationally are at least three dozen. It isn't as important to us."

Tabloid talkies

As for the recently heated-up tabloid-type talkshows, Riley says: "We've done some of those subjects over the years, but we handled it differently. The audience in the early morning is not looking for that kind of programming, that kind of visibility of the host or hostess, like Geraldo or Sally Jessy Raphael. That works better later in the day. That's directed at people who want to sit around.

"(Tabloid shows) usually take the same subject and put a new twist to it, like 'Cousins Who Sleep Together.' All three network morning shows have resisted any temptation to fall into that trap. Tabloid tv has a history of going up very fast, down very fast. They run out of subjects. People tire of the intensity. They're going to wear out their welcome very fast.

"When you get up in the morning, it's bad enough out there. The news is going to be bad enough. We're not Boy Scouts, but we want to help you into your day."

Two talkshows square off at 9 a.m. — "Geraldo" on WNBC and "Live With Regis & Kathie Lee" on WABC. The first represents the currently popular "confrontational" show, and the other the old-style "benign" program. February Nielsens give "Regis & Kathie Lee" a 6.1 rating and 25 share in New York, while "Geraldo" lags with a 3.84 rating and 16 share.

Per Jack Fentress, v.p. and director of programming for tv rep firm Petry National, the change to controversial began about five years ago when "Oprah came on the scene and hit the stage running. She was very hip, knew what could get an audience going and picked topics out of the normal stream of tv subjects."

(In New York, the Chicago-based "Oprah" and Gotham-based "Donahue" are opposite one another at 4 p.m. "Oprah" was enthroned in February Nielsens for New York with a 10.9 rating and 26 share, and "Donahue" had a 5.8 rating and 14 share. "Sally Jessy Raphaël," batting in the same league with many of the same kind of guests and topics, had a 4.9 rating and 20 share in the Big Apple, where she airs on WABC at 10 a.m.)

Oprah ahead

The Nielsen Syndication Service weekly national ranking for the period ending Feb. 26 listed "Oprah" in third place among top syndicated shows, with "Donahue" ninth and "Geraldo" 11th.

Delia Fine, senior producer of "Geraldo," says her host, with his news background and experience on "20/20," "in many ways takes a slightly more aggressive approach to a story" than either Donahue or Winfrey. That makes "Geraldo" almost the polar opposite of "Regis & Kathie Lee," which maintains an air of lighthearted domesticity; many viewers assume wrongly that the cohosts, who begin each program with a chat, are husband and wife.

When "Regis & Kathie Lee" went into national syndication in September, "we tried to keep the formula and provide the same kind of homey atmosphere," says senior producer Michael Gelman. "We continue to have chefs on our show, but now they may come from Seattle or San Francisco as well as New York."

The program is 95% live, which poses the usual anxieties. "Once we had a savant (what used to be called an idiot savant) on the show who had this incredible ability to play the piano," Gelman recalls. "Toward the end of the show, but not at the end, we discovered he didn't know how to stop. He finally completed the piece, but we had to rearrange the show."

The senior producer of "Donahue," Lorri Antosz Benson, says the veteran host (the show's been around since 1967) was the first to give the studio audience an opportunity to talk. "Today," she adds, "there are 10 shows like ours."

Asked how "Donahue" differs from the others, she points out he hosted "a panel of 15 or 16 top people in literature to do a reading of Salman Rushdie's book on Feb. 24, right in the middle of the ratings period. The others would probably think that was too political, and it wouldn't get big ratings. But he wanted to do it."

Benson has had her share of mishaps connected with the show, including the time she booked guests for the topic "People Who Stay Together But Are Not In Love." Watching the show at home, she spotted one of her supposedly alienated couples holding hands.

Sally first

When Sally Jessy Raphael started on tv Oct. 17, 1983, it was the first such nationally syndicated show to have a woman host, according to exec producer Bert Dubrow. "Here was Phil Donahue doing all these women's topics," he says. "It just never made sense to me."

He doesn't accept the "tabloid

tv" tag, and defines the show as "very personal, dealing with people's emotions, which I feel Sally does best. We have more closeups than the others. People are really expressing how they feel emotionally. The person out there who feels she's alone can say, 'I'm not.' "

Dubrow concedes some errors have been made during the show's five years, including one occasion that got a lot of press attention when one of Raphael's guests turned out to be a necrophiliac. "That was a mistake which, thank goodness, we could correct." The show never ran.

Viewer male

Some 35-45% of the topics discussed on the show come from viewer mail, with a letter-writer frequently invited to guest. Raphael also has an evening radio call in advice show Monday-Friday that helps give her an ear to what's on people's minds.

Although Raphael's show has many of the same guests and topics as the others, "We don't go out of our way to get guests yelling and screaming," Dubrow points out. "She's got her own barometer. Her show is much more quiet, more comfortable. She's been married over 20 years and has several children. Her frame of reference comes from that."

Three local tv talkshows round out the daytime list: "Best Talk In Town" at 9:30 a.m. on WPIX, "People Are Talking" a half-hour later on WWOR and "Live At Five" at 5 p.m. on WNBC.

In the evening, three WNET programs feature guests, although it's probably stretching it to call them talkshows. They are "Nightly Business Report" at 6:30, "Bill Moyers' World Of Ideas" at 7:30 and "MacNeil/Lehrer Newshour" at 8.

At 11 p.m., with "The Eleventh Hour," WNET really gets into the talkshow act. The program was launched Jan. 9. Executive producer Steven Weinstock says the station was seeking "a night presence in its hometown, which it hadn't had since 'The 51st State' in the early '70s. It was felt that the local news at 11 o'clock — made up essentially of rape, pillage and fire — was not the best of the local news broadcasts.

"A great number of people," he continues, "just want to know: Are we still in one piece? Do we need to wear a pair of galoshes tomorrow? Are the Knicks going to lose? We thought we'd give them something more substantial to go to bed with."

Per Weinstock, the aim is "to take all the disparate information and focus on one story. We are not looking to report the news but to put it in context, go a little deeper."

Gift of gab

Gift of gab still counts for something. "We have to pay some respect to the medium," Weinstock says. "Some people are terrific minds and terrible television. It's a job of casting the right people. You're looking for great talkers."

Heat more than light seems to be the aim of "Morton Downey Jr." at 11:30 on WWOR, if the show's critics are to be believed. The program has a 3 rating and a 10 share in New York and a tendency to grab topics by the jugular — a show for viewers who want their evenings to end not with a wimper but a bang. Downey stalks among his guests in shirtsleeves, a cigaret wedged between his fingers, acting the ringmaster to the antagonists his show attracts.

Just for laughs

"David Letterman," which starts an hour later on WNBC, is out for laughs, according to producer Robert Morton. "We structure the interviews with that in mind," he points out. "You wouldn't see too many guests talking about serious things."

It directly follows the California-based "Tonight Show With Johnny Carson," which it resembles, says Morton, except that Letterman "looks for guests who appeal to a younger demographic group." He adds the two shows have friendly relations and exchange information so they won't end up featuring the same guests. This is helped by their being based on opposite coasts, as an actor on promotional tour tends to make multiple appearances in one region before going on to another.

"Letterman" is followed on WNBC at 1:30 a.m. by "Later," hosted by Bob Costas, who is a sports reporter but has guests from all walks on his one-to-one show. Opposite him on WWOR is "Joe Franklin," which leans toward showbiz guests. Franklin has been on for 37 years, and he sometimes reruns tapes of some of his older programs to catch the attention of nostalgia buffs. He also hosts a radio show.

Franklin signs off at 2:30 a.m., giving the talk-addicted the chance for a brief sleep before "This Morning's Business" at daybreak.

News Faces of '88: Who Had Most Airtime

Variety 6/28/89

Top newswoman Rita Braver

Top newsman Bruce Morton

	Correspondent	Network	'87 Rank	Appearances '87	Appearances '88	% Change
1	Bruce Morton	CBS	9	69	146	+111%
2	Dean Reynolds	ABC	83	49	114	+132%
3	Bill Plante	CBS	3	155	111	-28%
4	Sam Donaldson	ABC	2	147	110	-25%
5	Bob Schieffer	CBS	41	56	107	+91%
6	Robert Hager	NBC	5	76	100	+31%
7	Brit Hume	ABC	8	79	100	+26%
8	Chris Wallace	NBC*	1	120	99	-17%
9	John McWethy	ABC	4	107	97	-9%
10	David Martin	CBS	6	95	96	+1%
11	Jim Wooten	ABC	66	55	94	+70%
12	Rita Braver	CBS	24	73	93	+27%
13	Lisa Myers	NBC	92	26	85	+226%
14	Martin Fletcher	NBC	65	45	82	+82%
15	Bob Simon	CBS	33	51	79	+54%

Source: J. Foote with Cary O'Dell, U. of Southern Illinois/Vanderbilt Archives

By VERNE GAY

New York Bruce Morton, CBS' senior political correspondent, was the most visible correspondent on network tv in 1988, according to recently compiled data.

Not only did the longtime CBS newsman appear at the top of a list of network tv's 50 reporters in terms of on-air exposure, but he also increased his appearances over 1987 by 111%, due mainly to his coverage of the presidential elections.

The list, based on Vanderbilt Archives data and compiled by Professor Joe Foote, chairman of the radio/tv department at U. of Southern Illinois, shows who the most utilized reporters on network tv were last year.

If you were a top Washington political correspondent in '88, chances are you made dramatically more appearances on the nightly newscasts than you had in at least the three previous years because of the election (Foote's information covers only individual appearances on nightly newscasts).

But there were other big stories, and a handful of correspondents benefited: ABC's Dean Reynolds, who was ranked 83d out of all network news correspondents in '87, jumped into second place last year. Reynolds was ABC's point man on the turmoil in the Mideast, and his network devoted more time to that story (an estimated 218 minutes, according to researcher Andrew Tyndall) than any other network.

The information is intriguing for other reasons. The networks, always in search of the next big star, found time to devote their precious minutes to a handful of up-and-comers.

Some, like NBC's Lisa Myers, were lucky to land a big story, like the elections. She covers the war on drugs this year, and her appearances are down substantially from last year. "1988 was a terrific year for me," says Myers, who joined the network in '81.

Other were dogged: Robert Hager, NBC's most visible reporter, boosted his appearances substantially last year because of his coverage of Pan Am Flight 101. A versatile reporter who covers a wide range of subjects, he specializes in aviation and regulatory agencies. Hager took two months in '87 to work on a documentary, lowering his '87 total.

Judd rising

ABC's Jackie Judd, one of that net's rising stars, took a page from Hager's book. She worked the political beat last year in D.C. and is currently handling the uprising in China.

Other reporters gained airtime because of national catastrophies. The Yellowstone fire ignited the appearances of CBS' Denver-based Bob McNamara (50). CBS' Miami-based Juan Vasquez (52) tangled with Hurricane Gilbert, the riots and other assorted problems.

There were curious turns of fate, as revealed by the data. Sam Donaldson, who over the past six years ranked as network's second-most visible correspondent, saw his number of pieces cut back last year.

A possible reason? All White House coverage appeared to wane during the election, and Donaldson, who covered Michael Dukakis, was one of several reporters on the campaign trail, says Foote.

New execs stress that some of the year-to-year differences in appearances are irrelevant. Some reporters, for example, find a dearth or plethora of news on their beats in a given year. Others change their beats. Other reporters, especially the investigative ones, take months on certain pieces. They showup least often on the nightly news. "With Brian Ross, who takes maybe two or three months before he hits with a big investigative report, his impact is great, even if his face doesn't appear every night," says NBC "Nightly News" senior producer Cheryl Gould.

Foote's numbers reveal some developing trends in network news:

- Last year, only seven women appeared on the list of the networks' 50 top correspondents. Of these, CBS' Rita Braver (93 appearances) was the topper. The others were Myers of NBC (85), CBS' Lesley Stahl (70), NBC's Andrea Mitchell (68), CBS' Susan Spencer (60), ABC's Judd (50) and CBS' Deborah Potter (50). Women had a significantly better showing in '88, but the numbers were still low. Only CBS' Braver, Stahl and Spencer have remained in the top 50 bracket over the past six years.
- 1988 was a year in which the rich got richer: Some of the nets' top correspondents generally got greater exposure as the news divisions boosted coverage of the political races. But not everyone gained — in some cases simply because the stories they covered cooled off. Correspondents with 50 or more appearances in 1987 who didn't make the '88 top 50 club: Mike Jensen (NBC), 38 appearances, down 38%; Doug Tunnel (CBS), 37, −30%; Frank Currier (CBS), 37, −33%; Mike Lee (ABC), 38, −30%; Carl Stern (NBC) 43, −33%; Robin Lloyd (NBC), 40, −25%; Allen Pizzey (CBS), 21, −60%, and Phil Jones (CBS), 41, −59%.
- Israel, a massive story in '88, gave great visibility to some, especially ABC's Reynolds, CBS' Bob Simon (79) and NBC's Martin Fletcher (82). China could do the same for some correspondents this year.
- Reports out of the capital continue to dominate the nightly news shows overwhelmingly. "The networks say they've backed off of Washington coverage, but the data say otherwise," per Foote.

FASTEST RISERS IN '88

Rank	Correspondent/beat	Network	Appearances '87	Appearances '88	% over '87
1	LISA MYERS (diplomatic)	NBC	26	85	+226%
2	DEBORAH POTTER (congressional)	CBS	16	50	+212%
3	JACKIE JUDD (general assignment)	ABC	20	50	+150%
4	DEAN REYNOLDS (Israel)	ABC	49	114	+132%
5	JUAN VASQUEZ (Central America)	CBS	24	52	+116%
6	BRUCE MORTON (chief political)	CBS	69	146	+111%
7	BOB SCHIEFFER (political)	CBS	56	107	+91%
8	MARTIN FLETCHER (Middle East)	ABC	45	82	+82%
9	JOHN COCHRAN (White House)	NBC	39	70	+79%
10	BOB McNAMARA (Denver)	CBS	28	50	+78%

Source: J. Foote with Cary O'Dell, U. of Southern Illinois/Vanderbilt Archives

TOP NETWORK CORRESPONDENTS

ABC

Rank	Correspondent/beat	Appearances	% vs. '87
1	DEAN REYNOLDS (Middle East)	114	+132%
2	SAM DONALDSON (chief White House)	110	-25%
3	BRIT HUME (Senate)	100	+26%
4	JOHN McWETHY (National Security)	97	-9%
5	JIM WOOTEN (Congress)	94	+70%
6	RAY GANDOLF (sports)	70	+204%
7	GARY SHEPHARD (Los Angeles)	68	+58%
8	BOB ZELNICK (Pentagon)	65	-22%
9	DENNIS TROUTE (general assignment)	62	+19%
10	BARRIE DUNSMORE (London)	60	+71%
11	WALTER RODGERS (Justice Dept.)	59	-13%
12	JOHN QUINONES (general assignment)	54	+14%
13	JACKIE JUDD (general assignment)	50	+150%
	BARRY SERAFIN (national)	50	+25%

CBS

Rank	Correspondent/beat	Appearances	% vs. '87
1	BRUCE MORTON (chief political)	146	+111%
2	BILL PLANTE (State Dept.)	111	-28%
3	BOB SCHIEFFER (chief Washington)	107	+91%
4	DAVID MARTIN (Pentagon)	96	+1%
5	RITA BRAVER (law)	93	+27%
6	BOB SIMON (Tel Aviv)	79	+54%
7	ERIC ENBERG (general assignment)	72	+44%
8	LESLEY STAHL (White House)	70	+59%
9	BARRY PETERSON (Moscow)	65	+66%
10	RAY BRADY (business)	64	-8%
11	WYATT ANDREWS (White House)	60	+25%
	SUSAN SPENCER (national)	60	0%
13	TOM FENTON (London)	59	+43%
14	TERRENCE SMITH (Washington)	57	+29%
15	RICHARD SCHLESINGER (New York)	54	+68%
	RICHARD ROTH (New York)	54	+22%
17	BRUCE HALL (Atlanta)	53	+12%
18	JUAN VASQUEZ (Miami)	52	+116%
19	DAVID DOW (Los Angeles)	51	+10%
20	BOB McNAMARA (Denver)	50	+78%
	DEBORAH POTTER (Congress)	50	+212%

NBC

Rank	Correspondent/beat	Appearances	% vs. '87
1	ROBERT HAGER (general assignment)	100	+31%
2	CHRIS WALLACE (White House)*	99	-17%
3	LISA MYERS (diplomatic)	85	+226%
4	MARTIN FLETCHER (Middle East)	82	+82%
5	JOHN COCHRAN (chief White House)	70	+79%
6	ANDREA MITCHELL (Congress)	68	+36%
7	GEORGE LEWIS (general assignment)	66	+53%
8	BOB KUR (national)	61	+64%
9	JOHN DANCY (diplomatic)	60	-36%
10	FRED FRANCIS (Pentagon)	58	+26%
11	DENNIS MURPHY (general assignment)	57	+67%
12	KEN BODE (national politics)**	54	+31%
13	JIM MIKLASZEWSKI (White House)	52	+26%
14	ROGER O'NEIL (general assignment)	51	+34%
	ED RABEL (general assignment)	51	+75%
16	DON OLIVER (general assignment)	50	+42%

*Now with ABC **Left network

Source: J. Foote with Cary O'Dell, U. of Southern Illinois/Vanderbilt Archives

Cable Bashing is the Major Activity at NAB Confab; Local Cable Sales a Threat

VARIETY 5/3/89

By BRIAN LOWRY

Las Vegas Despite the buzz surrounding High-D tv and threats from the telcos, cable appeared to remain public enemy No. 1 at the National Assn. of Broadcasters convention.

That was clearly the theme of a joint NAB-Television Bureau of Advertising (TVB) presentation, "Hooray For Television," and a Blair Television panel. Former outlined a study praising free tv vis-à-vis cable; latter recapped a "Television 1995" study by Wilkofsky Grue Associates.

Some in attendance took exception, saying TVB's cable bashing is too parochial. They asserted that resources would be put to better use by focusing on longterm strategies to expand the entire tv advertising pie.

TVB prez William Moll did cite the necessity for strategic positioning of television, saying the coming years will be "the most marketing-intensive era" in the history of the medium because of increased competition.

Hecht report

Still, the bulk of the TVB-NAB presentation dealt with a commissioned Norman Hecht Research study trumpeting free tv's advantages over cable in terms of reach, frequency and impact while seeking to debunk the perception that cable attracts a higher-quality audience.

Using four markets of varying size as examples, the study found that cable's claim to more upscale viewership has been mitigated by increased cable penetration and that no significant qualitative audience difference exist. Blair, meanwhile, found in an advertiser survey that audience quality emerged as the key factor in support of buying time on cable webs.

While the networks have been telling a similar story in their salvos against cable, the Hecht study brings it to a local level. It found that broadcast tv remains the dominant viewing choice (seven hours daily) in cable home.

In addition, Moll noted that the explosive days of growth in cable penetration are over and should top out — an assessment shared by Blair. Cable is currently at 55% penetration, or 48.5-million households, while TVB estimates tv households will climb to 90.4-million (or 98.2% of all homes) by 1990, with an average 1.9 tv sets per home.

While Blair's longterm forecast for the industry — initially outlined in November — is positive, its most recent data on spot tv sales for '89 can be described as unspectacular yet hopeful.

Not surprisingly, based on the tone of the day, Blair pointed to local cable advertising sales as one of the biggest threats on the horizon to the spot market. While 69% of cable systems do not sell local time, per TVB, McAuliff suggested this is the direction cable will take in the future and that local stations must "head them off at the pass" to protect their franchises. Ted Turner's TNT may be setting that trend in motion, allocating four minutes to systems for local sales, compared with two minutes from most competing cable webs.

Campaign to Ballyhoo Free TV Launched at NAB Conference

VARIETY 5/3/89

Las Vegas A campaign to remind the public, Congress and advertisers about the benefits of free television was launched here during the annual confab of the National Assn. of Broadcasters.

A variation on the industry's anti-cable rhetoric that has been building momentum in recent months, the campaign reportedly will involve every tv station in the country. "It is a true industrywide effort, with broad support coming from every segment of the industry, from local stations to the networks," claimed Milton Maltz, chairman of Cleveland's Malrite Communications Group.

Maltz, who is also chairman of the NAB Free Television Task Force, first proposed the campaign at last year's meeting and has been working on it since. He presented the plan during a luncheon with tv members April 30.

The effort, which will be heavily promoted with contests and billboards, among other things, will underscore the importance of free tv and the threats posed by cable, said Maltz.

Kickoff will be a primetime show emceed by Walter Cronkite that will be aired simultaneously on commercial stations throughout the nation. Date is unspecified, but it will be within the next two months, according to the NAB.

Promotional kits will be sent to all stations, which will be asked to produce local spots and campaigns with their tv peers.

Campaign is slated to last one year, but the possibilities of an extension are good, Maltz said. "We will do what it takes to present our message," he promised. Other trade groups, such as the Assn. of Independent Television Stations, will also participate.

Target of the campaign is the cable industry's ability to drop and reposition stations in the absence of the Federal Communications Commission's must-carry rules, said Maltz. He said cable has been guided "by the relentless forces of sheer greed," dropping stations at will and repositioning them on whim.

He reminded the luncheon audience that a cable multiple-system operater topper once warned him that "in due time, you'll have to pay for carriage on my systems."

Cable chagrined

The campaign's vigorous anti-cable message was greeted with chagrin by one cable tv industry representative here. The industry as a whole questions the wisdom of the effort, he suggested.

For example, stations risk antagonizing their local cable operators at a time when there are few must-carry infractions to report. It could be seen as overkill in the wake of the industry's efforts on Capitol Hill to paint cable as anti-competitive monsters, the cable rep said.

As has been mentioned before at broadcast and cable trade shows, the public is not overly concerned with how it receives its programming — whether it be over-the-air, direct broadcast satellite or fiber optics. The debate in recent years has not been whether viewers are willing to pay for video fare, but how much. *—Paul Harris*

Recapping the Freshman Class of 1988–89

VARIETY 5/3/89

By BOB KNIGHT

New York Successful new shows are the stuff of a commercial network's future. The 1988-89 season was not marked by an explosion of new product, but there was enough fresh material introduced, with varying degrees of acceptance, to provide a little hope.

A single smash hit does not provide enough impetus to turn a web from loser to winner, but two probably would. The last smash before the recent season was NBC's "The Golden Girls," and all three webs have been looking for a comedy breakthrough since.

ABC-TV came up with one in 1988-89 in "Roseanne," starring Roseanne Barr. The blue-collar comedy started out in advantageous position, Tuesdays at 8:30, but soon proved it didn't need a protected time period to pull large-scale audiences. In the spring, having lost Tuesday night to NBC, ABC did the unthinkable with "Roseanne," moving it back a half-hour, and it responded by beating "The Cosby Show" for the No. 1 weekly spot as often as not.

No web had ever moved a hit to a new time period during its first year on the air. Viewers establish habit patterns before a season is too old — and breaking that habit can be risky. That "Roseanne" flourished after the move underscores the skein's status as 1988-89's one legitimate smash hit.

NBC's "Dear John" and "Empty Nest" contributed hit

numbers, but the feeling is their strong shows were more the result of advantageous time periods than sheer programming strength. The numbers count, of course, but neither looks capable of drawing commensurate numbers in an unprotected timeslot as yet. Even so, they finished ninth and 11th, respectively.

'Mysteries' a mystery

More impressive (and unexpected) was the 17th-place ranking pulled off by NBC's "Unsolved Mysteries," which had no lead-in help on Wednesday yet managed to grow during the season, first beating ABC's "Head Of The by season's end, had split time Class" with some regularity, then taking the same web's "Growing Pains" frequently. The show opened to negative press but did the job all season long.

Another NBC September newcomer that made the grade was Tuesday's "Midnight Caller." Blessed with a good but not well-known lead (Gary Cole), this series soon began beating Emmy Award winner "thirtysomething" in the 10 o'clock time period and, period wins with the ABC entry. Overall it marked 41st, but it did well enough to be considered a success.

The only other successful September starter was CBS' "Murphy Brown," which ranked 36th for the season. CBS had trouble all season getting viewers on Monday night, so "Brown" seldom had much of a lead-in, but it did well enough to justify its return — and served as CBS' token 1988-89 new hit.

A number of series have surfaced since midseason with some success — at least enough to warrant consideration for next season. The most promising is "ABC Mystery Movie," which ranked 30th for the season. The Peter Falk "Columbo" event and the Burt Reynolds "B.L. Stryker" element fared well enough, but Louis Gossett Jr.'s "Gideon Oliver" may disappear before next season rolls around.

ABC's "Anything But Love" and "Have Faith" ran up good numbers behind "Roseanne," but what wouldn't? NBC's "One Of The Boys" and "Nearly Departed" have been moderately accept-(although not a newcomer) was ABC's "Full House." A struggler during its initial season last year, ed, but they were in choice time periods. Doing well enough in more extended tryouts have been "Father Dowling Mysteries" and "Nightingales" on NBC. The former did better than anything else NBC has tried on Friday at 8; the latter held its own on Wednesday at 10 but was threatened by various pressure groups because of its content.

'Jake' surprises

One of the big surprises of the season was the late-arriving "Jake & The Fatman" on CBS. Just strong enough last year to earn a backup order for this one, but not strong enough to make the 1988-89 sked at the start, this skein switched its locale to Hawaii — and ran a full rating point above its top 1987-88 score once it finally made the sked on Wednesday at 9. Ranking in a 33d-place tie for the season, "Jake" has earned a place on next season's sked, giving CBS the anchor it needs to build an all-action slate next year on the night.

Another surprise of the season "House" has emerged as the best ratings gatherer on the web's Friday sitcom block this year, enabling ABC to wrest control of the night from CBS. ABC could decide to move "House" to another night to build a new sitcom block. Such a move would open its current 8:30 period for use as a hammock for a new series next season.

Airing in November, "War & Remembrance" got decent numbers but wasn't the blockbuster called for by its $120-million cost

The miniseries genre did not have a great year. "War & Remembrance" was a very expensive disappointment (although hardly an abject flop). "Lonesome Dove," however, was a surprise hit. Motown Prods. and Qintex Entertainment had enough faith in its content to ignore conventional wisdom, which said audiences had abandoned the Western.

ABC's "The Women Of Brewster Place" was a success (24.0/37), but beyond that minis did not produce the hefty numbers of past years. NBC's "Brotherhood Of The Rose" (20.1/33), CBS' "Jack The Ripper" (17.6/29) and NBC's "Favorite Son" (16.0/26) were the best of the lot numberwise; CBS' "Dadah Is Death" hitting the low mark (10.7/18).

Saatchi Has Seen the Future of Network TV; It Looks Like More of the Same

VARIETY 5/3/89

By BOB KNIGHT

New York After analyzing the pilot crop for the 1989-90 primetime season, ad agency Saatchi & Saatchi DFS has concluded that tv of the '90s won't look too different from tv of the '80s.

The only difference, according to the agency, will be the competitive environment in which the networks operate. S&S opined that "to prosper throughout the '90s, the (networks) must acknowledge that they are competing not only with each other, but also with independent stations, cable, pay-television and videocassettes."

Pilots 'report'

These views are part of the agency's "Primetime Program Development 1989-90" annual report, which evaluates 117 new pilots (an alltime high, due to the inclusion of Fox projects in the survey). The agency's head count for pilots is 33 for ABC, 35 for CBS and 34 for NBC.

The pilots offer no major programming breakthroughs, Saatchi says. Noting that last year's crop emphasized drama, the report points out that more than half of all fall 1989 pilots are comedies, with the percentage even higher at NBC and Fox. S&S sees signs that sitcoms are losing popularity; it says comedy will probably account for 10 of the top 15 series when the '88-89 season ends.

According to the report, the pivotal issue this season has been program content. In the past, content boundaries were tested mostly by made-for-tv movies, but during the recent season it became clear that regular series — dramas and sitcoms — were not only testing boundaries but often crossing them.

Looking ahead, S&S sees a return of action drama, with more adventure-oriented, male-dominated fare expected. Men are back as series leads, according to S&S; so are policemen. Non-traditional families are big, with a wealth of single parents, inherited kids and merged families on tap.

Fantasy content is also hot; so are predominantly black casts. And after the success of stand-up comedienne Roseanne Barr this past season, stand-up comedy is also in demand. Reality programming seems on the upturn; such shows are relatively inexpensive to produce and seem to work as counterprogramming to "unbeatable" competition, the agency says.

S&S expects NBC's programming formula to be urban, upscale and innovative, as the pressure to find the next generation of hits remains strong. ABC's pilots "recall the network's past shows: urban, youthful comedies and action dramas." CBS will continue to try to attract younger viewers, with a focus on 8 p.m. shows. S&S feels the web has been able to attract a wealth of talent and "seems committed to turning things around."

Fox Broadcasting has established itself "as a force to be reckoned with, especially on Sunday night," the agency says, but cannot rest on its successes. Its challenge is to expand its scope as it builds Monday night and works to become a true "fourth network."

Syndie Faces Big Squeeze: Timeslots at a Minimum

Variety 7/19/89

By JOHN DEMPSEY

New York An airtight syndication marketplace has put a squeeze on big-budget series-development projects for the 1990-91 season.

"I don't see anything on the horizon that's exciting for the fall of 1990, at least if you go by what's in development right now" in firstrun syndication, says Jack Fentress, v.p. of programming for Petry National.

Mitchell Praver, v.p. of programming for Katz Continental, says, "You're not seeing as many deal-driven projects being launched because there's too much money at stake; the risks are too high."

Most industry experts predict that fall 1990 timeslot availabilities will be at a minimum in both prime access and early fringe, the two most lucrative dayparts for tv stations. "Because stations have been burned so often in the last few years," says Janeen Bjork, v.p. of programming for Seltel, "they've become gun-shy. Rather than taking a chance on an untried new program, they'll stick with what they have, even if it's performing only marginally in the ratings." (Two strips Bjork cites in the stations-got-burned category are GTG Marketing's "USA Today On TV," this season, and Lorimar's "Truth Or Consequences," the year before.)

Timeslots 'not there'

Phil Oldham, executive v.p. of domestic sales for Genesis Entertainment, says, "I don't see any niches for new programming" in the 1990-91 season. "The time periods just aren't there."

Various sources say the following 5-a-week series proposals are in active development for 1990-91: King World's "Monopoly" (from Merv Griffin Prods.), Orbis Communications' "Joker's Wild," Orion TV's "Name That Tune" and Worldvision Enterprises "Scruples" (this is just a preliminary list).

Except for "Monopoly," all are planned as updated versions of gameshows that ran successfully in the past. And, as Oldham puts it, these distributors are playing it safe because "gameshows are the cheapest form of programming to produce."

Gameshows may be the cheapest relative to other genres, but Bob Turner, president of Orbis Communications, says he'll be ponying up a strapping $1.5-million for the launch of "Joker's Wild" between now and January's National Assn. of Television Program Executives convention. That figure covers the construction of the sets and the production of the pilot, all of the research and marketing expenses, the ad-promo costs and overhead for Orbis' nine salespeople. If the pilot doesn't make it to series, that's $1.5-million down the drain.

The syndicators of all these gameshow pilots will make their initial pitches for access time period. But whether the pilots end up on the air as series will depend on the ratings of such returning access series as King World's "Inside Edition" and LBS Communications' "Family Feud."

Early fringe booked, too

But traditionally, when a series aimed at access gets shut out by no-vacancy signs on stations throughout the country, it goes after the next-best time period, early fringe. However, "there's nothing but wall-to-wall talkshows and sitcoms in early fringe," says Oldham. "All of the holes are plugged up."

The talkshows include King World's "Oprah Winfrey Show," Multimedia's "Donahue" and Paramount/Tribune's "Geraldo." The off-network sitcoms, all potential Nielsen winners, include Columbia's "Who's The Boss?" Warner Bros.' "Growing Pains," Buena Vista's "Golden Girls" and Warner's "Head Of The Class."

When the marketplace is unable to shoehorn a new series into access or early fringe, the show ends up in the much weaker daytime or late-fringe time periods. Hobbled with bad timeslots in many parts of the country, the series has almost no chance of surviving.

But not all syndication observers are convinced that dullness is going to be the watchword for new projects in 1990-91. "It's too early to write off the development year," says Matt Shapiro, v.p. of programing for MMT Sales. "The station that's No. 3 in the access time period in its market will be looking to buy new shows."

With overnight ratings available in 22 markets this fall, Shapiro says it'll be easier than ever before to get a quick fix on which shows are not well before the November books. "I think you'll see a number of new syndication projects surfacing in October," he says, many of them aimed at the time periods of failing shows.

But the consensus is that because of severe marketplace restraints the producers won't be funneling lavish budgets into any of these new program ideas.

Sunny Daytime Upfront Could Hit $900 Million

Variety 6/28/89

By VERNE GAY

New York Capping the most extraordinary upfront net tv sales period in history, advertisers are expected to buy their upfront daytime tv inventory this week.

As the big buyers continue to snap up primetime inventory in record amounts — some sources say the total pot could hit $3.8-billion (excluding Fox) — many have begun turning their sights on daytime. One top agency exec predicted the entire market, including primetime, could be over by the end of the week.

Sources said daytime will also be unusually strong, with total ad expenditures expected to exceed $900-million, or $100-million over last year's daytime upfront.

That should come as very welcome news to the networks, who have struggled with the daypart in recent years. In general, daytime ad budgets have declined and prices with them. Each of the Big Three have developed promotion and marketing programs to draw advertisers back in; those may be having some effect.

Prices are expected to go up 8-9% on a cost-per-thousand basis. That compares to average 3-network CPM increases in primetime of 13-15%. Pricing in the daytime upfront market was flat to slightly up last year. Daytime has ridden a severe roller-coaster for about three years in a row now: Because of the networks' seeming inability to predict daytime scatter-strength, they have either undersold or oversold their upfront inventory.

For example, daytime upfront prices plunged 20% in 1986, becoming the only daypart ever to suffer across-the-board price cuts in the upfront market. The networks cut prices to attract buyers, and it worked. But the webs ended up with a shortage of scatter, with the result that scatter prices went through the roof.

The next year, 1987, the networks boosted prices by 20%. They sold less inventory as a result, hoping for a resurgence in scatter that never materialized. They then ended up eating huge amounts of commercial spots.

Once again, in 1988, pricing was soft to lure buyers back. The Big Three sold a lot of inventory, but had little left for scatter and daytime scatter went through the roof last year.

"The networks have overreacted on pricing for three consecutive years," said Irwin Gotleib, senior v.p. and director of national broadcast for D'Arcy Masius Benton & Bowles, New York. "The pendulum keeps overswinging and at some point, they have to catch it midpoint."

Meanwhile, the networks have their hands full on the primetime upfront.

Sources said the networks, and in particular ABC, played extreme hardball with buyers this year. One source spoke of a call placed by an ABC salesman to a buyer at 3:30 in the morning, warning the buyer that unless he bought then and there, the price would go up the next morning.

Some buying execs were believed to be infuriated with ABC's hardball tactics. As one top exec noted, "They came in with clubs in both hands."

ABC set market pace

The network also dictated the pace of the entire market. While NBC sat back, ABC was vigorously cutting deals. Indeed, ABC panicked the entire market in the middle of last week when it told buyers that it was "shutting down" — that is, accepting no more orders.

ABC did that to find out exactly how much inventory it had left to sell — but the networks also shut down in the heat of battle to frighten buyers and boost prices.

Early this week, top dog NBC was closing in on $1-billion; ABC was believed to be right behind it, while CBS was in the $500-600-million range.

Some execs disputed that ABC had unfairly manipulated the market. "They were hard but fair," said John Sisk, senior v.p., J. Walter Thompson. "ABC has determined that they are a network with growth potential."

Others noted a particular aggressiveness on the part of each of the networks. As Sisk noted, "All three have owners who are profit driven."

Including Fox, the total primetime upfront is expected to exceed $4-billion this year.

Ad Agencies Eclipse Daytime; Early, Late Fringe Growing

VARIETY 3/1/19

By VERNE GAY

New York Recently released data indicates that advertiser spending habits changed dramatically in 1988, in terms of where the money was spent and who spent it.

Spending was way up in late-night and daytime. Packaged goods advertising fell slightly while automotive spending increased significantly. The Big Three networks also had to tangle with generally weaker ad spending across the board, which resulted in less than enthralling bottom lines for both ABC and CBS.

The so-called fringe dayparts of early morning and late night are by far the hottest growth areas in network tv. Advertisers ponied up $1.25-billion on fringe dayparts last year, up 26%. Indeed, late night is becoming so popular it may soon surpass daytime in total spending.

Poor selling environment

Once the leader in total ad revenues among network dayparts, daytime has suffered declines for several years now. Many advertisers perceive daytime as a poor selling environment, and have migrated to late night where they believe the best and youngest sales prospects reside.

Daytime spending last year was $1.668-billion flat compared to 1987. Primetime spending was $5.211-billion, up 5%, while Saturday/Sunday spending (all sports) was $1.453-billion, up 19%. Gains in prime and weekend were Olympic-related.

A change for the better was in the number of advertisers spending $20-million or more. About 105 advertisers each spent $20-million or more on network tv last year compared to 84 in 1987. The Olympics are partly accountable for this growth. So is the 15-second commercial unit, which has drawn a handful of smaller advertisers to network tv.

One new spender

Only one major new network tv spender emerged last year, shoemaker Reebok, which spent $20-million. Apple Computer brought some happy news to the webs. A former $50-million spender which virtually disappeared from the airwaves two years ago, Apple spent $30-million last year.

While no new advertising category emerged last year, several grew significantly. Freight advertisers spent $68-million last year, more than double the 1987 figure. Household furnishings ballooned into an $80-million category. Travel advertising spurted ahead 36% to $169-million while computers chipped in another $186.5-million, up 17%. The Olympics probably accounted for most of this.

Unfortunately for the networks, four of their most important categories were flat or down in 1988. Beers and wines fell 7% to $479.8-million. Two years ago, wines were the fastest growing category on net tv. Soaps fell 1% to $320-million. Food advertising rose 2% to $1.6-billion. Toiletries were even at $869-million and pet foods plummeted 22% to $126-million.

There are a myriad of reasons behind the shifting spending patterns. Corporate mergers and consolidations accounted for some cutbacks. Barter syndication also siphoned significant portions of packaged goods' budget. And of course, there is the Olympics.

Packaged goods, once by far the most important advertising category on network tv, has been spending less money in recent years. Yet while packaged goods are defecting from network tv, automotives are embracing it with renewed passion.

P&G and GM

The examples of Procter & Gamble and General Motors are most telling. P&G was once not only the largest programmer of network tv but its largest advertiser as well. In 1978, P&G spent an estimated $262-million on net tv, or 6.3% of all net tv revenues. In 1988, P&G spent $370-million, or 3.7% of the total. Last year the company had 37 brands on net tv each spending $4-million or more, but 14 of those brands clocked budget cuts. (Two years ago, P&G chopped $100-million off of its tv budget.)

GM's experience is the exact opposite. Ten years ago, GM's $107-million network budget represented 2.6% of total network revenues. Last year it represented 4.6%, and spent $443-million, becoming the single-most freewheeling spender for ad time.

In 1988, the company had a total of 35 auto brands on network spending $2-million or more each. Twenty-seven of these had big increases last year. Oldsmobile Cutlass Supreme and Grand Prix got $73-million worth of blurbs last year. GM spent nothing on these cars in 1987.

GM's colossal spurt is partly attributable to new car introductions as well as an all-out attempt to win back market share lost to Ford in recent years. GM has also stayed away from barter syndication, thus funneling even more money into net tv.

A trunk-full of revenue

The auto company's massive buy was no doubt spurred in part by the Olympics. But GM spent only $55-million on both Winter and Summer Games which means that even without the Olympics, GM still spent $338-million, or 43% more than the year before.

GM's big push galvanized the rest of the auto industry. Ford and Chrysler boosted network budgets and a host of imports also spent more. Automotive advertisers spent $1.3-billion last year and, after foods, have become net tv's biggest buyers.

P&G has strayed from network tv because of cost. High prices have forced it into syndication, unwired networks, and Fox. Weakness in daytime pricing also accounts for lower P&G numbers. BAR counts P&G spots on P&G-owned shows. Because advertisers have evacuated daytime, prices have softened in all shows. As a result, P&G's daytime rates have declined with the rest of the market. P&G spent $164.9-million on daytime last year, down 17%.

Top 10 Network Advertisers in 1988

Rank	Company	Amount spent	Percentage change
1.	General Motors	$443.4-million	+62
2.	Philip Morris	$388.6-million	+3
3.	Procter & Gamble	$370.2-million	−2
4.	Kellogg	$297.7-million	+25
5.	McDonald's	$245.4-million	+14
6.	Anheuser-Busch	$207.3-million	+11
7.	Lever Bros.	$189.7-million	−11
8.	Ford	$175.7-million	+9
9.	Johnson & Johnson	$174.5-million	−4
10.	RJR Nabisco	$174-million	−17

Top 5 Syndication Advertisers

Rank	Company	Amount spent	Percentage change
1.	Philip Morris	$88.6-million	+38
2.	Procter & Gamble	$60.9-million	+25
3.	Bristol-Myers	$32.6-million	+19
4.	Lever Bros.	$27-million	−13
5.	Nestle	$26.3-million	+49

Top 5 National Spot Advertisers

Rank	Company	Amount spent	Percentage change
1.	General Motors	$312.2-million	+21
2.	Procter & Gamble	$222.9-million	−63
3.	Ford	$165.7-million	+20
4.	Philip Morris	$155.5-million	−11
5.	Chrysler	$146.8-million	+34

How Syndie's Big Spenders See '89

By ROBERT MARICH

Hollywood Executives in charge of programming for some of the nation's largest groups of television stations are approaching the new year with reserve.

As a result, the outlook for the $2.5-billion program syndication business generally continues to be subdued. The outlook is not surprising since the quadrennial Olympics/election advertising fell short of expectations in 1988.

Stations continue to scale back their acquisitions of syndicated programs and many distributors still are smarting from program deals that fell through after the syndication boom three years ago.

There was considerable turnover last year in the highest echelon of group program execs. Of the three leading groups of independent stations, two have new execs at the programming helm. Two of the group programming execs at the three networks' owned-stations divisions are gone.

Among independents, Fox Stations promoted research exec Steven Leblang in November to veep-programming to replace David Simon, who left for Walt Disney Co. in London. At Tribune Broadcasting, company veteran Marc Schacher was promoted in October to director of program services, succeeding Mel Smith, who went to the company's syndication arm.

At network-owned station groups, Eugene Lothery, who has worked at CBS since 1968, became v.p. of programming for the CBS Television Stations division in December, succeeding Alan Shaklan, who became g.m. of the company's new Miami station. At NBC, Weston J. Harris retired in September and now is a consultant after being in charge of NBC group-programming projects since 1971. A successor has not been named.

Here's how some of the group broadcasting execs see the trends in syndication:

Marc Schacher

Television stations are becoming stingy in acquiring movies, said Marc Schacher, director of program services for Tribune Broadcasting.

''Stations used to pride themselves on having large inventories of movies of all types — the classics, martial arts films, whatever,'' said Schacher. ''Now, that's changed because stations are streamlining their inventories. You can't afford as many as you did before.''

Schacher, who is the top program exec for Tribune's six independents in major markets, all heavily into movies, said he's inclined to focus on A titles that can stand up in the sweeps periods, rather than hunt for bargain B titles. Movie packages generally are bought by Tribune stations individually.

This year, Schacher said, Tribune will not bring a big checkbook to NATPE because there are no gaps to be plugged. ''In most cases, the needs are specific to our individual stations, as opposed to the group,'' he explained. Tribune, of course, is one of the three most potent chains of independents because it operates stations in the nation's three largest markets.

Aside from the obvious tilt to tabloid tv, he said there's a minitrend to firstrun programming loosely related to the wrestling genre. It takes the form of hybrid gameshow/action series: Samuel Goldwyn's ''American Gladiator'' and ''Roller Games,'' a 1-hour weekly high tech/action series from Qintex Entertainment. ''There certainly has been an audience for wrestling if you look at how well NBC has done with its 'Main Event' that runs once a month on Saturday night,'' he said.

Schacher took his post in October, replacing Mel Smith, who was appointed v.p./programming for Tribune Entertainment. Schacher held program director posts at Tribune stations in Chicago and Denver, and jobs in ratings research.

John A. Trinder

TVX Broadcast Group prez/CEO John A. Trinder said he'll seek programming ''with a keen eye and a sharp pencil'' for his indie stations.

''I wouldn't say we'll be at NATPE with a bag full of cash, but we will be making selective buys,'' said Trinder. He became TVX boss last May, replacing Tim McDonald, who left the company after its investment banker, Salomon Bros., was forced to meet interest obligations from the company's debt. That problem has been eased through a recapitalization effort.

Many in the industry felt TVX was headed for the trash heap, but Trinder, who joined the company's Norfolk, Va., station as an account exec in 1975, bristles at the suggestion that TVX is troubled. ''I'm going to remind syndicators that we never, never were late on payments'' for programs, he said.

He added that tv stations in general, which had pulled back from syndication buying last year as they digested excess programming bought in 1984-86, seem to have again become active in acquisitions. ''We all had bought too much programming,'' he said.

George Moynihan

George Moynihan, senior v.p. of Group W Television Station Group, is a man with early fringe on his mind.

He is seeking a replacement for ''Hour Magazine,'' the daytime talkshow that holds down early fringe spots on Group W's five affiliates in major markets. ''Hour'' was canceled by Group W's syndication arm in December.

After the Group W and NBC-owned station groups announced in December a joint venture to develop syndicated programming, Moynihan said early fringe was a much-discussed daypart in his first meeting with NBC brass. ''I anticipate that we'll be working together closely,'' he said.

Moynihan said he envisions NBC Prods., run by NBC Entertainment prez Brandon Tartikoff, will produce shows that initially would air on a few of the partners' stations. Eventually, Group W Prods. can syndicate such programming to stations outside NBC and Group W, he added.

In other syndication developments, Moynihan said Lorimar's ''The People's Court,'' which airs on three Group W stations, is being renewed, although some other court and gameshows may get the ax.

What incident in the past season stands out in Moynihan's memory? He found the tepid audience interest in ''USA Today: The Television Show'' spoke volumes about America's tastes.

''With all the publicity and glowing stories about it, it wasn't even sampled,'' he said. That reception indicates that upbeat journalism-of-hope takes a back seat to tabloid in the hearts of the masses, he said.

Kevin O'Brien

Kevin O'Brien, exec v.p. for the three independent tv stations of Cox Enterprises, said he thought six months ago that syndication was headed for a year marked by little program development.

''I must say that I'm pleasantly surprised by the number of pilots and different projects'' that popped up since then, said O'Brien. ''There's a lot of action out there. I think that's a good omen for our business.''

O'Brien, also v.p./g.m. of Cox' KTVU San Francisco, said he is rooting for LBS Communications' ''Family Feud,'' which is off to an uneven start nationally. O'Brien said ''Feud'' is holding its own on KTVU.

O'Brien, who joined Cox in June 1986 after 18 years with Metromedia stations (now the backbone of the Fox-owned stations), said his division is autonomous from the parent company's other tv interests. Cox also is involved in Al Masini's Television Program Enterprises and Tribune Co.'s Teletrib, the syndication concern.

G. Gregory Miller

For C. Gregory Miller, v.p.-television operations at Great American Broadcasting (formerly Taft), this is a year to tread water.

''I don't see us buying many programs,'' said Miller, who joined the company in 1974 and took his present job in 1985.

''In our case, we're pretty much standing pat because our schedules are doing reasonably well,'' he said, referring to the company's five affiliate stations.

''Based on everything we've seen in screening groups, there is nothing that's really major or a hot product that's new,'' he said.

His company holds interests in ''Entertainment Tonight,'' ''Star Search'' and ''Lifestyles Of The Rich And Famous'' as a result of deals by Miller's predecessor, Lucille Salhany, now prez of domestic tv for Paramount Pictures.

Steve Leblang

Steve Leblang, v.p.-programming for the seven Fox Television Stations, says time periods for firstrun syndication series are becoming scarcer.

''Much of the growth in firstrun earlier in the 1980s was a result of stations not having a lot of fresh off-network series,'' he said. Only one major off-network sitcom package became available in 1985-86, Leblang said, while now there are 2-3 sitcom packages every season.

Facing scarcity of off-network sitcoms in the mid-1980s, stations embraced firstrun, but that trend is waning, judging by what's taking the place of the flunkouts from syndication's class of 1988-89. "In the case of the medical shows, 'USA Todays,' and 'Wipe Outs,' they will be replaced with off-network product," he said.

Leblang is the point man for syndication programming of a truly heavyweight group since Fox owns independent stations in seven of the top 10 markets. The group, of course, is the backbone of the Fox Broadcasting network.

He assumed the job in November, replacing David Simon. Leblang's background is research, which he calls excellent training because it focuses him on the audience rather than dealmaking. "I learn more from reading 'American Demographics' than I do from reading the television trades."

Leblang said he also will eagerly follow the toy industry's Feb. 13-22 convention to see if it spawns products that will spark slumping toy advertising. If the toy business remains in the doldrums, kidvid will fizzle for stations, he said.

Peter Kohler

Television stations should build longterm franchises around the information programs they acquire and not look upon such programming as easily expendable, said Peter Kohler, v.p. at Gannett Broadcasting.

Tv stations should focus on "longer-term appeal, at demographics and building a franchise, not a quick fix" from information programs that can complement their local news, Kohler said.

Gannett's 10 stations, of which nine are network affiliates, do not have a group-level exec for programming because it makes few group deals. However, eight of the 10 stations carry "USA Today: The Television Program," the news/information daily series syndicated by another Gannett unit.

Kohler said heavy promotion, good time periods and compatibility with adjacent shows helped "USA Today" perform well on Gannett affiliates. The strip has been disappointing in national ratings, however. Gannett's Boston station, the only independent of the group, just dropped "USA Today."

Kohler, a former print newsman who held posts at CBS before joining Gannett in 1986, said tv stations should emphasize localness in their news, and not let satellite feeds of distant events or technology overwhelm them. For stations, "the fundamental element is local news," he said.

Upfront TV Sales Eye Record

VARIETY 5/24/89

By VERNE GAY

New York The 1989-90 network tv upfront market could soar to $3.7-billion in primetime sales, or $400-million over last year's record haul.

Network and agency sources said the biggest market could start rolling as early as next week, and conclude by the last week in June. In recent years, primetime upfront has wrapped up just after July 4.

Execs say the market will move a little earlier this year because of the projected strength. Also, there are no apparent obstacles to an orderly market; the writers' strike and a changeover in ratings methodologies slightly delayed upfront markets in 1987 and '88.

Pricing, also expected to be strong, could have significant impact on the scope of the market. If the networks hold out for extremely high prices in anticipation of a bull market, advertisers may be inclined to hold back budgets for scatter, syndication or cable. "The networks will dictate their own destiny," said one exec.

Nevertheless, some agencies are telling their clients to expect cost-per-thousand increases in the 9-10% range. There will, of course, be significant differences among networks. Per one tv source, CBS, with weak audience members among adults 18-49, will settle at 0 to 5% increases. NBC is expected to nail 9-10% increases; ABC may end up with increases of 11-13%.

As usual, there are disagreements over the market's total size. While no one expects the market to decline, there are significant differences among some forecasts. Lintas New York, a General Motors agency, said in a recent inhouse newsletter that the market would hit $3.5-billion, or 8% over last year. But the market "could exceed that," said Lou Schultz, exec v.p.-media services. However, Grey Advertising said primetime upfront will register only 2-3% increases.

Network sources said the market could hit $3.7-billion if Fox' potential revenues are included in the total. There are no figures on Fox' 1988 upfront performance, although some sources have pegged the weblet's upfront sales at about $90-million. Levels could exceed $150-million this year, and may top out at $190-million.

Fox' expected gains are due to massive ratings growth this past year; but sources said Fox could also draw money from the Big Three networks which would slightly dampen their own projected gains. Others expect Fox to siphon money out of the barter marketplace.

There is another big unknown in the overall equation: Olympics money. In 1988, advertisers spent up to $800-million on the 1988 Winter and Summer Olympics. No one knows for certain how much of that money will resurface in 1989-90 (advertisers create additional funds for network Olympic buys). The networks expect about $200-million to reappear.

In upfront buys, advertisers secure commercial inventory for the fourth quarter and the first three quarters of the folloiwng year. They do this to get guarantees on purchased ratings levels and to get lower prices than they might otherwise get during the scatter, or short-term, market.

Conventional wisdom says advertisers will load up on longterm buys if scatter has been strong, as it has been this year. If scatter has been weak, advertisers hold back during the upfront, anticipating lower scatter rates in the future.

The 1988 market demolished conventional wisdom, however. Many experts anticipated a weak upfront following a lackluster '87-88 scatter. In addition, upfront budgets were expected to be weak because of the uncertainty caused by the writers' strike.

Instead, the opposite occurred: primetime upfront buys exploded, as advertisers snapped up a record $3.3-billion worth of commercials — or $300-million more than the year before.

Carmakers' surge

Sources now attribute the 1988 bull market to an unexpected buying splurge by major U.S. automotive advertisers. Led by GM, the Big Three out of Detroit injected an estimated $500-million in upfront buys.

For 1989-90, sources say the autos will be back in force. It is unclear, however, whether they will exceed last year's spending levels.

This year's network market is also expected to get a boost from two unlikely sources: promotion and barter syndication. One agency exec says "the pendulum has swung in the other direction" with promotion. Promotion budgets have grown rapidly in recent years, but there has been growing disenchantment among some major advertisers over promotion.

Likewise, barter syndication may be in for some tough sledding because of a dearth of hit shows.

Despite the projected bull market in primetime, the networks could still end up with modest gains overall. Robert Coen, senior v.p. director of forecasting for McCann-Erickson, said network would have "nearly a zero gain" for 1989, although he has made no projections for 1990. If advertisers take money from scatter to fund upfront buys, the short-term market in primetime could be a weak one. Also, the daytime marketplace could even fall below last year's levels, according to some forecasts.

Right now, agency execs are putting estimates on individual shows' performances. ABC unveiled its fall schedule May 23. Execs will now send client buy specifications to the networks. The networks then send those back with initial pricing estimates. Following intense negotiations, the market could be over within three weeks.

Unit Costs Soar 25–30% For November Prime

VARIETY 8/30/89

By VERNE GAY

New York The average 30-second unit for the three nets during primetime in November will cost advertisers $148,788, according to a VARIETY survey. While similar numbers were not available from last year, an estimate based on A.C. Nielsen cost supplement reports determined that the November-to-November 30-second price increase was roughly 25-30%.

The VARIETY unit cost survey was based on estimates obtained from several advertising agencies and reflect costs only during primetime in November. These costs do not represent average rates for the fall or for any other months. Prices vary dramatically from month to month and from season to season. November also is one of the highest-priced periods of the year (July is usually the cheapest).

These prices also vary somewhat with prices that the networks are currently commanding for November inventory. Scatter prices are much higher for many of these shows.

ABC, for example, is asking a whopping $500,000 per 30-second unit in the scatter market for "Roseanne" in November. If the network somehow manages to unload a unit at that price, it would set a record for highest-priced commercial unit (excluding the Super Bowl). NBC sold a few 30-second units for "The Cosby Show" in the high-$400,000 range last season.

As expected, NBC has the highest average commercial cost, $165,341 per 30-second unit. ABC's average went for $158,750, while CBS' average for the 30-second spots was $122,273.

Weblet Fox made a strong showing at $76,670. Top dollar on Fox goes to "Married . . . With Children" and "America's Most Wanted," both with $140,000.

NBC's top-dollar show for the second year in a row is "Cheers," at $330,000 per 30-second unit. The cheapest show is "Baywatch," at $90,000.

ABC's "Roseanne" wins the crown for the most expensive network show, at $375,000 per. ABC's economy models are "Life Goes On" and "Homeroom," both at $55,000.

CBS' winner is "Murder, She Wrote," at $240,000. "Paradise" sits at the bottom of the heap, at $60,000.

Pricing on a show is typically a good reflection of advertiser demand for the property: the stronger the demand, the stronger the price.

Thus, the accompanying chart can give clues about which shows advertisers — and the networks — believe will be big hits as well as which shows are expected to be big losers.

CBS' Monday night, for example, gets a ringing endorsement, with the exception of bookend shows "Major Dad" and "Newhart." Weak unit costs suggest weak advertiser confidence in their chances for success.

Likewise, ABC's early Sunday lineup was virtually written off. But the web's Tuesday night sked is expected to emerge once again as a big winner.

NBC has strength across the board with the exception of Friday night. Relatively anemic prices suggest that the ad community has little confidence in the web's attempts to rebuild that once potent night.

PRIMETIME NETWORK TELEVISION SCHEDULE 1989-1990

SUNDAY

Time	ABC	CBS	NBC	FOX
7:00	Life Goes On* 55K	60 Minutes 215K	Disney 110K	Booker* 80K
7:30				
8:00	Free Spirit* 65K	Murder, She Wrote 240K	Sister Kate* 150K	America's Most Wanted 140K
8:30	Home-room* 55K		My Two Dads** 160K	Hidden Video* 115K
9:00	Movie 140K	Movie 175K	Movie 180K	Married... With Children** 140K
9:30				Open House* 80K
10:00				Tracy Ullman Show** 60K
10:30				Garry Shandling 55K

MONDAY

Time	ABC	CBS	NBC	FOX
8:00	MacGyver 95K	Major Dad* 115K	ALF 200K	21 Jump Street** 85K
8:30		People Next Door* 125K	Hogan Family 180K	
9:00	Football 235K	Murphy Brown 195K	Movie 160K	Alien Nation* 55K
9:30		Famous Teddy Z 150K		
10:00		Designing Women** 180K		Movie 75K
10:30		Newhart ** 135K		

TUESDAY

Time	ABC	CBS	NBC
8:00	Who's the Boss? 260K	Rescue 911* 95K	Matlock 125K
8:30	Wonder Years 250K		
9:00	Rosanne 375K	Wolf* 100K	Heat of the Night 145K
9:30	Chicken Soup* 230K		
10:00	thirty-something 200K	Island Son 65K	Midnight Caller 140K

WEDNESDAY

Time	ABC	CBS	NBC
8:00	Growing Pains 200K	Peaceable King-dom* 75K	Unsolved Mys-teries 155K
8:30	Head of the Class 175K		
9:00	Anything But Love** 150K	Jake and the Fatman 100K	Night Court 195K
9:30	Doggie Howser, M.D.* 115K		Nutt House* 130K
10:00	China Beach 150K	Wiseguy 145K	Quantum Leap 110K

THURSDAY

Time	ABC	CBS	NBC
8:00	Mission Impos-sible** 60K	48 Hours 95K	Cosby 300K
8:30			Different World 280K
9:00	The Young Riders 90K	Top of the Hill* 105K	Cheers 330K
9:30			Dear John 250K
10:00	Prime Time Live* 110K	Knots Landing 155K	L.A. Law 260K

FRIDAY

Time	ABC	CBS	NBC
8:00	Full House** 150K	Snoops* 105K	Bay-watch* 90K
8:30	*Family Matters 125K		
9:00	**Perfect Strangers 145K	Dallas 100K	Hard-ball* 120K
9:30	Just the Ten of Us 115K		
10:00	20/20 140K	Falcon Crest 95K	Mancuso FBI* 125K

SATURDAY

Time	ABC	CBS	NBC	FOX
8:00	Mr. Bel-vedere** 95K	Paradise 60K	227 115K	Cops** 55K
8:30	Living Dolls* 60K		Amen 135K	**The Reporters 50K
9:00	Mystery Movie 80K	Tour of Duty** 70K	Golden Girls 210K	
9:30			Empty Nest 200K	Beyond Tomor-row** 45K
10:00		Saturday Night with Connie Chung 70K	Hunter 160K	Local

*New Program **Time Period Change

TV's a White Man's World: New Study Supports Claims by Women, Minority Orgs

VARIETY 8/30/89

By PAUL HARRIS

Washington Minorities and women seeking to change the way these groups are portrayed on primetime tv received some ammunition last week from the National Commission on Working Women.

A new report from the group supports complaints that tv inaccurately paints a world of little racial bias, with minority characters comfortable in roles subservient to whites. Hispanics, Asians and Native Americans virtually are invisible during primetime, and so is poverty, the group points out.

The portrait is outlined in "Unequal Picture: Black, Hispanic, Asian and Native American Characters On Television." Authoring group is the National Commission on Working Women of Wider Opportunities For Women.

Study contends that although tv's images of blacks and women have improved, the white male-dominated tv production industry is largely ignoring reality. "Tv denies the reality of racism in America,'' writes author Sally Steenland. ''Racial tension is commonplace in the real world but virtually invisible among white and minority characters on entertainment television.''

The report focuses on primetime programming by the three networks and Fox this spring. Among its findings: Of the 162 producers working on 30 shows with minority characters, only 12 (7%) are members of a minority group. Of those dozen, only four are female.

Rest of the story

Among its other findings:

- On tv, almost all minority characters are black. Of the 78 minority characters on primetime, 65 (83%) are black, nine (12%) are Hispanic and three (4%) are Asian. Only one is Native American.
- Nine out of 10 minority characters are middle-class or wealthy. Less than 10% are working-class, and none are poor.
- Almost three-quarters of all minority females on tv are featured in sitcoms, the only genre in which they have leading roles.
- Tv's world is one of racial harmony. The workplace is an egalitarian one with no hint of bias, even though whites are almost always in charge.
- Most programs reduce injustice to individual conflict, denying the reality of oppressive social structures. Only a few shows present a multicultural world.

The study said a few network shows, such as ''A Man Called Hawk'' and ''In The Heat Of The Night,'' consistently touch on issues of racial injustice and cultural differences. It praised ''A Different World'' for its high percentage of black producers.

It was funded by a grant from the Ford Foundation.

CBS's Olympics Wager: Riding Sports to the Top

VARIETY 8/30/89

By RICHARD HUFF

New York CBS' decision to pay $300-million for the 1994 Winter Olympics is part of its plan to acquire top sporting events at any cost as a way to bring the network back into contention in the ratings race.

It was a strategy discussed publicly last year when the network outbid NBC by nearly $70-million for the rights to the 1992 Winter Olympics. With its winning bid of $243-million, the third-place network gave notice it was mounting a comeback and would let sports lead the way.

Later in 1988, CBS snared broadcast rights to a Major League Baseball package, with a staggering $1.1-billion bid. Package includes regular-season games, playoffs and championship series.

And with each purchase, the industry has responded with negative comments about the network's approach to sports.

NBC sports prez Dick Ebersol called the '94 Games' price ''substantially in excess of what their value would be to NBC and its affiliates.''

''I don't know what they're trying to do over there,'' said another NBC executive. He called CBS' $300-million bid for the '94 Games, to be held in Lillehammer, Norway, too high, adding ''we thought $243-million (for the 1992 games) was high too.''

The Intl. Olympic Committee had set a $300-million minimum for the games. According to the IOC, there was no bid from either of the other networks. NBC, which bid $175-million for the 1992 Winter Games, dropped out of the race early when it heard what CBS was offering. ABC, suffering from about a $75-million loss from the 1988 Winter Olympics, sat out both auctions. ABC was protesting the IOC's rejection of its offer to buy the rights to the 1992 Summer and Winter Games for $500-million and forgo the bidding process.

The 1994 pricetag falls just short of that for the '88 Winter Olympics which cost ABC $309-million. Last year NBC agreed to pay over $400-million for the '92 Summer Olympics.

In addition to the Olympics and MLB, CBS has full or partial rights to the National Football League, including the Super Bowl, the National Basketball Assn., the National Collegiate Athletic Assn., golf, tennis, auto racing and several secondary sports.

The network has taken the approach that paying big dollars for high-profile sporting events makes sense. The idea is that the sports department can ''rent'' the network an audience to watch promos for its entertainment programming. CBS will be getting up to 51 hours of normally high-rated Olympic programming during the heart of February sweeps. In the past, programs that have debuted after such events as the Super Bowl have gotten good sampling from audiences that otherwise may not have watched.

Emblem of endeavor

More importantly, big sporting events give the network a trophy to hold before its affiliates as a symbol of its commitment, especially at a time when CBS' entertainment programming has been less than spectacular. Some industry execs cite the ABC of the '70s as an example of a network that used sports as a calling card; NBC tried to distinguish itself with sports in the '60s.

''Sports is the only area where you can go out and buy a rating point,'' said Robert Wussler, a former Turner Broadcasting senior executive and the newly named president of Comsat Video Enterprises. ''You can't do that with producers, you can't do that with writers, you can't do that with stars. Sports is the only place that you can predict ratings with a relatively high degree of accuracy.''

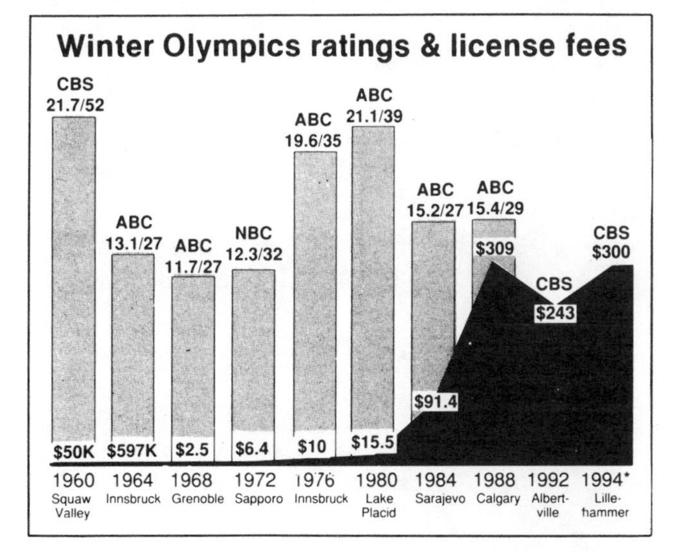

The gathering of the network's sports properties has been expensive, and some executives wonder when the money will run out. ''Eventually they'll break the bank and not be able to afford some of the smaller events,'' said CNBC president Michael Eskridge, who ran NBC's Olympic operations.

There's a tradeoff

''There are limitations with everything,'' said Wussler. ''The more you spend in one area, the less you're going to spend in another.''

Added another executive: ''Having these sports are great, but the (entertainment) shows have to work on their own.''

While CBS says it expects the two Olympics to be successful, executives outside the network have their doubts. Typically, the Winter Olympics get lower ratings than their summer counterparts because the Winter Games have fewer events and the U.S. does not fare as well in them. All of the executives contacted said they do not expect CBS to come out of its Olympic purchases financially ahead.

When planning a bid for such events as the Olympics or NBA games, networks at times cut the numbers close to get the event. However, it's not easy to predict a soft advertising market or lower-than-average ratings — when planning 3-6 years in advance.

CBS might be able to offset any potential losses by selling a portion of the two Olympics to a cable service. When NBC bid for the 1992 Summer Olympics in Barcelona, the network included a provision enabling it to sell some coverage to cable for about $75-million. NBC has since aligned itself with Cablevision Systems, and the two will offer a package of

Olympic events on pay-per-view. CBS executives says they have not decided if they will offer cable a package.

Turner Broadcasting, which was discussed when NBC was shopping its Olympic coverage around, and the USA Network are possible buyers. A USA executive said his network would love to talk about buying Olympic coverage from CBS. Turner's superstation, TBS, and cable net TNT have large appetites for sport. ESPN, ABC's all-sports cable service, also could be in the running, although it's unlikely CBS would offer it a package.

"Obviously, CBS doesn't think it's spending too much," said Wussler. "For (CBS prez) Larry Tisch and (CBS sports prez) Neal Pilson, it might be a smart gamble."

Acts of Violence & Crime Tallied in Variety Study

VARIETY 8/16/89

By VERNE GAY

New York Television can be a very treacherous place.

Take the week of March 27. During the primetime hours on the Big Three networks alone, a total of 56 people bit the dust. That works out to just under one death per hour.

These are just some of the key findings of a study commissioned last spring by VARIETY to count the total number of acts of violence and crime featured on the 3-network primetime schedule during one week.

Injuries were also tabulated. Counting those who escaped death during the week, the figures are daunting: 132 people were involved in some sort of violent mishap or altercation. While most of these were just beaten, pushed or grabbed, a few others were shot, clubbed or stabbed.

As in real life, violence on the networks was remarkable in its diversity. At least four poor souls suffered from animal bites; two were frozen to death; there was one attempted suicide and one rape; a drug sale resulted in one injury, while various gun battles left six wounded.

For its study, VARIETY commissioned Harry Kovsky Inc., an Irvington-N.Y.-based firm specializing in tv program content and format analysis. The researcher counted instances of crime or violence during "primetime," which entailed only the 8-11 time period, Monday-Sunday. Overruns past 11 were not part of the study and Sunday, 7-8, was also excluded.

Kovsky counted a total of 382 "acts of crime or violence" during 63 hours of primetime programming March 27-April 2, for an average of 6.06 acts per hour.

A crime was defined as "any act in which there is a visible or implied violation of law or in which

The 25 shows with the most crime and violence

Rank	Network	Show	No. of Acts
1	ABC	Beverly Hills Cop (M)	36
2	ABC	A Man Called Hawk	26
3	ABC	B.L. Stryker	23
4	NBC	Hillside Strangler (M)	16
5	NBC	In The Heat of the Night	15
6	CBS	Wildcats (M)	14
7	CBS	The Equalizer	13
	CBS	Stroker Ace (M)	13
	CBS	Murder, She Wrote	13
10	ABC	MacGyver	12
	NBC	Hunter	12
	CBS	Hard Time On Planet Earth	12
13	ABC	The Wonder Years	11
	NBC	Unsolved Mysteries	11
15	CBS	The Morning After (M)	10
	ABC	20/20**	10
17	ABC	thirtysomething	9
18	ABC	Academy Awards*	8
	NBC	Matlock	8
	ABC	The Shaggy Dog (M)	8
21	NBC	Midnight Caller	7
	ABC	Moonlighting	7
23	CBS	Tour of Duty	6
24	CBS	West 57th**	5
	CBS	48 Hours**	5
		Total acts	**310**

*Film clips **News magazine M-Movie

The study reports a total of 51 (of a total of 67) tv shows with at least one or more instances of c/v. (Seven shows had only one instance). ABC's "Beverly Hills Cop" was king of the hill in the c/v count. The show had one murder, one attempted murder, eight beatings, pushings or grabbings, one burglary, two threats-to-injure, six threats-to-kill, one gun battle, one tie-up, one break-in, three reckless driving episodes, one scam, an instance of property damage, and three people killed while escaping (among other things).

The top-25 shows in terms of criminal or violent activity accounted for 81.2% of all instances of c/v on net tv.

Death and injury by network

	ABC		CBS		NBC	
	Dead	Injured	Dead	Injured	Dead	Injured
Animal bite	-	3	-	-	-	1
Animal injury	2	-	2	9	3	-
Assault with object	-	-	-	2	-	-
Attempted murder	-	2	-	-	-	-
Attempted suicide	-	-	-	1	-	-
Beating/pushing/grabbing	-	31	-	32	-	11
Drug use/sale	-	-	-	-	-	1
Fist fight	-	3	-	-	-	-
Freezing to death/exposure	2	-	-	-	-	-
Gun battle	-	6	9	-	-	-
Kidnap/missing person	-	-	-	1	-	-
Killing/self-defense	2	-	-	-	-	-
Movie/tv within program	1	3	-	-	-	-
Murder	4	-	7	-	14	-
Natural death	-	-	-	-	1	-
Nightmare/fantasy	-	-	-	-	1	-
Property damage	-	-	-	1	-	-
Rape	-	-	1	-	-	-
Reckless driving	-	-	-	-	1	-
Restraint/tying	-	1	-	-	-	-
Shot/killed/inj. during escape	2	2	-	1	-	2
Sports injury/accident	-	-	-	1	-	-
Spousal/child abuse	-	-	-	3	-	-
Threat to injure	-	1	-	-	-	-
Threat to kill	-	1	-	-	-	-
Other/injury death	1	12	3	-	-	1
Total	**14**	**65**	**22**	**51**	**20**	**16**

There was little rhyme or reason to death patterns on network tv that week. True, murder is a favorite way of dispatching characters, but CBS lost nine during a gun battle on "Tour Of Duty" while NBC had three killed in various animal attacks. The 3-network average death rate was 18.7, or 0.85 deaths per hour. Average injuries per web were 44, or 2.1 injuries per hour.

characters give a verbal description of a violation of law."

Violence was defined as "an act in which there is visible or implied harm of abuse to a person, animal, or property, or in which the tv actors give a verbal description of harm or abuse to a person, animal or property."

There were some gray areas, such as an attempt to kill accompanied by beating, pushing or shoving. These incidents were counted as crimes.

A non-violent crime could be more innocuous, but a crime nonetheless. One instance involved "fixing a race." Another involved a fraud, and so on.

The range of crime and violence was varied. For example, the VARIETY study revealed instances of "bullying" on "The Wonder Years" which were counted as violent acts. (See separate story.) But on other shows gun battles, car chases, murders, rapes and many other violent acts were counted.

The violence on tv controversy is a complex one. Earlier this month, the House passed a bill which essentially said members of the tv industry would not be in violation of antitrust laws if they got together to devise guidelines that would "alleviate the negative impact of violence," according to the bill. The Senate had earlier passed a slightly different bill after several died.

While physicians testifying before the House favored such a bill, the tv networks did not. Representatives from NBC and CBS said the bill was unnecessary. ABC said it would be almost impossible to implement.

This study does not presume to provide an answer to the controversy, but a "snapshot" of violence and crime on network tv during one week last spring.

Weapons of choice

Type	No. of incidents 3-web total
Pistol	14
Needle/drugs	4
Rope/necktie	4
Gun heard or not seen	3
Hands	2
Knife/razor	2
Rifle	2
Animal trap	1
Axe/hatchet/pick-ax	1
Blow torch	1
Blunt object	1
Book	1
Cannon	1
Car/truck	1
Electrocution	1
Explosive	1
Foot	1
Gun-automatic	1
Guns-varied	1
Stick/club	1
Tin can	1
Vase/crockery	1
Total	**46**

There was a wide array of weapons used to beat, blast, kill, maim, demolish or terrorize. The all-time favorite, of course, was the gun. But cannons, needles, electricity, hands, neckties and blowtorches were also found to be useful, if not always tasteful.

How the Networks Set Standards for Portraying Violent Behavior

VARIETY 8/16/89

New York Does television cause violent behavior or does it simply reflect it?

As far as the Big Three networks are concerned, the answer is obvious: tv reflects and mimics; it doesn't set standards of behavior.

While that may come as news to advertisers who spend billions each year on the medium to influence the public's behavior, the networks use this as their basic line of defense in the violence debate.

"Network standards are an attempt to gauge what the sensitivities of the public are and to stay in tune with them," says Alan Gerson, NBC's v.p. of program standards and marketing policy. "As far as the network is concerned, we are a respectful three paces behind" those sensitivities. "As the public's sensibilities change, so do the network's."

A CBS spokesman uses another metaphor: "We reflect societal norms; we're never ahead of that curve and will never be."

"Essentially, and this may sound self-serving, we've got a real handle" on tv violence, said Alfred Schneider, v.p. of policy and standards, Capital Cities/ABC. "ABC is always vigilant and vigorous about portrayal of violence."

The three webs have longstanding policies on violence in tv shows which essentially bar the depiction of "gratuitous" violence. In other words, violence may be depicted if it is "reasonably" related to plot development or character delineation.

ABC established "incident classification and analysis forms" in the mid-'70s that were designed to analyze closely how violence fits into the action, and to determine whether it was gratuitous.

Likewise, NBC's policy allows violence only if "sensitively" portrayed and not gratuitous or excessive. Also, "it shall not be depicted ... as a solution to problems," read the standards guidelines.

"We take a long and hard and careful look at violence," said Gerson. "We show the consequences of violence rather than the physical impact ... Sometimes we don't get credit for the care we take."

The networks also are particularly sensitive about being singled out in studies on tv violence. They argue that such studies fail to include the "universe" of viewing options, including cable and homevideo, which often feature far more violence than network shows.

Schneider says paycable and homevideo have a far greater incidence of violence, partly because their windows for theatricals open years before the networks' window opens. In that respect, he says the networks are years behind what the public already sees elsewhere today. *—Verne Gay*

The Top Ten Rated Shows: How Violent Were They?

VARIETY 8/16/89

ABC

The top 10 shows on ABC accounted for 63.3% of the ratings and 58.3% of the c/v acts. The reason: "Beverly Hills Cop." The boxoffice hit was the most violent show on primetime tv that week.

Without "Cop" on the sked that week, ABC's top 10 would have accounted for 72.3% of total ratings, but only 47% of all acts of c/v. The network had a total of 175 instances of c/v.

ABC's "Academy Awards" had eight c/v acts (not counting the Rob Lowe/Snow White song-and-dance routine at the telecast's outset). They occurred in movie clips.

CBS

The top 10 shows on CBS accounted for 58.9% of the web's total ratings that week, and had 49.5% of the c/v acts. Three shows had high counts, including top-rated "Murder, She Wrote," but the rest had fairly low c/v counts.

Ratings bombs like "Hard Time On Planet Earth," "Stroker Ace" and "The Morning After" accounted for a total of 30.6 ratings points and 35 acts of c/v. The net had a total of 111 acts of c/v.

NBC

The top 10 shows on NBC accounted for 50.4% of the network's ratings that week, but only 44.8% of all acts of crime or violence occurred in those shows. The average was jacked up by the telepic "Hillside Strangler" and "Hunter." But three of the top 10 shows had no crime or violence.

"The Cosby Show" is an example of a top-rated show with no recorded instances of either crime or violence. The network had a total of 96 incidences of crime and/or violence (c/v.)

ABC's 'A Man Called Hawk'

Top 10 rated shows by network

ABC

Rank	Show	Rating	No. of Acts
1	Academy Awards	29.8	8
2	Roseanne	25.5	3
3	The Wonder Years	20.9	11
4	Barbara Walters	20.3	3
5	Who's The Boss?	19.7	1
6	Beverly Hills Cop (M)	19.4	36
7	Anything But Love	17.6	5
8	B.L. Stryker	15.2	23
9	MacGyver	15.1	12
10	Full House	14.9	0
		Total	102*

CBS

Rank	Show	Rating	No. of Acts
1	Murder, She Wrote	21.2	13
2	Knots Landing	16.6	1
3	Dallas	14.8	5
4	Designing Women	14.6	0
5	Murphy Brown	13.8	1
6	NCAA Championships	13.5	0
7	Newhart	12.4	3
8	Falcon Crest	12.2	5
9	The Equalizer	11.9	13
10	Wildcats (M)	11.5	14
		Total	55

NBC

Rank	Show	Rating	No. of Acts
1	The Cosby Show	26.0	0
2	A Different World	24.8	3
3	Cheers	24.4	0
4	Hillside Strangler (M)	23.1	16
5	The Golden Girls	21.1	0
6	Dear John	20.0	0
7	Empty Nest	19.5	1
8	Hunter	19.0	12
9	Matlock	17.8	8
10	L.A. Law	17.8	3
		Total	43

M-Movie *-Filmclips **-News magazine shows

Top 10 most violent shows by network

ABC

Rank	Show	Rating	No. of Acts
1	Beverly Hills Cop (M)	19.4	36
2	A Man Called Hawk	7.8	26
3	B.L. Stryker	15.2	23
4	MacGyver	15.1	12
5	The Wonder Years	20.9	11
6	20/20 **	13.6	10
7	thirtysomething	11.0	9
8	The Shaggy Dog (M)	7.9	8
9	Academy Awards *	29.8	8
10	Moonlighting	6.9	7
		Total	150

CBS

Rank	Show	Rating	No. of Acts
1	Wildcats (M)	11.5	14
2	Stroker Ace (M)	10.6	13
3	The Equalizer	11.9	13
4	Murder, She Wrote	21.2	13
5	Hard Time On Planet Earth	9.8	12
6	The Morning After (M)	10.2	10
7	Tour Of Duty	9.2	6
8	48 Hours **	10.0	5
9	West 57th **	7.9	5
10	Dallas	14.8	5
		Total	96

NBC

Rank	Show	Rating	No. of Acts
1	Hillside Stranger (M)	23.1	16
2	In The Heat Of The Night	16.6	15
3	Hunter	19.0	12
4	Unsolved Mysteries	15.2	11
5	Matlock	17.8	8
6	Midnight Caller	13.5	7
7	Quantum Leap	10.2	5
8	Unsub	9.2	4
9	Nightingales	7.7	4
10	Friday Night Surprise	8.7	3
		Total	85

The Three Networks' 10 Most Violent Programs

Variety 8/16/89

ABC

ABC's top 10 accounted for 85.7% of all recorded instances of c/v on the web, but 47.1% of the ratings. The latter figure would have been much smaller without "Beverly Hills Cop."

CBS

CBS' top 10 shows counted for 86.5% of all c/v acts, but 48.4% of the ratings.

NBC

The top most violent shows on each network were not necessarily tops in terms of ratings. NBC's top 10 c/v shows accounted for 88.5% of all c/v acts on the network. But those shows accounted for only 33.3% of total ratings.

Top 10 shows without crime or violence

Show	Network	Rating
The Cosby Show	NBC	26.0
Cheers	NBC	24.4
The Golden Girls	NBC	21.1
Dear John	NBC	20.0
Alf	NBC	17.1
The Hogan Family	NBC	16.9
Amen	NBC	16.3
Full House	ABC	14.9
227	NBC	14.9
Designing Women	CBS	14.6

Believe it or not, there are plenty of net tv shows without even the slightest hint of crime or violence. They are certainly members of a minority group, but their clout is great. ABC's violence-free trio accounted for 42.4 rating points that week (out of a total of 313.5 rating points). CBS' trio accounted for 38.5 rating points, out of 241.8. NBC's 11 shows accounted for a whopping 192 rating points, out of a total of 423.7.

Network Evening News Shares Plummet From 76 to 59 Since 1979–80

VARIETY 9/20/89

New York Network evening news 3-web shares have dropped from 76 to 59 since 1979-80. CBS-TV has lost the most shares (8), followed by NBC-TV (5) and ABC-TV (4). It's more a case of CBS falling than of NBC and ABC catching up.

Erosion on national network news numbers has followed a different pattern than that of network primetime numbers. First of all, it began later for news than it did for primetime (about two season), and it hasn't followed the same relentless pattern of decrease.

For the past 10 years, the network newscast scores are as follows:

Season	ABC Rtg.	ABC Sh.	CBS Rtg.	CBS Sh.	NBC Rtg.	NBC Sh.	3-Net Rtg.	3-Net Sh.
'79-'80	12.1	24	14.0	28	12.1	24	38.2	76
'80-'81	11.8	23	13.5	26	11.8	23	37.1	72
'81-'82	11.7	21	12.8	24	11.7	21	35.9	68
'82-'83	10.7	20	13.1	25	10.6	20	34.4	66
'83-'84	10.2	20	12.4	23	10.1	19	32.7	62
'84-'85	10.4	21	12.4	24	10.4	20	33.2	64
'85-'86	11.0	19	12.4	23	11.4	22	34.8	66
'86-'87	10.0	20	10.8	21	11.2	22	32.0	62
'87-'88	10.2	20	10.7	21	9.7	19	30.6	60
'88-'89	10.7	20	10.8	20	10.1	19	31.6	59

Perusal of the above numbers shows a steady 4-share decline in 3-web shares for each of the first three seasons, then a 2-share drop in '82-'83, then another 4-share drop — and then a 2-share rise. Since that time the range between the last three seasons is in the 3-share range, suggesting the erosion slowed.

An overall appraisal of the 10-year action is that CBS has come back to the pack in that period. Although '86-'87 is the only season won by a web other than CBS (NBC), CBS is off eight share points since '79-'80, while NBC is down five and ABC four.

In individual web numbers, there are some incongruities. NBC's score for '88-'89 (when it ran third) is equal to its score in '83-'84. Its winning '86-'87 score (11.2/22) is followed in '87-'88 by 9.7/19 — the most drastic season-to-season change (three points) on the chart.

ABC's '88-'89 score (10.7/20) equals its score for '82-'83 and is higher than any season score for the web since '82-'83 with one exception — '85-'86, when it drew an 11.0/19. *—Bob Knight*

Regular Season Ratings

VARIETY 5/17/89

The following ratings cover the 1988-89 regular season, from Sept. 19, 1988, to April 16, 1989. Some 126 series qualify for the report, of which 34 have had five or fewer airings during the season (and are so marked).

Rank	Series	Rtg.
1.	The Cosby Show (NBC)	25.5
2.	Roseanne (ABC)	23.6
3.	A Different World (NBC)	22.9
4.	Cheers (NBC)	22.5
5.	The Golden Girls (NBC)	21.3
6.	Who's The Boss? (ABC)	20.9
7.	60 Minutes (CBS)	20.7
8.	Murder, She Wrote (CBS)	19.3
9.	Empty Nest (NBC)	19.1
10.	Anything But Love (ABC)	19.0
11.	Dear John (NBC)	18.5
12.	Alf (NBC)	17.6
	L.A. Law (NBC)	17.6
14.	Matlock (NBC)	17.5
15.	Unsolved Mysteries (NBC)	17.3
	In The Heat Of The Night (NBC)	17.3
17.	Growing Pains (ABC)	17.2
18.	Hunter (NBC)	17.0
19.	NFL Monday Night Football (ABC)	16.9
	Head Of The Class (ABC)	16.9
21.	Night Court (NBC)	16.7
22.	NBC Monday Movie	16.5
23.	The Wonder Years (ABC)	16.4
	The Hogan Family (NBC)	16.4
25.	One Of The Boys (NBC)*	16.3
26.	NBC Sunday Movie	16.2
27.	Knots Landing (CBS)	16.1
28.	CBS Sunday Movie	15.8
29.	ABC Tuesday Movie#	15.6
30.	Amen (NBC)	15.5
31.	ABC Mystery Movie	15.4
	Dallas (CBS)	15.4
33.	Jake & The Fatman (CBS)*	15.2
	Designing Women (CBS)	15.2
35.	Nearly Departed (NBC)*	15.1
	ABC Sunday Movie	15.1
37.	Head Of The Class (Thurs.) (R) (ABC)*#	15.0
38.	Murphy Brown (CBS)	14.9
39.	Full House (ABC)	14.7
	ABC Monday Movie*#	14.7
41.	227 (NBC)	14.6
42.	Family Ties (NBC)	14.3
	Head Of The Class (Tues.) (R) (ABC)*#	14.3
	NBC Movie Of The Week (Tues.)*#	14.3
45.	Midnight Caller (NBC)	14.1
	Growing Pains (Thurs.) (R) (ABC)*#	14.1
	CBS Monday Movie#	14.1
48.	My Two Dads (NBC)	14.0
49.	20/20 (ABC)	13.9
50.	thirtysomething (ABC)	13.8
51.	Growing Pains (Tues.) (R) (ABC)*#	13.7
52.	Perfect Strangers (ABC)	13.6
53.	MacGyver (ABC)	13.1
	Mr. Belvedere (ABC)	13.1
	Kate & Allie (CBS)	13.1
56.	Moonlighting (ABC)	12.9
	Dream Street (NBC)#	12.9
58.	Day By Day (NBC)	12.7
	Nightingales (NBC)	12.7
	Father Dowling Mysteries (NBC)	12.7
61.	Newhart (CBS)	12.6
62.	Just The Ten Of Us (ABC)	12.5

Rank	Program	Rating
	Wiseguy (CBS)	12.5
	ABC Thursday Movie*#	12.5
	Falcon Crest (CBS)	12.5
66.	CBS Tuesday Movie	12.3
67.	Highway To Heaven (NBC)*	12.2
68.	Hooperman (ABC)	12.1
69.	Coach (ABC)	11.9
	Baby Boom (NBC)	11.9
71.	China Beach (ABC)	11.7
72.	Robert Guillaume Show (ABC)*	11.6
73.	CBS Thursday Movie*#	11.4
74.	Coming Of Age (CBS)*	11.1
	Miami Vice (NBC)	11.1
76.	The Equalizer (CBS)	11.0
77.	CBS Friday Movie*#	10.9
78.	Paradise (CBS)	10.6
	Almost Grown (CBS)	10.6
80.	Dynasty (ABC)	10.5
	The Cavanaughs (CBS)*	10.5
82.	Who's The Boss? (Sun.) (R) (ABC)*#	10.4
83.	Live-In (CBS)	10.3
	Unsub (NBC)	10.3
85.	Perfect Strangers (Sun.) (R) (ABC)*#	10.2
86.	Beauty & The Beast (CBS)	10.1
87.	48 Hours (CBS)	10.0
	Magical World Of Disney (NBC)	10.0
	Heartland (CBS)	10.0
90.	Who's The Boss? (Sat.) (R) (ABC)*#	9.8
91.	ABC Saturday Movie *#	9.6
	Mission: Impossible (ABC)	9.6
	Perfect Strangers (Sat.) (R) (ABC)*#	9.6
94.	Tour Of Duty (CBS)	9.3
	Tattinger's (NBC)#	9.3
96.	High Risk (CBS)*#	9.2
97.	A Man Called Hawk (ABC)	9.0
	Quantum Leap (NBC)*	9.0
	Incredible Sunday (ABC)	9.0
	Sonny Spoon (NBC)#	9.0
101.	Hard Time On Planet Earth (CBS)	8.9
102.	Something Is Out There (NBC)#	8.8
103.	Dolphin Cove (CBS)	8.7
104.	Police Story (ABC)*#	8.1
	Smothers Brothers Comedy Hour (CBS)#	8.1
106.	ABC Family Classics (ABC)	7.9
	Live! Dick Clark Presents (CBS)#	7.9
	Mr. Belvedere (Sun.) (R) (ABC)*#	7.9
109.	Annie McGuire (CBS)	7.8
	The Van Dyke Show (CBS)	7.8
111.	Jim Henson Hour (NBC)*	7.7
112.	Dirty Dancing (CBS)	7.4
113.	West 57th (CBS)	7.2
	Burning Questions (ABC)*	7.2
115.	Simon & Simon (CBS)	6.9
116.	Knightwatch (ABC)	6.6
117.	Great Circuses Of The World (ABC)	6.5
118.	Heartbeat (ABC)	6.4
119.	Murphy's Law (ABC)	6.3
120.	TV 101 (CBS)	6.1
121.	Raising Miranda (CBS)	6.0
122.	A Fine Romance (ABC)	5.4
123.	First Impressions (CBS)*#	5.0
	Frank's Place (R) (CBS)*#	5.0
125.	Studio 5B (ABC)*#	4.9
	Men (ABC)*#	4.9

*5 episodes or less # Canceled

All-Time Top 75 Programs

VARIETY 5/17/89

New York The American viewer's program preference has evolved from series such as "Gunsmoke," "Bonanza," "Wagon Train" and "Beverly Hillbillies" (which rank in the lower half of the listing) to Super Bowl games.

Regular series make up 45% of the programs ranked, with sporting events second with 25%, followed by miniseries (16%), movies (8%) and specials (6%).

(From Jan. 30, 1960 through April 17, 1989) Nielsen Media Research

RANK	PROGRAM	TELECAST DATE	NETWORK	MINS.	AVERAGE AUDIENCE (%)	SHARE	AVERAGE AUDIENCE (000)
1	MASH Special	Feb. 28, 1983	CBS	150	60.2	77	50,150
2	Dallas	Nov. 21, 1980	CBS	60	53.3	76	41,470
3	Roots-Pt. VIII	Jan. 30, 1977	ABC	115	51.1	71	36,380
4	Super Bowl XVI	Jan. 24, 1982	CBS	213	49.1	73	40,020
5	Super Bowl XVII	Jan. 30, 1983	NBC	204	48.6	69	40,480
6	Super Bowl XX	Jan. 26, 1986	NBC	231	48.3	70	41,490
7	Gone With The Wind-Pt. 1	Nov. 7, 1976	NBC	179	47.7	65	33,960
8	Gone With The Wind-Pt. 2	Nov. 8, 1976	NBC	119	47.4	64	33,750
9	Super Bowl XII	Jan. 15, 1978	CBS	218	47.2	67	34,410
10	Super Bowl XIII	Jan. 21, 1979	NBC	230	47.1	74	35,090
11	Bob Hope Christmas Show	Jan. 15, 1970	NBC	90	46.6	64	27,260
12	Super Bowl XVIII	Jan. 22, 1984	CBS	218	46.4	71	38,800
	Super Bowl XIX	Jan. 20, 1985	ABC	218	46.4	63	39,390
14	Super Bowl XIV	Jan. 20, 1980	CBS	178	46.3	67	35,330
15	The Day After	Nov. 20, 1983	ABC	144	46.0	62	38,550
16	Roots Pt. VI	Jan. 28, 1977	ABC	120	45.9	66	32,680
16	The Fugitive	Aug. 29, 1967	ABC	60	45.9	72	25,700
18	Super Bowl XXI	Jan. 25, 1987	CBS	206	45.8	66	40,030
19	Roots Pt. V	Jan. 27, 1977	ABC	60	45.7	71	32,540

20	Ed Sullivan	Feb. 9, 1964	CBS	60	45.3	60	23,240
21	Bob Hope Christmas Show	Jan. 14, 1971	NBC	90	45.0	61	27,050
22	Roots Pt. III	Jan. 25, 1977	ABC	60	44.8	68	31,900
23	Super Bowl XI	Jan. 9, 1977	NBC	204	44.4	73	31,610
	Super Bowl XV	Jan. 25, 1981	NBC	220	44.4	63	34,540
25	Super Bowl VI	Jan. 16, 1972	CBS	170	44.2	74	27,540
26	Roots Pt. II	Jan. 24, 1977	ABC	120	44.1	62	31,400
27	Beverly Hillbillies	Jan. 8, 1964	CBS	30	44.0	65	22,570
28	Roots Pt. IV	Jan. 26, 1977	ABC	60	43.8	66	31,190
	Ed Sullivan	Feb. 16, 1964	CBS	60	43.8	60	22,445
30	Super Bowl XXIII	Jan. 22, 1989	NBC	213	43.5	68	39,320
31	Academy Awards	April 7, 1970	ABC	145	43.4	78	25,390
32	Thorn Birds Pt. III	March 29, 1983	ABC	120	43.2	62	35,990
33	Thorn Birds Pt. IV	March 30, 1983	ABC	180	43.1	62	35,900
34	CBS NFC Championship	Jan. 10, 1982	CBS	195	42.9	62	34,960
35	Beverly Hillbillies	Jan. 15, 1964	CBS	30	42.8	62	21,960
36	Super Bowl VII	Jan. 14, 1973	NBC	185	42.7	72	27,670
37	Thorn Birds Pt. II	March 28, 1983	ABC	120	42.5	59	35,400
38	Super Bowl IX	Jan. 12, 1975	NBC	190	42.4	72	29,040
	Beverly Hillbillies	Feb. 26, 1964	CBS	30	42.4	60	21,750
40	Super Bowl X	Jan. 18, 1976	CBS	200	42.3	78	29,440
	Airport	Nov. 11, 1973	ABC	170	42.3	63	28,000
40	Love Story	Oct. 1, 1972	ABC	120	42.3	62	27,410
	Cinderella	Feb. 22, 1965	CBS	90	42.3	59	22,250
	Roots Pt. VII	Jan. 29, 1977	ABC	60	42.3	65	30,120
45	Beverly Hillbillies	March 25, 1964	CBS	30	42.2	59	21,650
46	Beverly Hillbillies	Feb. 5, 1964	CBS	30	42.0	61	21,550
47	Beverly Hillbillies	Jan. 29, 1964	CBS	30	41.9	62	21,490
	Super Bowl XXII	Jan. 31,1988	ABC	229	41.9	62	37,120
49	Miss America Pageant	Sept. 9, 1961	CBS	150	41.8	75	19,600
	Beverly Hillbillies	Jan. 1, 1964	CBS	30	41.8	59	21,440
	Wagon Train	Jan. 27, 1960	NBC	30	41.8	62	18,890
52	Super Bowl VIII	Jan. 13, 1974	CBS	160	41.6	73	27,540
	Bonanza	March 8, 1964	NBC	60	41.6	62	21,340
54	Beverly Hillbillies	Jan. 22, 1964	CBS	30	41.5	61	21,290
55	Gunsmoke	April 2, 1960	CBS	30	41.4	69	18,710
	Bonanza	Feb. 16, 1964	NBC	60	41.4	60	21,240
57	Gunsmoke	Mach 5, 1960	CBS	30	41.3	66	18,670
	The Cosby Show	Jan. 22, 1987	NBC	30	41.3	56	36,100
59	Academy Awards	April 10, 1967	ABC	150	41.2	75	22,620
60	Bonanza	Feb. 9, 1964	NBC	60	41.0	58	21,030
	Winds Of War-Pt.7	Feb. 13, 1983	ABC	177	41.0	56	34,150
62	Gunsmoke	Jan. 28, 1961	CBS	30	40.9	65	19,180
63	Bonanza	March 28, 1965	NBC	60	40.8	63	21,460
	Gunsmoke	Jan. 30, 1960	CBS	30	40.8	63	18,440
65	All In The Family	Jan. 8, 1972	CBS	30	40.7	62	25,270
	Bonanza	March 7, 1965	NBC	60	40.7	61	21,410
67	Gunsmoke	March 26, 1960	CBS	30	40.6	66	18,350
	Beverly Hillbillies	Feb. 20, 1963	CBS	30	40.6	59	20,220
69	Gunsmoke	Feb. 25, 1961	CBS	30	40.5	64	19,000
	Beverly Hillbillies	May 1, 1963	CBS	30	40.5	62	20,170
	Roots Pt. I	Jan. 23, 1977	ABC	120	40.5	61	28,840
	Wagon Train	Feb. 10, 1960	NBC	30	40.5	61	18,310
	Bonanza	Feb. 2, 1964	NBC	60	40.5	58	20,780
74	Gunsmoke	Dec. 17, 1960	CBS	30	40.4	68	18,260
75	Gunsmoke	May 7, 1960	CBS	30	40.4	67	18,260
	Bonanza	Feb. 21, 1965	NBC	60	40.4	61	21,250

PLEASE NOTE:

- Average audience % rankings based on reports from Jan. 30, 1960, through April 17, 1989.
- Above data represent sponsored programs telecast on individual networks, i.e., no unsponsored or joint network telecasts are reflected in the above listings.
- Programs under 30 minutes scheduled duration are excluded.

Theatrical Movie Rankings, 1988–89

Sept. 1, 1988 — Aug. 31, 1989

Rank	Title	Web	Date	Rtg.	Share
1.	Top Gun	NBC	5- 8	20.3	32
2.	Beverly Hills Cop (R) (S)	ABC	4- 2	19.4	30
3.	Witness (S)	CBS	5- 7	19.1	31
4.	Romancing The Stone (R) (S)	ABC	4- 9	18.2	28
5.	The Wizard Of Oz (R) (S)	CBS	3-19	18.1	27
6.	Back To The Future (S)	NBC	11-13	18.0	26
7.	Ferris Bueller's Day Off	NBC	5-14	17.7	29
8.	Raiders Of The Lost Ark (R) (S)	ABC	2-12	17.5	28
9.	Sudden Impact (R) (S)	ABC	1- 8	17.4	26
10.	Cocoon (S)	CBS	10- 2	17.1	31
	Spies Like Us	NBC	12-11	17.1	27
12.	Raw Deal	ABC	1-15	16.7	26
13.	Manhunter (S)	NBC	3- 5	15.7	26
14.	Down And Out In Beverly Hills (S)	ABC	11- 6	15.5	25
	The Karate Kid Part II (S)	NBC	11-15	15.5	24
16.	Commando	ABC	10-30	15.4	25
17.	Return Of The Jedi (S)	NBC	3-19	15.1	23
18.	The Delta Force (S)	NBC	11-29	15.0	23
19.	Ghostbusters (R) (S)	ABC	1-29	14.7	23
	The Ten Commandments (R) (S)	ABC	3-26	14.7	26
21.	Pretty In Pink	NBC	4- 3	14.6	22
22.	Rambo: First Blood Part II (R)	NBC	3-12	14.2	22
23.	Stand By Me (S)	ABC	9-27	13.8	22
24.	Night Shift (R) (S)	ABC	9-28	13.5	22
25.	Tough Guys	NBC	4-24	13.4	22
26.	Runaway (S)	CBS	11-30	13.3	22
	Gung Ho	NBC	1-15	13.3	21
28.	Irreconcilable Differences (R)	NBC	8-20	13.1	23
	Agnes Of God	CBS	9-29	13-1	21
30.	The Sound Of Music (R) (S)	NBC	12-30	13.0	23
31.	Goonies (S)	NBC	10-23	12.8	20
	Gremlins (S)	NBC	1- 1	12.8	20
33.	Out Of Africa (Part 2) (S)	CBS	10-17	12.7	20
	Star Wars (R) (S)	CBS	11-23	12.7	21
	One Magic Christmas (S)	NBC	12-23	12.7	23
36.	Tank	CBS	9- 1	12.6	22
	A View To A Kill (R) (S)	ABC	1- 1	12.6	20
38.	Ruthless People	ABC	2- 5	12.4	18
39.	Runaway Train	NBC	1-30	12.3	19
	Police Academy 3: Back In Training (S)	CBS	11-15	12.3	18
41.	Out Of Africa (Part 1) (S)	CBS	10-16	12.1	20
	The Man With One Red Shoe (S)	ABC	9-20	12.1	20
43.	For Your Eyes Only (R)	ABC	6-11	12.0	21
44.	Twice In A Lifetime (S)	NBC	10-11	11.9	19
	Winnie The Pooh & The Blustery Day (R) (S)	NBC	3- 5	11.9	19
46.	Starman	CBS	1-10	11.8	19
47.	Wildcats	CBS	3-28	11.5	19
48.	The Man With The Golden Gun (R) (S)	ABC	7- 9	11.4	20
49.	Aliens (S)	CBS	3-14	11.3	18
	Baby It's You	NBC	9- 5	11.3	19
	The Karate Kid Part II (R) (S)	NBC	2-24	11.3	19
52.	Police Academy (R) (S)	NBC	11-11	11.2	19
	Armed And Dangerous (S)	CBS	1- 7	11.2	18
54.	Murphy's Romance (R) (S)	CBS	1-30	11.1	17
	Heaven Can Wait (R)	ABC	1-22	11.1	18
	Cannonball Run II (R)	ABC	12-25	11.1	24
	Star Trek II: The Wrath Of Khan (R) (S)	ABC	6-18	11.1	20
58.	All Of Me (R) (S)	ABC	9-22	10.9	18
	Superman III (R) (S)	ABC	12-31	10.9	22
	Never Say Never Again (R) (S)	ABC	3-12	10.9	17
	Police Academy 3: Back In Training (R) (S)	CBS	8-29	10.9	19
62.	Mad Max Beyond Thunderdome (R)	NBC	7- 2	10.8	21
	Rocky IV (R)	CBS	4- 4	10.8	17
64.	You Only Live Twice (R) (S)	ABC	7-23	10.7	19
	Winnie The Pooh & A Day For Eeyore (S)	NBC	1-29	10.7	17
66.	Dr. No (R)	ABC	7-10	10.6	18
	Stroker Ace (R) (S)	CBS	3-29	10.6	17
	St. Elmo's Fire	CBS	12-27	10.6	18
69.	The Spy Who Loved Me (R) (S)	ABC	8-20	10.5	19
	Starting Over (R)	ABC	7-30	10.5	19
71.	Magnum Force (R) (S)	ABC	11-19	10.4	18
	Places In The Heart (R) (S)	NBC	5-28	10.4	21
73.	Mary Poppins (R) (S)	ABC	11-24	10.3	19
	The Spy Who Loved Me (R) (S)	ABC	9-15	10.2	17
74.	The Morning After	CBS	4-2	10.2	16
76.	Dinosaur . . . Secret Of The Lost Legend (S)	NBC	1- 8	10.0	15
	Jaws III (R)	CBS	6-23	10.0	20
78.	Star Trek — The Motion Picture (R) (S)	ABC	8-13	9.9	18
79.	Swiss Family Robinson (Part 2) (R) (S)	NBC	3-19	9.8	16
80.	Goldfinger (R) (S)	ABC	1-14	9.7	17
81.	Benji The Hunted (S)	NBC	2-19	9.6	15
82.	Rocky II (R)	CBS	5-30	9.5	15
	Scarface (S)	ABC	1- 7	9.5	16
	Pee-wee's Big Adventure (R) (S)	CBS	9-16	9.5	17
85.	Live And Let Die (R) (S)	ABC	6-25	9.4	17
	Oh God! You Devil (R) (S)	NBC	1- 6	9.4	15
87.	Club Paradise (S)	ABC	10- 1	9.3	17
	Stripes (R)	CBS	11-29	9.3	14
	The Toy (R) (S)	ABC	12-17	9.3	17
	Swiss Family Robinson (Part 1) (R) (S)	NBC	3-12	9.3	16
91.	The Journey Of Natty Gann (R) (S)	ABC	9- 3	9.2	19
	Home Is Where The Heart Is (R)	NBC	6- 5	9.2	16
93.	Paternity (R)	NBC	6-30	9.0	18
94.	Poltergeist (R) (S)	ABC	9- 4	8.8	17
	Racing With The Moon	ABC	9- 8	8.8	15
	Gladiator (R)	ABC	8-27	8.8	15
	National Lampoon's European Vacation (R) (S)	CBS	4- 8	8.8	16
98.	Twilight Zone — The Movie (R)	CBS	6-30	8.6	18
99.	Silkwood (R) (S)	ABC	7- 2	8.5	17
	White Nights (S)	CBS	9-20	8.5	14
	'night Mother	ABC	6-12	8.5	15
	Earth's Final Fury (R) (S)	NBC	5-26	8.5	17
	Purple Hearts	CBS	3-21	8.5	14
104.	Ladyhawke (S)	CBS	7- 7	8.3	17
105.	Little Treasure	CBS	7-21	8.2	16
	The Shaggy Dog (Part 1) (R) (S)	ABC	3-23	8.2	14
107.	Santa Claus: The Movie (S)	ABC	12-14	8.1	19
108.	High Road To China (R) (S)	CBS	5-26	8.0	16
109.	Escape To Freedom	ABC	7-24	7.9	14
110.	Alien (R) (S)	ABC	7-16	7.8	14
	Bedknobs And Broomsticks (R) (S)	ABC	2-12	7.8	12
	Parent Trap (R) (S)	ABC	5-18	7.8	13
	The Shaggy Dog (Part 2) (R) (S)	ABC	3-30	7-8	13
114.	Bad Medicine	ABC	7-31	7.6	13
	Pete's Dragon (R) (S)	ABC	2-19	7.6	12
116.	The Man Who Wasn't There	NBC	9- 4	7.5	14
	Protocol (R) (S)	NBC	5-19	7.5	14
118.	20,000 Leagues Under The Sea (R) (S)	ABC	8- 5	7.3	16
119.	The Apple Dumpling Gang (Part 2) (R) (S)	NBC	8-15	7.2	15
	A Fine Mess (S)	CBS	1-14	7.2	12
121.	Lucas (S)	CBS	6-10	7.1	15
122.	Deal Of The Century (S)	CBS	10- 1	7.0	13
123.	King David	NBC	7-21	6.8	14
	The Keep	NBC	7-28	6.8	14
125.	The Apple Dumpling Gang (Part 1) (R) (S)	NBC	8- 6	6.7	14
126.	On Her Majesty's Secret Service (R) (S)	ABC	5-28	6.6	13
127.	The Barefoot Executive (Part 1) (R) (S)	NBC	8-27	6.1	13
128.	Real Genius (S)	CBS	6- 3	6.0	12
	The Sting II (R) (S)	ABC	7-15	6.0	13
130.	Herbie Goes Bananas (R) (S)	ABC	1-22	5.9	9
131.	Supergirl (R) (S)	ABC	7- 1	5.8	14
132.	Alice In Wonderland (R) (S)	ABC	12-24	5.3	13
133.	The Devil & Max Devlin (R) (S)	ABC	8-12	5.0	10

(R) - Repeat (S) - Special

1988–89 Miniseries Ratings

VARIETY 9/13/89

Sept. 1, 1988 — Aug. 31, 1989

Rank	Title	Rtg.	Share
1.	Lonesome Dove (CBS)	26.1	39
2.	I Know My First Name Is Steven (NBC)	24.5	39
3.	The Women Of Brewster Place (ABC)	24.0	37
4.	Brotherhood Of The Rose (NBC)	20.1	33
5.	War & Remembrance I (ABC)	18.6	29
6.	Jack The Ripper (CBS)	17.6	29
7.	Favorite Son (NBC)	16.0	26
8.	Internal Affairs (CBS)	14.9	24
9.	War & Remembrance II (ABC)	14.8	23
	Twist Of Fate (NBC)	14.8	23
11.	From The Dead Of Night (NBC)	14.4	22
12.	Around The World In 80 Days (NBC)	13.9	22
13.	The Great Escape II: The Untold Story (NBC)	13.4	21
14.	North & South (ABC) (R)	13.1	21
15.	If Tomorrow Comes (CBS) (R)	12.4	21
16.	Passion & Paradise (ABC)	12.2	20
17.	Guts & Glory: The Rise & Fall Of Oliver North (CBS)	11.8	19
18.	Hands Of A Stranger (NBC) (R)	11.5	20
19.	Dadah Is Death (CBS)	10.7	18
20.	Billionaire Boys Club (NBC) (R)	10.1	18
21.	This Is America, Charlie Brown (CBS)	9.5	17
22.	On Wings Of Eagles (NBC) (R)	8.3	16
23.	Monte Carlo (CBS) (R)	7.9	15
24.	Sins (CBS) (R)	7.7	15
25.	Earth*Star Voyager (ABC) (R)	6.4	12
26.	North & South, Book II (ABC) (R)	6.3	13
27.	Fresno (CBS) (R)	5.5	11

(R) - Repeat

Primetime Track Record

VARIETY 9/13/89

At start of 1989-90 season

ABC	CBS	NBC
10 years or more		
NFL Football (19) 20/20* (11)	60 Minutes* (14) Dallas (12) Knots Landing* (10)	Monday Movies (13)
5 to 9 years		
Who's The Boss? (5) Mr. Belvedere* (5)	Falcon Crest (8) Newhart (7) Murder, She Wrote (5)	Sunday Movies (8) Cheers (7) Night Court* (6) Cosby Show (5) Hunter (5)
4 years		
MacGyver Growing Pains Perfect Strangers*	Sunday Movie* Saturday Night with Connie Chung *#	Alf Golden Girls 227 Hogan Family*
3 years		
Head Of The Class	Designing Women	Matlock L.A. Law Amen
2 years		
thirtysomething China Beach* Wonder Years* Just The Ten Of Us*	Tour Of Duty Jake And The Fatman Wiseguy 48 Hours	A Different World My Two Dads In The Heat Of The Night*
1 year		
Roseanne Anything But Love* Mission: Impossible ABC Saturday Mystery*	Murphy Brown Paradise	Midnight Caller Unsolved Mysteries Quantum Leap* Dear John Empty Nest

**Midseason replacement*
#First 4 years as "West 57th"

Top-Rated Syndication Specials

VARIETY 8/23/89

Rank	Title	Date	AA%
1.	The Mystery Of Al Capone's Vaults	4/86	33.4*
2.	Return Of The Titanic	11/87	24.7*
3.	Nixon's Interview No. 1	5/77	24.3
4.	Andrea Doria: The Final Chapter	8/84	19.3*
5.	Geraldo: Murder Live From Death Row	4/88	18.7*
6.	It Came Upon A Midnight Clear	12/84	17.3*
7.	Star Trek: The Next Generation	10/87	16.3*
8.	American Vice: The Doping Of A Nation	12/86	16.2*
9.	Solid Gold Special	3/82	15.6*
10.	TV Net: Death Wish III	3/88	15.4*
11.	Miracle On 34th Street (colorized 1947)	12/85	14.7
	Solid Gold Christmas Special	12/82	14.7*
	The Yearling	9/73	14.7
14.	Solid Gold Countdown '83	3/84	14.5*
15.	Countdown '82: A Solid Gold Special	3/83	14.2*
16.	Universal Pictures Debut Net: Weird Science	10/88	14.1*
	A Cosmic Christmas	12/77	14.1
18.	White Christmas	12/86	13.7*
	14th Annual Music City News Awards	7/80	13.7
20.	A Cosmic Christmas	12/78	13.5
21.	TV Net Movie: Murphy's Law	9/88	13.4*
22.	Universal Pictures Debut Net: Sixteen Candles	12/86	13.3*
23.	Sons Of Scarface: The New Mafia	8/87	13.2*
24.	Mysteries Of The Pyramids . . . Live	5/88	13.1*
	Solid Gold Christmas Special	12/83	13.1*
	Solid Gold '79	3/80	13.1

*Includes multiple airings.
Source: Nielsen Media Research.

MONDAY-FRIDAY NIELSEN RATINGS

(Arithmetic averages, not weighted by market size)

PRIME ACCESS

Network Affiliates

(TOP 100 MARKETS)

Household Share/Rank	Program	No. Of Stations	Nov. 87 Shr	Nov. 87 +/–	Feb. 88 Shr	Feb. 88 +/–	May 88 Shr	May 88 +/–	NOVEMBER 1988 Lead-In Shr	Lead-In +/–	H'hold Shr	H'hold Rtg	W25-54 Shr	W25-54 Rank	M25-54 Shr	M25-54 Rank	Kids Shr	Kids Rank
1	Wheel Of Fortune	93	33	–3	32	–2	32	–2	26	+4	30	18	24	2	20	5	17	5
2	Jeopardy	54	28	NC	28	NC	27	+1	27	+1	28	16	23	3	20	5	13	8
3	Cosby Show	30	19	+5	18	+6	19	+5	22	+2	24	14	28	1	24	2	37	1
4	PM Magazine	19	23	–1	22	NC	21	+1	24	–2	22	13	23	3	20	5	13	8
5	Cheers	10	17	+3	19	+1	20	NC	20	NC	20	11	22	6	23	3	17	5
	Night Court	5	16	+4	15	+5	16	+4	22	–2	20	12	23	3	25	1	24	3
7	Mash	4	20	–2	19	–1	20	–2	16	+2	18	10	18	8	21	4	13	8
8	Entertainment Tonight	40	15	+2	16	+1	16	+1	17	NC	17	10	19	7	16	9	9	13
	Family Ties	4	18	–1	19	–2	20	–3	23	–6	17	10	18	8	16	9	32	2
10	Hollywood Squares	4	21	–5	23	–7	23	–7	19	–3	16	10	15	13	13	11	13	8
11	Current Affair	24	12	+3	13	+2	14	+1	13	+2	15	9	17	10	17	8	7	14
	Family Feud	28	16	–1	15	NC	14	+1	15	NC	15	9	16	11	13	11	19	4
13	Win, Lose Or Draw	64	17	–3	18	–4	18	–4	16	–2	14	8	16	11	12	15	17	5
14	Newhart	3	19	–6	18	–5	17	–4	22	–9	13	7	12	14	13	11	11	12
15	USA Today	67	16	–4	17	–5	16	–4	19	–7	12	7	12	14	13	11	4	15

DAYTIME

Network Affiliates

(TOP 100 MARKETS)

Household Share/Rank	Program	No. Of Stations	Nov. 87 Shr	Nov. 87 +/–	Feb. 88 Shr	Feb. 88 +/–	May 88 Shr	May 88 +/–	NOVEMBER 1988 Lead-In Shr	Lead-In +/–	H'hold Shr	H'hold Rtg	W25-54 Shr	W25-54 Rank	M25-54 Shr	M25-54 Rank	Kids Shr	Kids Rank
1	Donahue	55	34	–1	35	–2	33	NC	25	+8	33	7	35	2	33	2	6	11
	Oprah Winfrey Show	8	30	+3	29	+4	30	+3	21	+12	33	10	43	1	29	3	6	11
3	Geraldo	52	21	+8	22	+7	24	+5	22	+7	29	6	35	2	36	1	8	7
4	Sally Jessy Raphael	61	23	+2	25	NC	25	NC	28	–3	25	5	30	4	20	4	5	15
5	Jeopardy	4	19	+1	19	+1	20	NC	20	NC	20	5	18	5	18	6	10	3
6	Divorce Court	13	18	–1	19	–2	19	–2	20	–3	17	3	18	5	18	6	9	6
	Peoples Court	10	18	–1	18	–1	16	+1	16	+1	17	3	14	13	16	8	8	7
8	Hour Magazine	32	20	–4	20	–4	18	–2	20	–4	16	3	17	8	11	17	3	23
	Love Connection	8	17	–1	16	NC	17	–1	17	–1	16	3	17	8	15	11	6	11
10	Hollywood Squares	9	15	NC	15	NC	15	NC	18	–3	15	3	14	13	13	13	8	7
	Judge	12	16	–1	17	–2	18	–3	18	–3	15	3	16	10	16	8	7	10
	Live w/Regis & Kathie Lee	71	16	–1	18	–3	17	–2	18	–3	15	3	18	5	13	13	5	15
	Magnum, P.I.	8	16	–1	18	–3	18	–3	22	–7	15	3	15	11	20	4	10	3
14	Superior Court	18	15	–1	17	–3	16	–2	17	–3	14	3	14	13	13	13	5	15
15	Facts of Life	4	20	–8	14	–2	14	–2	19	–7	12	3	13	16	16	8	17	2
	Family Medical Center	34	16	–4	15	–3	14	–2	18	–6	12	2	15	11	11	17	5	15
	Sweethearts	19	17	–5	17	–5	16	–4	21	–9	12	2	12	17	13	13	5	15
18	Family Ties	3	27	–17	23	–13	22	–12	19	–9	10	2	7	23	10	19	6	11
	Gimme A Break	3	16	–6	11	–1	12	–2	15	–5	10	2	12	17	5	23	25	1
	Group One Medical	28	16	–6	15	–5	14	–4	18	–8	10	2	11	19	9	20	4	20
	New Newlywed Game	4	15	–5	14	–4	10	NC	9	+1	10	2	9	22	14	12	3	23
	On Trial	17	16	–6	17	–7	17	–7	14	–4	10	2	11	19	9	20	4	20
	Wipeout	19	13	–3	15	–5	14	–4	12	–2	10	2	10	21	8	22	4	20
24	Love Boat	3	9	–2	10	–3	15	–8	12	–5	7	1	7	23	4	24	10	3

EARLY FRINGE

Network Affiliates

(TOP 100 MARKETS)

Household Share/Rank	Program	No. Of Stations	Nov. 87 Shr	Nov. 87 +/–	Feb. 88 Shr	Feb. 88 +/–	May 88 Shr	May 88 +/–	NOVEMBER 1988 Lead-In Shr	Lead-In +/–	H'hold Shr	H'hold Rtg	W25-54 Shr	W25-54 Rank	M25-54 Shr	M25-54 Rank	Kids Shr	Kids Rank
1	Oprah Winfrey Show	91	31	+5	33	+3	33	+3	24	+12	36	12	48	1	29	1	6	25
2	Wheel Of Fortune	4	28	+4	28	+4	29	+3	30	+2	32	15	27	3	19	10	13	11
3	Cosby Show	51	21	+5	21	+5	20	+6	18	+8	26	11	31	2	27	3	31	2
	Jeopardy	42	27	–1	27	–1	27	–1	24	+2	26	10	25	6	23	7	5	26
5	Andy Griffith Show	7	22	NC	22	NC	21	+1	25	–3	22	8	25	6	29	1	20	5
	Donahue	43	22	NC	22	NC	22	NC	20	+2	22	8	26	5	20	8	3	37
	Geraldo	35	17	+5	17	+5	18	+4	21	+1	22	8	27	3	26	4	4	29

MONDAY-FRIDAY NIELSEN RATINGS

(Arithmetic averages, not weighted by market size)

EARLY FRINGE Affiliates

Continued from page 131

Household Share/Rank	Program	No. Of Stations	Nov. 87 Shr	Nov. 87 +/−	Feb. 88 Shr	Feb. 88 +/−	May 88 Shr	May 88 +/−	NOVEMBER 1988 Lead-In Shr	Lead-In +/−	H'hold Shr	H'hold Rtg	W25-54 Shr	W25-54 Rank	M25-54 Shr	M25-54 Rank	Kids Shr	Kids Rank
8	Cheers	14	17	+2	20	−1	20	−1	20	−1	19	8	23	8	24	6	10	16
	Mash	20	19	NC	19	NC	20	−1	17	+2	19	7	18	11	25	5	9	19
10	People's Court	82	20	−2	20	−2	20	−2	19	−1	18	7	17	13	17	14	4	29
11	Divorce Court	21	17	NC	18	−1	17	NC	21	−4	17	6	15	17	14	20	3	37
	Family Feud	40	18	−1	18	−1	17	NC	18	−1	17	6	17	13	15	18	8	22
	Hour Magazine	7	26	−9	20	−3	22	−5	22	−5	17	5	18	11	8	34	1	46
14	Entertainment Tonight	5	19	−3	20	−4	19	−3	17	−1	16	7	19	10	18	12	5	26
	Jeffersons	5	21	−5	19	−3	21	−5	16	NC	16	6	15	17	17	14	15	8
	Judge	41	17	−1	17	−1	17	−1	19	−3	16	5	14	22	14	20	4	29
	Superior Court	31	17	−1	18	−2	17	−1	17	−1	16	5	14	22	16	16	4	29
	Win, Lose Or Draw	31	18	−2	19	−3	17	−1	21	−5	16	5	20	9	13	25	7	24
19	A-Team	4	19	−4	16	−1	17	−2	17	−2	15	5	11	35	18	12	17	7
	Current Affair	26	17	−2	16	−1	15	NC	19	−4	15	5	17	13	14	20	4	29
	Different Strokes	10	20	−5	18	−3	20	−5	22	−7	15	4	14	22	11	29	18	6
	Night Court	17	16	−1	17	−2	16	−1	15	NC	15	6	17	13	20	8	10	16
	Simon & Simon	5	18	−3	20	−5	17	−2	24	−9	15	5	13	26	15	18	3	37
24	Love Connection	17	15	−1	17	−3	17	−3	18	−4	14	4	15	17	13	25	3	37
	Newhart	5	15	−1	15	−1	16	−2	17	−3	14	5	15	17	16	16	8	22
	On Trial	4	18	−4	15	−1	16	−2	24	−10	14	5	12	32	19	10	5	26
	Three's Company	10	16	−2	16	−2	13	+1	12	+2	14	5	13	26	14	20	12	14
28	Facts Of Life	17	15	−2	15	−2	15	−2	13	NC	13	4	13	26	8	34	15	8
	Family Ties	33	16	−3	16	−3	14	−1	16	−3	13	5	13	26	12	28	14	10
	Gimme A Break	7	17	−4	19	−6	18	−5	16	−3	13	4	12	32	10	32	13	11
	Kate & Allie	13	17	−4	16	−3	16	−3	15	−2	13	5	14	22	13	25	9	19
	Little House On The Prairie	9	19	−6	17	−4	17	−4	18	−5	13	4	15	17	11	29	11	15
	Silver Spoons	4	17	−4	15	−2	14	−1	14	−1	13	5	13	26	7	37	21	4
	Duck Tales	5	14	−1	16	−3	16	−3	20	−7	13	4	7	39	8	34	42	1
35	USA Today	10	18	−7	19	−8	18	−7	16	−5	11	5	10	37	9	33	2	44
	Webster	7	18	−7	17	−6	15	−4	19	−8	11	4	12	32	7	37	13	11
37	Magnum P.I.	8	17	−7	16	−6	14	−4	19	−9	10	3	11	35	14	20	3	37
	St. Elsewhere	3	13	−3	12	−2	10	NC	12	−2	10	3	13	26	11	29	4	29
39	Hollywood Squares	13	13	−4	12	−3	12	−3	14	−5	9	3	8	38	7	37	4	29
	Double Dare	5	13	−4	12	−3	11	−2	13	−4	9	3	5	44	5	43	23	3
41	Benson	3	10	−2	8	NC	8	NC	8	NC	8	3	6	42	7	37	10	16
	Family Medical Center	5	12	−4	11	−3	12	−4	16	−8	8	3	7	39	6	42	3	37
43	Sweethearts	4	17	−10	18	−11	13	−6	13	−6	7	2	6	42	7	37	2	44
44	Group One Medical	13	15	−9	15	−9	15	−9	18	−12	6	2	7	39	5	43	1	46
	Wipeout	5	9	−3	9	−3	8	−2	9	−3	6	2	4	45	5	43	3	37
46	C.O.P.S.	3	5	−2	4	−1	2	+1	3	NC	3	1	1	48	2	46	9	19
47	New Dating Game	3	3	−1	2	NC	3	−1	2	NC	2	1	2	46	2	46	4	29
48	New Newlywed Game	4	2	−1	2	−1	2	−1	4	−3	1	1	2	46	1	48	1	46

LATE NIGHT

Network Affiliates

(TOP 100 MARKETS)

Household Share/Rank	Program	No. Of Stations	Nov. 87 Shr	Nov. 87 +/−	Feb. 88 Shr	Feb. 88 +/−	May 88 Shr	May 88 +/−	NOVEMBER 1988 Lead-In Shr	Lead-In +/−	H'hold Shr	H'hold Rtg	W25-54 Shr	W25-54 Rank	M25-54 Shr	M25-54 Rank	Kids Shr	Kids Rank
1	Sanford & Son	3	29	NC	29	NC	36	−7	24	+5	29	8	32	1	28	1	57	1
2	Tonight Show	96	27	−2	27	−2	26	−1	25	NC	25	6	23	4	21	3	11	19
	Newhart	5	23	+2	23	+2	20	+5	35	−10	25	8	27	2	20	4	22	7
4	Cheers	31	20	+3	22	+1	23	NC	24	−1	23	7	26	3	22	2	24	6
5	Mash	28	21	−1	21	−1	22	−2	22	−2	20	6	21	5	20	4	17	13
	Night Court	13	19	+1	20	NC	18	+2	26	−6	20	5	20	6	19	6	15	15
7	Hill Street Blues	4	17	+2	20	−1	18	+1	22	−3	19	5	19	8	19	6	22	7
	Jeffersons	5	14	+5	17	+2	16	+3	23	−4	19	5	20	6	17	8	30	4
9	Barney Miller	6	22	−4	22	−4	23	−5	23	−5	18	4	15	13	15	11	28	5
10	Benson	7	19	−2	20	−3	20	−3	20	−3	17	4	19	8	14	13	40	2
	Morton Downey Jr.	11	10	+7	11	+6	10	+7	18	−1	17	2	17	12	17	8	20	11
12	Late Night 1	62	15	+1	15	+1	17	−1	19	−3	16	2	19	8	13	15	16	14
	Current Affair	11	16	NC	16	NC	16	NC	20	−4	16	3	14	15	11	17	14	16
	Magnum, P.I.	16	15	+1	18	−2	19	−3	21	−5	16	3	18	11	16	10	20	11
15	Entertainment Tonight	22	18	−3	16	−1	16	−1	23	−8	15	3	14	15	10	19	9	22
	USA Today	9	17	−2	19	−4	19	−4	27	−12	15	4	14	15	14	13	11	19
	Three's Company	4	21	−6	20	−5	19	−4	22	−7	15	4	13	18	15	11	33	3
18	ABC-Niteline	79	15	−2	16	−3	15	−2	20	−7	13	3	11	19	12	16	11	19
	Love Connection	11	12	+1	13	NC	13	NC	15	−2	13	4	11	19	11	17	22	7
20	St. Elsewhere	3	11	+1	14	−2	12	NC	18	−6	12	3	15	13	8	22	14	16
	Taxi	7	17	−5	19	−7	17	−5	21	−9	12	3	11	19	10	19	14	16
	WKRP In Cincinnati	3	13	−1	15	−3	14	−2	21	−9	12	3	9	22	8	22	22	7
23	On Trial	3	15	−7	13	−5	11	−3	18	−10	8	1	6	23	10	19	3	23
24	Sweethearts	3	11	−8	8	−5	9	−6	13	−10	3	—	2	24	1	24	—	24

MONDAY-FRIDAY NIELSEN RATINGS

(Arithmetic averages, not weighted by market size)

PRIME ACCESS

Independent Stations

(TOP 100 MARKETS)

Household Share/Rank	Program	No. Of Stations	Nov. 87		Feb. 88		May 88		NOVEMBER 1988 Lead-In		H'hold		W25-54		M25-54		Kids	
			Shr	+/−	Shr	+/−	Shr	+/−	Shr	+/−	Shr	Rtg	Shr	Rank	Shr	Rank	Shr	Rank
1	Cosby Show	12	12	+5	12	+5	12	+5	12	+5	17	10	21	1	17	1	41	1
2	Night Court	28	9	+3	9	+3	8	+4	11	+1	12	7	14	2	16	2	20	3
3	Current Affair	7	10	+1	11	NC	12	−1	11	NC	11	7	12	3	10	6	8	15
4	Cheers	27	8	+2	9	+1	9	+1	9	+1	10	6	12	3	13	3	13	7
	Family Ties	21	11	−1	11	−1	10	NC	10	NC	10	6	11	5	9	7	27	2
6	Love Connection	3	9	−1	8	NC	9	−1	7	+1	8	5	11	5	9	7	6	17
	Mash	19	9	−1	8	NC	8	NC	7	+1	8	5	9	7	11	5	10	12
8	Newhart	15	9	−2	8	−1	8	−1	9	−2	7	4	8	8	8	9	12	10
	Star Trek	10	6	+1	6	+1	6	+1	8	−1	7	4	7	9	12	4	12	10
10	A-Team	10	6	NC	6	NC	6	NC	7	−1	6	4	5	15	7	12	16	5
	Facts Of Life	7	8	−2	7	−1	7	−1	7	−1	6	3	6	11	4	16	18	4
	Kate & Allie	8	6	NC	7	−1	6	NC	8	−2	6	4	6	11	5	14	14	6
	Star Trek-Next Generation	4	4	+2	6	NC	4	+2	6	NC	6	3	6	11	8	9	13	7
	WKRP In Cincinnati	10	7	−1	7	−1	7	−1	7	−1	6	3	7	9	8	9	9	14
	Three's Company	8	7	−1	9	−3	8	−2	7	−1	6	4	6	11	6	13	13	7
16	It's A Living	5	4	NC	4	NC	4	NC	7	−3	4	2	4	16	4	16	6	17
	New Newlywed Game	7	4	NC	4	NC	4	NC	4	NC	4	2	4	16	3	19	5	20
	Taxi	4	5	−1	6	−2	6	−2	7	−3	4	2	4	16	5	14	6	17
19	Gidget	3	2	+1	1	+2	2	+1	4	−1	3	1	2	21	2	20	8	15
	Gong Show	6	3	NC	4	−1	4	−1	4	−1	3	2	4	16	4	16	10	12
21	New Dating Game	3	5	−3	3	−1	3	−1	4	−2	2	1	3	20	2	20	2	21

DAYTIME

Independent Stations

(TOP 100 MARKETS)

Household Share/Rank	Program	No. Of Stations	Nov. 87		Feb. 88		May 88		NOVEMBER 1988 Lead-In		H'hold		W25-54		M25-54		Kids	
			Shr	+/−	Shr	+/−	Shr	+/−	Shr	+/−	Shr	Rtg	Shr	Rank	Shr	Rank	Shr	Rank
1	Geraldo	3	9	+4	10	+3	11	+2	7	+6	13	3	17	1	22	1	5	48
2	Little House On The Prairie	6	11	+1	10	+2	9	+3	12	NC	12	2	14	2	15	3	22	15
3	Andy Griffith Show	21	9	NC	10	−1	10	−1	9	NC	9	2	10	3	17	2	14	27
4	Barnaby Jones	4	10	−2	11	−3	11	−3	7	+1	8	2	4	23	9	16	3	60
	Fall Guy	4	8	NC	5	+3	6	+2	7	+1	8	2	6	11	13	6	17	23
	Live w/Regis & Kathie Lee	5	10	−2	10	−2	8	NC	11	−3	8	2	8	6	11	11	6	41
	Alvin & Chipmunks	18	7	+1	7	+1	7	+1	5	+3	8	2	3	35	5	33	33	3
	Dennis The Menace Cartoon	10	7	+1	8	NC	8	NC	5	+3	8	2	4	23	3	50	35	2
	Tom & Jerry	9	7	+1	7	+1	7	+1	7	+1	8	2	3	35	7	22	29	9
10	Happy Days	11	9	−2	9	−2	7	NC	7	NC	7	2	6	11	12	7	12	33
	Honeymooners	3	11	−4	7	NC	9	−2	14	−7	7	2	5	18	14	5	4	52
	I Love Lucy	38	8	−1	7	NC	7	NC	6	+1	7	2	9	4	9	16	13	30
	Leave It To Beaver	16	7	NC	7	NC	7	NC	7	NC	7	1	6	11	12	7	18	21
	Love Boat	4	9	−2	8	−1	8	−1	8	−1	7	2	9	4	5	33	8	37
	Flintstones	11	7	NC	7	NC	7	NC	6	+1	7	2	4	23	5	33	27	12
	Jetsons	6	5	+2	7	NC	6	+1	5	+2	7	2	3	35	5	33	33	3
	Real Ghostbusters	9	7	NC	7	NC	7	NC	5	+2	7	2	3	35	7	22	33	3
18	Divorce Court	23	5	+1	6	NC	6	NC	6	NC	6	1	7	8	7	22	4	52
	Judge	8	6	NC	7	−1	6	NC	6	NC	6	1	7	8	7	22	6	41
	Laverne & Shirley	6	6	NC	6	NC	5	+1	7	−1	6	1	6	11	5	33	11	35
	Magnum, P.I.	4	8	−2	9	−3	9	−3	8	−2	6	2	4	23	11	11	6	41
	On Trial	8	6	NC	9	−3	8	−2	5	+1	6	1	6	11	5	33	6	41
	Quincy	12	5	+1	7	−1	6	NC	6	NC	6	1	6	11	11	11	4	52
	Rockford Files	7	5	+1	5	+1	6	NC	5	+1	6	1	5	18	15	3	5	48
	Sally Jessy Raphael	4	1	−5	13	−7	13	−7	6	NC	6	2	8	6	8	18	1	67
	Superior Court	7	6	NC	5	+1	5	+1	5	+1	6	1	6	11	12	7	5	48
	Three's Company	3	8	−2	7	−1	9	−3	4	+2	6	1	4	23	8	18	16	24
	Double Dare	4	9	−3	9	−3	11	−5	10	−4	6	1	2	46	3	50	16	24
	Gumby	7	5	+1	5	+1	5	+1	5	+1	6	1	2	46	4	46	40	1
	My Little Pony	17	8	−2	8	−2	6	NC	7	−1	6	1	2	46	3	50	33	3
	Popeye	8	6	NC	5	+1	5	+1	5	+1	6	2	2	46	6	29	28	11
	Woody Woodpecker & Friends	12	6	NC	7	−1	6	NC	5	−1	6	2	2	46	6	29	32	7
33	Benson	4	8	−3	8	−3	4	+1	10	−5	5	1	7	8	12	7	7	38
	Brady Bunch	7	6	−1	7	−2	7	−2	6	−1	5	1	3	35	5	33	15	26
	Gilligan's Island	8	4	+1	5	NC	5	NC	5	NC	5	1	3	35	10	14	13	30
	Hour Magazine	16	6	−1	5	NC	6	−1	8	−3	5	1	5	18	4	46	4	52

MONDAY-FRIDAY NIELSEN RATINGS

(Arithmetic averages, not weighted by market size)

DAYTIME Independents *Continued from page 136*

Household Share/Rank	Program	No. Of Stations	Nov. 87 Shr	Nov. 87 +/–	Feb. 88 Shr	Feb. 88 +/–	May 88 Shr	May 88 +/–	NOVEMBER 1988 Lead-In Shr	Lead-In +/–	H'hold Shr	H'hold Rtg	W25-54 Shr	W25-54 Rank	M25-54 Shr	M25-54 Rank	Kids Shr	Kids Rank
	I Dream Of Jeannie	46	5	NC	5	NC	4	+1	5	NC	5	1	4	23	7	22	10	36
	Love Connection	7	5	NC	5	NC	5	NC	3	+2	5	1	4	23	10	14	4	52
	New Newlywed Game	33	4	+1	5	NC	4	+1	5	NC	5	1	4	23	5	33	5	48
	Trapper John, M.D.	7	6	–1	5	NC	6	–1	6	–1	5	1	5	18	6	29	7	38
	Too Close For Comfort	7	6	–1	7	–2	5	NC	6	–1	5	1	4	23	8	18	12	33
	Care Bears	10	5	NC	4	+1	4	+1	5	NC	5	1	2	46	3	50	26	13
	JEM	10	5	NC	5	NC	5	NC	6	–1	5	1	2	46	2	60	23	14
	Smurfs	17	5	NC	5	NC	5	NC	5	NC	5	1	2	46	3	50	22	15
	Teddy Ruxpin	5	5	NC	6	–1	5	NC	4	+1	5	1	1	59	1	64	30	8
46	Family Medical Center	8	6	–2	8	–4	8	–4	5	–1	4	1	5	18	8	18	3	60
	Gimme A Break	4	5	–1	8	–4	5	–1	6	–2	4	1	3	35	3	50	4	52
	Group One Medical	5	10	–6	8	–4	7	–3	5	–1	4	1	3	35	5	33	2	64
	Hollywood Squares	23	4	NC	4	NC	4	NC	4	NC	4	1	3	35	5	33	6	41
	Mork & Mindy	8	4	NC	5	–1	5	–1	5	–1	4	1	4	23	7	22	13	30
	New Dating Game	38	5	–1	5	–1	4	NC	4	NC	4	1	3	35	6	29	4	52
	Sanford & Son	3	3	+1	2	+2	2	+2	2	+2	4	1	2	46	5	33	6	41
	Simon & Simon	3	8	–4	8	–4	7	–3	7	–3	4	1	4	23	7	22	3	60
	Sweethearts	6	5	–1	6	–2	5	–1	6	–2	4	1	4	23	5	33	2	64
	Bugs Bunny	4	4	NC	4	NC	3	+1	3	+1	4	1	1	59	2	60	14	27
	Scooby Doo	14	4	NC	4	NC	4	NC	3	+1	4	1	1	59	3	50	29	9
	Thundercats	8	3	+1	3	+1	2	+2	3	+1	4	1	1	59	3	50	21	17
58	Diff'rent Strokes	3	6	–3	6	–3	6	–3	5	–2	3	1	2	46	4	46	6	41
	Gidget	35	3	NC	4	–1	3	NC	4	–1	3	1	2	46	3	50	7	38
	Gong Show	16	3	NC	4	–1	4	–1	4	–1	3	1	2	46	5	33	3	60
	Wipeout	11	4	–1	4	–1	4	–1	3	NC	3	1	3	35	4	46	4	52
	Finders Keepers	4	5	–2	4	–1	4	–1	5	–2	3	1	1	59	3	50	14	27
	Ghostbusters	6	3	NC	3	NC	3	NC	3	NC	3	1	1	59	2	60	19	19
	G.I. Joe	4	6	–3	6	–3	5	–2	3	NC	3	1	1	59	1	64	19	19
	Snorks	11	3	NC	3	NC	3	NC	3	NC	3	1	1	59	1	64	20	18
66	Jeffersons	3	2	NC	2	NC	2	NC	2	NC	2	1	2	46	2	60	2	64
	Comic Strip	3	2	NC	3	–1	2	NC	4	–2	2	1	1	59	1	64	18	21

EARLY FRINGE

Independent Stations

(TOP 100 MARKETS)

Household Share/Rank	Program	No. Of Stations	Nov. 87 Shr	Nov. 87 +/–	Feb. 88 Shr	Feb. 88 +/–	May 88 Shr	May 88 +/–	NOVEMBER 1988 Lead-In Shr	Lead-In +/–	H'hold Shr	H'hold Rtg	W25-54 Shr	W25-54 Rank	M25-54 Shr	M25-54 Rank	Kids Shr	Kids Rank
1	Duck Tales	88	10	+1	12	–1	11	NC	9	+2	11	4	5	28	6	20	39	1
2	Cheers	9	11	–1	11	–1	11	–1	9	+1	10	6	12	1	12	1	16	36
	Family Ties	31	10	NC	9	+1	8	+2	8	+2	10	5	11	3	8	8	23	17
	Gimme A Break	8	12	–2	12	–2	12	–2	10	NC	10	5	11	3	7	12	26	8
	Night Court	4	9	+1	9	+1	8	+2	9	+1	10	5	11	3	11	3	22	21
6	Diff'rent Strokes	12	10	–1	10	–1	10	–1	8	+1	9	4	7	15	5	29	25	12
	Facts Of Life	32	10	–1	10	–1	10	–1	8	+1	9	4	9	9	5	29	22	21
	Kate & Allie	7	7	+2	8	+1	7	+2	8	+1	9	5	12	1	7	12	22	21
	Love Connection	3	10	–1	9	NC	9	NC	9	NC	9	4	10	6	11	3	7	62
	Mash	3	14	–5	14	–5	12	–3	11	–2	9	5	10	6	9	5	10	55
	Silver Spoons	37	9	NC	10	–1	10	–1	8	+1	9	4	7	15	6	20	26	8
	Three's Company	21	11	–2	11	–2	10	–1	9	NC	9	5	9	9	8	8	20	27
13	Brady Bunch	21	8	NC	9	–1	9	–1	8	NC	8	3	6	23	4	40	24	14
	Good Times	4	6	+2	7	+1	9	–1	7	+1	8	4	8	12	6	20	21	26
	Newhart	5	10	–2	10	–2	10	–2	8	NC	8	4	10	6	9	5	10	55
	Punky Brewster	18	8	NC	8	NC	8	NC	7	+1	8	4	5	28	3	49	30	6
	Webster	42	9	–1	9	–1	8	NC	8	NC	8	4	7	15	4	40	27	7
	Disney/World Of Disney	6	9	–1	9	–1	9	–1	11	–3	8	4	8	12	7	12	33	2
	Alvin & Chipmunks	54	8	NC	8	NC	7	+1	6	+2	8	2	3	44	4	40	33	2
	Dennis The Menace Cartoon	24	8	NC	8	NC	8	NC	6	+2	8	2	4	36	6	20	33	2
21	Andy Griffith Show	14	6	+1	8	–1	7	NC	6	+1	7	3	9	9	9	5	16	36
	Magnum, P.I.	4	11	–4	11	–4	9	–2	6	+1	7	4	8	12	7	12	8	60
	Fun House	54	7	NC	7	NC	7	NC	8	–1	7	3	4	36	3	49	26	8
	G.I. Joe	6	7	NC	6	+1	7	NC	7	NC	7	2	2	54	5	29	23	17
	Jetsons	34	6	+1	7	NC	7	NC	7	NC	7	2	3	44	4	40	24	14
	Real Ghostbusters	46	8	–1	8	–1	8	–1	7	NC	7	2	3	44	5	29	26	8
	Scooby Doo	13	7	NC	8	–1	8	–1	7	NC	7	2	4	36	6	20	33	2
	Tom & Jerry	12	6	+1	7	NC	7	NC	6	+1	7	2	3	44	5	29	24	14
29	A-Team	15	7	–1	7	–1	7	–1	5	+1	6	3	5	28	7	12	14	44
	Happy Days	18	7	–1	7	–1	7	–1	7	–1	6	3	5	28	5	29	18	32
	Jeffersons	5	6	NC	7	–1	7	–1	7	–1	6	3	6	23	5	29	15	40
	Knight Rider	9	6	NC	6	NC	6	NC	7	–1	6	3	5	28	6	20	17	33
	Laverne & Shirley	4	6	NC	6	NC	6	NC	5	+1	6	3	5	28	4	40	15	40

MONDAY-FRIDAY NIELSEN RATINGS

(Arithmetic averages, not weighted by market size)

EARLY FRINGE Independents *Continued from page 138*

Household Share/Rank	Program	No. Of Stations	Nov. 87		Feb. 88		May 88		NOVEMBER 1988: Lead-In		H'hold		W25-54		M25-54		Kids	
			Shr	+/-	Shr	+/-	Shr	+/-	Shr	+/-	Shr	Rtg	Shr	Rank	Shr	Rank	Shr	Rank
	Leave It To Beaver	10	7	-1	6	NC	7	-1	6	NC	6	2	6	23	6	20	11	51
	9 To 5	3	8	-2	7	-1	7	-1	7	-1	6	3	6	23	5	29	10	55
	New Leave It To Beaver	11	7	-1	7	-1	6	NC	7	-1	6	3	6	23	6	20	22	21
	Sanford & Son	4	5	+1	6	NC	6	NC	5	+1	6	4	7	15	8	8	16	36
	Star Trek	6	5	+1	6	NC	6	NC	5	+1	6	3	7	15	12	1	10	55
	Bugs Bunny	7	4	+2	5	+1	6	NC	5	+1	6	2	3	44	5	29	22	21
	C.O.P.S.	32	6	NC	6	NC	7	-1	6	NC	6	2	2	54	3	49	19	29
	Double Dare	66	8	-2	8	-2	8	-2	8	-2	6	2	3	44	3	49	23	17
	Flintstones	15	6	NC	6	NC	6	NC	5	+1	6	2	3	44	4	40	19	29
	Thundercats	4	8	-2	6	NC	6	NC	8	-2	6	2	2	54	2	58	15	40
	Woody Woodpecker & Friends	25	6	NC	7	-1	7	-1	6	NC	6	2	3	44	5	29	25	12
45	Benson	3	5	NC	5	NC	5	NC	4	+1	5	3	7	15	6	20	12	48
	Gilligan's Island	7	4	+1	5	NC	4	+1	4	+1	5	2	4	36	4	40	12	48
	Gong Show	7	5	NC	6	-1	5	NC	6	-1	5	2	5	28	8	8	14	44
	Little House On The Prairie	9	5	NC	6	-1	7	-2	5	NC	5	3	7	15	3	49	14	44
	Taxi	4	6	-1	7	-2	6	-1	5	NC	5	3	7	15	7	12	12	48
	WKRP In Cincinnati	6	6	-1	7	-2	7	-2	5	NC	5	3	5	28	7	12	11	51
51	I Dream Of Jeannie	3	5	-1	4	NC	3	+1	4	NC	4	2	4	36	7	12	17	33
	It's A Living	4	7	-3	8	-4	6	-2	5	-1	4	2	4	36	3	49	7	62
	New Dating Game	6	4	NC	5	-1	4	NC	3	+1	4	2	3	44	3	49	5	64
	New Newlywed Game	5	5	-1	5	-1	5	-1	4	NC	4	2	4	36	4	40	5	64
	Finders Keepers	38	5	-1	6	-2	6	-2	5	-1	4	1	2	54	2	58	15	40
	Ghostbusters	5	5	-1	5	-1	6	-2	6	-2	4	1	2	54	4	40	19	29
	My Little Pony	6	5	-1	4	NC	5	-1	5	-1	4	1	1	61	2	58	23	17
	Popeye	10	5	-1	4	NC	5	-1	4	NC	4	1	2	54	3	49	13	47
	Smurfs	6	5	-1	5	-1	6	-2	4	NC	4	1	1	61	2	58	17	33
	Teenage Mutant Ninja Turtles	3	4	NC	4	NC	4	NC	5	-1	4	2	1	61	5	29	11	51
61	Gidget	8	4	-1	4	-1	4	-1	4	-1	3	1	3	44	3	49	11	51
	Hollywood Squares	4	6	-3	6	-3	5	-2	4	-1	3	2	4	36	2	58	5	64
	Gumby	7	3	NC	3	NC	4	-1	4	-1	3	1	1	61	2	58	20	27
	Snorks	5	5	-2	4	-1	4	-1	4	-1	3	1	1	61	2	58	16	36
65	Care Bears	4	1	+1	3	-1	2	NC	2	NC	2	1	1	61	2	58	10	55
	JEM	5	5	-3	2	NC	2	NC	4	-2	2	1	2	54	2	58	8	60

LATE NIGHT

Independent Stations

(TOP 100 MARKETS)

Household Share/Rank	Program	No. Of Stations	Nov. 87		Feb. 88		May 88		NOVEMBER 1988: Lead-In		H'hold		W25-54		M25-54		Kids	
			Shr	+/-	Shr	+/-	Shr	+/-	Shr	+/-	Shr	Rtg	Shr	Rank	Shr	Rank	Shr	Rank
1	Hill Street Blues	9	11	-2	9	NC	10	-1	8	+1	9	2	8	3	9	3	4	22
	Sanford & Son	7	8	+1	7	+2	10	-1	8	+1	9	4	9	1	10	2	25	2
3	Cheers	15	7	+1	7	+1	8	NC	7	+1	8	4	8	3	8	5	10	7
	Mash	8	9	-1	9	-1	10	-2	7	+1	8	4	8	3	9	3	8	10
	Quincy	4	9	-1	9	-1	6	+2	8	NC	8	2	9	1	7	7	30	1
6	Tonight Show	3	9	-2	6	+1	7	NC	5	+2	7	2	6	12	5	15	8	10
	Morton Downey Jr.	25	5	+2	4	+3	4	+3	5	+2	7	2	7	8	8	5	9	8
	Night Court	8	6	+1	5	+2	6	+1	8	-1	7	4	7	8	7	7	15	6
	Star Trek	9	6	+1	7	NC	6	+1	6	+1	7	2	7	8	11	1	6	18
10	Barney Miller	9	6	NC	6	NC	6	NC	5	+1	6	2	5	17	5	15	6	18
	Benson	4	9	-3	8	-2	8	-2	6	NC	6	2	6	12	4	19	8	10
	Fall Guy	4	7	-1	9	-3	8	-2	7	-1	6	2	8	3	6	12	19	4
	Friday The 13th: TV Series	7	7	-1	4	+2	4	+2	5	+1	6	1	4	21	7	7	9	8
	Honeymooners	10	7	-1	6	NC	7	-1	7	-1	6	2	6	12	7	7	8	10
	I Love Lucy	3	6	NC	7	-1	6	NC	5	+1	6	1	8	3	5	15	25	2
	Love Connection	18	5	+1	4	+2	4	+2	6	NC	6	2	7	8	4	19	7	15
	Taxi	12	7	-1	7	-1	7	-1	8	-2	6	2	6	12	7	7	8	10
18	Late Night 1	4	3	+2	2	+3	4	+1	3	+2	5	1	5	17	4	19	7	15
	Matt Houston	3	6	-1	3	+2	2	+3	5	NC	5	1	3	26	1	29	1	26
	New Newlywed Game	10	5	NC	4	+1	5	NC	6	-1	5	2	6	12	3	22	5	20
	Simon & Simon	6	5	NC	4	+1	4	+1	5	NC	5	2	5	17	5	15	2	23
	Three's Company	3	4	+1	3	+2	4	+1	7	-2	5	2	4	21	3	22	16	5
23	New Dating Game	10	5	-1	4	NC	4	NC	5	-1	4	1	4	21	2	26	7	15
	Rockford Files	4	5	-1	4	NC	4	NC	5	-1	4	1	4	21	3	22	2	23
	Twilight Zone	5	5	-1	6	-2	6	-2	5	-1	4	1	5	17	6	12	1	26
	WKRP In Cincinnati	6	5	-1	4	NC	4	NC	5	-1	4	2	4	21	6	12	1	26
27	Gong Show	14	5	-2	4	-1	4	-1	4	-1	3	1	3	26	2	26	5	20
	On Trial	5	4	-1	5	-2	4	-1	5	-2	3	1	3	26	2	26	1	26
	Tales From The Darkside	4	3	NC	3	NC	2	+1	4	-1	3	1	2	29	3	22	2	23

(Effective Sept. 18, 1989)

(New Shows In CAPS)

MONDAY		8:00	8:30	9:00	9:30	10:00	10:30	
	(Local)	MacGyver		NFL Football				ABC
	(Local)	MAJOR DAD	PEOPLE NEXT DOOR	Murphy Brown	FAMOUS TEDDY Z.	Designing Women	Newhart	CBS
	(Local)	Alf	Hogan Family	Movie				NBC
	(Local)	21 Jump Street		ALIEN NATION				Fox

TUESDAY		8:00	8:30	9:00	9:30	10:00	10:30	
	(Local)	Who's The Boss?	Wonder Years	Roseanne	CHICKEN SOUP	thirtysomething		ABC
	(Local)	RESCUE: 911		WOLF		THE HAWAIIAN		CBS
	(Local)	Matlock		In The Heat Of The Night		Midnight Caller		NBC

WEDNESDAY		8:00	8:30	9:00	9:30	10:00	10:30	
	(Local)	Growing Pains	Head Of The Class	Anything But Love	DOOGIE HOWSER, M.D.	China Beach		ABC
	(Local)	A PEACEABLE KINGDOM		Jake & The Fatman		Wiseguy		CBS
	(Local)	Unsolved Mysteries		Night Court	NUTT HOUSE	Quantum Leap		NBC

THURSDAY		8:00	8:30	9:00	9:30	10:00	10:30	
	(Local)	Mission: Impossible		THE KID		PRIME TIME		ABC
	(Local)	48 Hours		TOP OF THE HILL		Knots Landing		CBS
	(Local)	The Cosby Show	A Different World	Cheers	Dear John	L.A. Law		NBC

FRIDAY		8:00	8:30	9:00	9:30	10:00	10:30	
	(Local)	Full House	FAMILY MATTERS	Perfect Strangers	Just The Ten Of Us	20/20		ABC
	(Local)	SNOOPS		Dallas		Falcon Crest		CBS
	(Local)	BAYWATCH		HARDBALL		MANCUSO, FBI		NBC

SATURDAY		8:00	8:30	9:00	9:30	10:00	10:30	
	(Local)	Mr. Belvedere	LIVING DOLLS	ABC Mystery Movie				ABC
	(Local)	Paradise		Tour Of Duty		West 57th		CBS
	(Local)	227	Amen	Golden Girls	Empty Nest	Hunter		NBC
	(Local)	Cops	The Reporters		Beyond Tomorrow			Fox

SUNDAY	7:00	7:30	8:00	8:30	9:00	9:30	10:00	10:30	
	LIFE GOES ON		FREE SPIRIT	HOMEROOM	Movie				ABC
	60 Minutes		Murder, She Wrote		Movie				CBS
	Magical World Of Disney		SISTER KATE	My Two Dads	Movie				NBC
	BOOKER		America's Most Wanted	HIDDEN VIDEO	Married/ Children	OPEN HOUSE	Tracey Ullman	Garry Shandling	Fox

(Effective January, 1990)

(New Shows In CAPS)

MONDAY		8:00	8:30	9:00	9:30	10:00	10:30	
	(Local)	MacGyver		MOVIE				ABC
	(Local)	Major Dad	Famous Teddy Z	Murphy Brown	Designing Women	Newhart	DOCTOR, DOCTOR	CBS
	(Local)	Alf	Hogan Family	Movie				NBC
	(Local)	21 Jump Street		Alien Nation		(Local)		Fox

TUESDAY		8:00	8:30	9:00	9:30	10:00	10:30	
	(Local)	Who's The Boss?	Wonder Years	Roseanne	Coach	thirtysomething		ABC
	(Local)	Rescue 911		MOVIE				CBS
	(Local)	Matlock		In The Heat Of The Night		Midnight Caller		NBC

WEDNESDAY		8:00	8:30	9:00	9:30	10:00	10:30	
	(Local)	Growing Pains	Head Of The Class	Doogie Howser, M.D.	Anything But Love	China Beach		ABC
	(Local)	GRAND SLAM		Jake & The Fatman		Wiseguy		CBS
	(Local)	Unsolved Mysteries		Night Court	Dear John	Quantum Leap		NBC

THURSDAY		8:00	8:30	9:00	9:30	10:00	10:30	
	(Local)	FATHER DOWLING MYSTERIES		The Young Riders		Primetime Live		ABC
	(Local)	48 Hours		Island Son		Knots Landing		CBS
	(Local)	The Cosby Show	A Different World	Cheers	GRAND	L.A. Law		NBC

FRIDAY		8:00	8:30	9:00	9:30	10:00	10:30	
	(Local)	Full House	Family Matters	Perfect Strangers	Just The Ten Of Us	20/20		ABC
	(Local)	MAX MONROE: LOOSE CANNON		Dallas		Falcon Crest		CBS
	(Local)	Baywatch		TRUE BLUE		Mancuso FBI		NBC

SATURDAY		8:00	8:30	9:00	9:30	10:00	10:30	
	(Local)	Mission: Impossible		ABC Saturday Mystery				ABC
	(Local)	Paradise		Tour Of Duty		Saturday Night With Connie Chung		CBS
	(Local)	227	Amen	Golden Girls	Empty Nest	Hunter		NBC
	(Local)	Cops	Totally Hidden Video	The Reporters		(Local)		Fox

SUNDAY	7:00	7:30	8:00	8:30	9:00	9:30	10:00	10:30	
	Life Goes On		FUNNIEST VIDEOS	Free Spirit	Movie				ABC
	60 Minutes		Murder, She Wrote		Movie				CBS
	Magical World Of Disney		ANN JILLIAN	Sister Kate	Movie				NBC
	Booker		America's Most Wanted	THE SIMPSONS	Married/ Children	Open House	Tracey Ullman	Garry Shandling	Fox

Includes series titles listed alphabetically by networks, timeslots, suppliers, production staff heads, cast regulars and semiregulars and estimated network license fee per segment (Costs do not include time charges or commercials)

ABC-TV

Series Title	Day	Hr.	Mins.	Supplier	Production principals	Cast regulars & semiregulars	Estimated prod. fee per episode
ABC Saturday Mystery, The	Sat.	9:00	120	Universal TV	SEP: William Link		$2.5-million
B.L. Stryker				Blue Period Prods.-TWS Prods.-Universal TV	EP: Tom Selleck. Burt Reynolds. Chas. Floyd Johnson. Chris Abbott SP: Tom Donnelly P: Alan Barnette	Burt Reynolds. Ossie Davis. Kristi Swanson. Dana Kaminski. Michael O. Smith. Alfie Wise. Rita Moreno	
Christine Cromwell				Universal TV	EP: Dick Wolf P: Lynn Guthrie. Michael Dugan CP: Robert Palm. Dan Sackheim	Jaclyn Smith. Celeste Holm. Ralph Bellamy	
Columbo				Universal TV	EP: Richard Alan Simmons SP: Philip Saltzman CP: Peter Ware	Peter Falk	
Kojak				Universal TV	EP: James McAdams SP: Stuart Cohen P: Marc Laub CP: Judith Stevens	Telly Savalas	
ABC Sunday Night Movie, The	Sun.	9:00	120	Various			2.5-million
Anything But Love	Wed.	9:00	30	Adam Prods.-20th Fox TV	EP: Robert Myman. Peter Noah SP: Janis Hirsch P: Peter Schindler CP: Bruce Rasmussen	Jamie Lee Curtis. Richard Lewis. Ann Magnuson. Joseph Maher. Richard Frank. Holly Fulger	450.000
Chicken Soup	Tue.	9:30	30	Carsey-Werner Co.	EP: Marcy Carsey. Tom Werner. Saul Turteltaub & Bernie Orenstein SP: Paul Perlove P: Faye Oshima D: Alan Rafkin	Jackie Mason. Lynn Redgrave. Rita Karin. Kathryn Erbe. Johnny Pinto. Alisan Porter	450.000
China Beach	Wed.	10:00	60	Sacret Inc. Prods.-Warner Bros. TV	EP: John Sacret Young SP: John Wells. Georgia Jeffries P: Mimi Leder CP: Geno Escareega. Fred Gerber	Dana Delany. Brian Wimmer. Michael Boatman. Marg Helgenberger. Concetta Tomei. Robert Picardo. Jeff Kober. Ned Vaughn. Nancy Giles	950.000
Doogie Howser, M.D.	Wed.	9:30	30	Steven Bochco Prods.	EP: Steven Bochco SP: Stephen Cragg P: Scott Goldstein. Phil Kellard. Tom Moore. Jill Gordon. Vic Rauseo. Linda Morris	Neil Patrick Harris. James B. Sikking. Belinda Montgomery. Lawrence Pressman. Max Casella. Mitchell Anderson. Kathryn Layng	450.000
Family Matters	Fri.	8:30	30	Miller/Boyett Prods.-Lorimar-Telepictures	EP: Thomas L. Miller. Robert L. Boyett. William Bickley & Michael Warren SP: Alan Eisenstock & Larry Mintz P: Robert Blair CP: Harriette Ames-Regan	JoMarie Payton-France. Reggie VelJohnson. Rosetta LeNoire. Darius McCrary. Kellie Williams. Jamie Foxworth. Telma Hopkins. Joseph Loyal Wright. Julius Royal Wright	450.000
Free Spirit	Sun.	9:00	30	Columbia Pictures TV	EP: Richard Gurman. Phil Doran SP: Howard Meyers P: Mark Fink CP: Bob Rosenfarb. Jon Spector D: Art Dielhenn	Corinne Bohrer. Franc Luz. Edan Gross. Paul Scherrer. Alyson Hannigan	450.000
Full House	Fri.	8:00	30	Jeff Franklin Prods.-Miller/Boyett Prods.-Lorimar-Telepictures	EP: Jeff Franklin. Thomas L. Miller. Robert L. Boyett SP: Rob Dames P: Don Van Atta. Lenny Ripps. Marc Warren. Dennis Rinsler CP: Kim Weiskopf	John Stamos. Bob Saget. David Coulier. Candace Cameron. Jodie Sweetin. Mary Kate Olsen. Ashley Fuller Olsen. Lori Loughlin	475.000
Growing Pains	Wed.	8:00	30	Warner Bros. TV	EP: Michael Sullivan. Dan Guntzelman. Steve Marshall. David Kendall P: Henry Johnson. Tim O'Donnell	Alan Thicke. Joanna Kerns. Kirk Cameron. Tracey Gold. Jeremy Miller	500.000
Head Of The Class	Wed.	8:30	30	Eustis Elias Prods.-Warner Bros. TV	EP: Richard Eustis. Michael Elias P: Alan Rosen. Frank Pace CP: Ray Jessel. Steve Kreinberg. Andy Guerdat. Jonathan Roberts	Howard Hesseman. Jeannette Arnette. William G. Schilling. Dan Frischman. Robin Givens. Khrystyne Haje. Tony O'Dell. Kimberly Russell. Brian Robbins. Daniel Schneider. Rain Pryor. Michael DeLorenzo. Lara Piper. De'Voreaux White	475.000
Homeroom	Sun.	8:30	30	Castle Rock Entertainment	EP: Gary Gilbert. Andrew Scheinman. Topper Carew	Darryl Sivad. Penny Johnson. Bill Cobbs. Jahary Bennett. Trent Cameron. Billy Dee Wil-	450.000

Series Title	Day	Hr.	Mins.	Supplier	Production principals	Cast regulars & semiregulars	Estimated prod. fee per episode
					P: Jan Siegelman, David Cohen, Roger Schulman, Trish Soodik D: Linda Day	liams	
Just The Ten Of Us	Fri.	9:30	30	GSM Prods.-Warner Bros. TV	EP: Michael Sullivan, Dan Guntzelman, Steve Marshall P: Rich Reinhart, Nick Lerose, Henry Johnson	Bill Kirchenbauer, Deborah Harmon, Heather Langenkamp, Brooke Theiss, Jamie Luner, JoAnn Willette, Matt Shakman, Heidi Zeigler	450,000
Life Goes On	Sun.	7:00	60	Toots Prods.-Warner Bros. TV	EP: Michael Braverman SP-D: Ron Rubin P: Phillips Wylly Sr.	Bill Smitrovich, Patti LuPone, Monique Lanier, Christopher Burke, Kellie Martin	900,000
Living Dolls	Sat.	8:30	30	Columbia Pictures TV-ELP Communications	EP: Ross Brown, Phyllis Glick P: Bob Colleary, Martha Williamson CP: Valri Bromfield D: John Sgueglia	Michael Learned, Leah Remini, Alison Elliott, Deborah Tucker, Halle Berry, David Moscow	450,000
MacGyver	Mon.	8:00	60	Henry Winkler/John Rich Prods.-Paramount Network TV	EP: Henry Winkler, John Rich, Stephen Downing SP: Michael Greenburg	Richard Dean Anderson, Dana Elcar	900,000
Mission: Impossible	Thu.	8:00	60	Jeffrey Hayes Prods.-Paramount Network TV	EP: Jeffrey Hayes SP: Frank Abatemarco P: Ted Roberts, Darryl Sheen	Peter Graves, Thaao Penghlis, Tony Hamilton, Phil Morris, Jane Badler	700,000
Mr. Belvedere	Sat.	8:00	30	Lazy B/FOB Prods.-20th Fox TV	EP: Frank Dungan & Jeff Stein, Liz Sage SP: Jeff Ferro, Ric Weiss P: Patricia Rickey CP: Geri Maddern D: Don Corvan	Christopher Hewett, Ilene Graff, Rob Stone, Tracy Wells, Brice Beckham, Bob Uecker	475,000
NFL Monday Night Football	Mon.	9:00	120	ABC Sports	EP: Geoffrey Mason P: Ken Wolfe D: Craig Janoff	Al Michaels, Frank Gifford, Dan Dierdorf, Lynn Swann	2.5-million
Perfect Strangers	Fri.	9:00	30	Miller/Boyett Prods.-Lorimar TV	EP: Thomas L. Miller, Robert L. Boyett, William Bickley & Michael Warren SP: Paula A. Roth P: James O'Keefe, Bob Griffard, Howard Adler, Alan Plotkin	Bronson Pinchot, Mark Linn-Baker, Melanie Wilson, Rebeca Arthur, Belita Moreno, Sam Anderson	475,000
Primetime Live	Thu.	10:00	60	ABC News	EP: Richard N. Kaplan Sr. P: Ira Rosen, Amy Sacks	Diane Sawyer, Sam Donaldson, Chris Wallace	500,000
Roseanne	Tue.	9:00	30	Carsey-Werner Co.	EP: Marcy Carsey, Tom Werner, Jeff Harris, Allan Katz SP: Danny Jacobson P: Al Lowenstein CP: Norma Safford Vela D: John Pasquin	Roseanne Barr, John Goodman, Laurie Metcalf, Lecy Goranson, Sara Gilbert, Michael Fishman, Natalie West	500,000
thirtysomething	Tue.	10:00	60	Bedford Falls Co.-MGM/UA TV	EP; Marshall Herskovitz, Edward Zwick SP: Scott Winant P: Richard Kramer CP: Ellen S. Pressman	Timothy Busfield, Polly Draper, Mel Harris, Peter Horton, Melanie Mayron, Ken Olin, Patricia Wettig	900,000
20/20	Fri.	10:00	60	ABC News	EP: Victor Neufeld Sr. P: Jeff Diamond D: Jerry Paul	Hugh Downs, Barbara Walters, Bob Brown, John Stossel, Tom Jarriel	500,000
Who's The Boss?	Tue.	8:00	30	Columbia/Embassy TV	EP: Martin Cohan, Blake Hunter, Karen Wengrod, Ken Cinnamon SP: Danny Kallis P: John Anderson, Joe Fisch, Asaad Kelada D: Asaad Kelada	Tony Danza, Judith Light, Alyssa Milano, Danny Pintauro, Katherine Helmond	475,000
Wonder Years, The	Tue.	8:30	30	Black/Marlens Co.-New World TV	EP: Bob Brush, Bob Stevens, P: Ken Topolsky CP: Matthew Carlson	Fred Savage, Dan Lauria, Alley Mills, Jason Hervey, Olivia d'Abo, Danica McKellar, Josh Saviano, Daniel Stern	475,000
Young Riders, The	Thu.	9:00	60	Ogiens/Kane Co.-MGM/UA TV	EP: Michael Ogiens, Josh Kane, Jonas McCord SP: Ed Spielman, Dennis Cooper P: Harvey Frand	Ty Miller, Josh Brolin, Stephen Baldwin, Gregg Rainwater, Yvonne Suhor, Anthony Zerbe, Melissa Leo, Brett Cullen	900,000

CBS-TV

Series Title	Day	Hr.	Mins.	Supplier	Production principals	Cast regulars & semiregulars	Estimated prod. fee per episode
CBS Sunday Movie	Sun.	9:00	120	Various			2,500,000
Dallas	Fri.	9:00	60	Lorimar TV	EP: Leonard Katzman, Larry Hagman, Ken Horton SP: Howard Lakin P: Cliff Fenneman, Mitchell Wayne Katzman	Barbara Bel Geddes, Patrick Duffy, Larry Hagman, Howard Keel, George Kennedy, Ken Kercheval, Cathy Podewell, Charlene Tilton, Sheree J. Wilson, Kimberly Foster, Michael Wilding, Sasha Mitchell	1,000,000
Designing Women	Mon.	10:00	30	Bloodworth/Thomason Mozark Prods.-Columbia Pictures TV	EP: Harry Thomason, Linda Bloodworth Thomason SP: Pam Norris P: Tommy Thompson, Douglas Jackson CP: David Trainer	Delta Burke, Dixie Carter, Annie Potts, Jean Smart, Meshach Taylor	475,000
Falcon Crest	Fri.	10:00	60	Amanda/MF Prods.-Lorimar TV	EP: Jerry Thorpe, Michael Filerman, Joel Surnow P: Philip L. Parslow	Jane Wyman, David Selby, Gregory Harrison, Lorenzo Lamas, Kristian Alfonson, Margaret Ladd, Chao-Li Chi	950,000
Famous Teddy Z, The	Mon.	9:30	30	Hugh Wilson Prods.-Columbia Pictures TV	EP: Hugh Wilson SP: Richard Dubin	Jon Cryer, Alex Rocco, Milton Selzer, Erica Yohn, Jane Sibbett, Tom LaGrua, Josh Blake	450,000
48 Hours	Thu.	8:00	60	CBS News	EP: Andrew Heyward Sr. P: Catherine Lasiewicz, Steve Glauber, Al Briganti	Dan Rather, Bernard Goldberg	500,000
Island Son	Tue.	10:00	60	Maili Point Enterprises-Lorimar TV	EP: Nigel & Carol Evan McKeand, Richard Chamberlain, Martin Rabbett SP: Les Carter, Susan Sisko P: Christopher Chulack	Richard Chamberlain, Ray Bumatai, Timothy Carhart, Betty Carvalho, Clyde Kusatsu, William McNamara, Brynn Thayer	900,000
Jake And The Fatman	Wed.	9:00	60	Fred Silverman Co.-Dean Hargrove Prods.-Viacom Prods.	EP: Fred Silverman, Dean Hargrove, David Moessinger, Jeri Taylor, Bernie Kowalski P: Fred McKnight	William Conrad, Joe Penny, Alan Campbell	925,000
Knots Landing	Thu.	10:00	60	Roundelay/MF Prods.-Lorimar TV	EP: David Jacobs, Michael Filerman, Lawrence Kasha P: Mary-Catherine Harold, Lynn Marie Latham, Bernard Lechowick	William Devane, Kevin Dobson, Michele Lee, Donna Mills, Ted Shackelford, Joan Van Ark, Nicollette Sheridan, Lynne Moody, Larry Riley, Tonya Crowe, Pat Petersen	950,000
Major Dad	Mon.	8:00	30	SBB Prods.-Spanish Trail Prods.-Universal	EP: Earl Pomerantz, Richard C. Okie, Gerald McRaney, John C. Stephens	Gerald McRaney, Shanna Reed, Matt Mulhern, Marlon Archey, Whitney Kershaw, Marisa Ryan, Nicole Dubuc, Chelsea Hertford	450,000
Murder, She Wrote	Sun.	8:00	60	Universal TV	EP: Peter S. Fischer SP: Robert F. O'Neill P: Robert E. Swanson, Robert Van Scoyk	Angela Lansbury, William Windom, Ron Masak	1,000,000
Murphy Brown	Mon.	9:00	30	Shukovsky/English Prods.-Warner Bros. TV	EP: Diane English, Joel Shukovsky P: Tom Seeley, Norm Gunzenhauser, Russ Woody, Gary Dontzig, Steven Peterman, Barnet Kellerman CP: Deborah Smith	Candice Bergen, Pat Croley, Faith Ford, Charles Kimbrough, Joe Regalbuto, Robert Pastorelli, Grant Shaud	475,000
Newhart	Mon.	10:30	30	MTM Enterprises	EP: Mark Egan, Mark Solomon SP: Bob Bendetson P: Stephen C. Grossman	Bob Newhart, Mary Frann, Peter Scolari, Julia Duffy, Tom Poston, William Sanderson, Tony Papenfuss, John Voldstad	500,000
Paradise	Sat.	8:00	60	Roundelay Prods.-Lorimar TV	EP: David Jacobs SP: James H. Brown CP: Robert Porter, Joel J. Feigenbaum	Lee Horsley, Jenny Beck, Matthew Newmark, Brian Lando, Michael Carter, Dehl Berti, Sigrid Thornton	900,000
Peaceable Kingdom	Wed.	8:00	60	Columbia Pictures TV	EP: Mark Waxman, Michael Vittes, Karen Harris P: Michael Vittes CP: Parke Perine	Lindsay Wagner, Tom Wopat, David Ackroyd, David Renan, Michael Manasseri, Melissa Clayton, Victor DiMattia, Conchata Ferrell	900,000
People Next Door, The	Mon.	8:30	30	The Sunshines Inc.-Wes Craven Films-Lorimar TV	EP: Steve & Madeline Sunshine, Wes Craven, Bruce Johnson SP: Robert Tischler P: Mark Masuoka, Robert Tischler CP: Lee Aronsohn D: J.D. Lobue	Jeffrey Jones, Mary Gross, Jaclyn Bernstein, Chance Quinn, Leslie Jordan, Christine Pickles	450,000

Series Title	Day	Hr.	Mins.	Supplier	Production principals	Cast regulars & semiregulars	Estimated prod. fee per episode
Rescue 911	Tue.	8:00	60	Arnold Shapiro Prods.-CBS Entertainment Prods.	EP: Arnold Shapiro SP: Jean O'Neill P: Nancy Platt Jacoby D: Chris Pechin	William Shatner	500,000
Saturday Night With Connie Chung	Sat.	10:00	60	CBS News	EP: Andrew Lack SP: Maurice Murad D: Don Roy King	Connie Chung	500,000
60 Minutes	Sun.	7:00	60	CBS News	EP: Don Hewitt Sr. P: Philip Scheffler D: Arthur Bloom	Mike Wallace, Morley Safer, Harry Reasoner, Ed Bradley, Meredith Vieira, Steve Kroft, Andy Rooney	650,000
Snoops	Fri.	8:00	60	Timalove Prods.-Viacom Prods.-Solt/Egan Co.	EP: Tim Reid, Hal Sitowitz, Sam Egan SP: Jo & Tom Perry P: David Auerbach CP: Lee Shaldon	Tim Reid, Daphne Maxwell Reid, John Karlen, Troy Curvey Jr.	900,000
Top Of The Hill	Thu.	9:00	60	Stephen J. Cannell Prods.	EP: Stephen J. Cannell SP: Jo Swerling Jr. P: Henry Colman	William Katt, Dick O'Neill, Jordan Baker	900,000
Tour Of Duty	Sat.	9:00	60	Zev Braun Prods.-New World TV	EP: Zev Braun SP: Steven Philip Smith P: Vahan Moosekian, Jim Westman CP: Jerry Patrick Brown, Robert Bielak, Carol Mendelsohn	Terence Knox, Stephen Caffrey, Tony Becker, Stan Foster, Ramon Franco, Miguel A. Nunez Jr., Dan Gauthier, Kim Delaney	950,000
Wiseguy	Wed.	10:00	60	Stephen J. Cannell Prods.	EP: Stephen J. Cannell, Les Sheldon SP: David A. Burke, Stephen Kronish, Jo Swerling Jr. P: Alfonse Ruggiero Jr., John Shulian CP: Clifton Campbell	Ken Wahl, Jonathan Banks, Jim Byrnes	925,000
Wolf	Tues.	9:00	60	CBS Entertainment Prods.	EP: David Peckinpah, Rod Holcomb SP: Garner Simmons P: Ken Swor CP: Tom Del Ruth	Jack Scalia, Nicholas Surovy, Joseph Sirola, J.C. Brandy, Mimi Kuzyk	900,000
					NBC-TV		
A Different World	Thu.	8:30	30	Carsey-Werner Co.-Bill Cosby	EP: Marcy Carsey, Tom Werner, Thad Mumford, Margie Peters SP: Susan Fales P: Joanne Curley Kerner, Debbie Allen, Cheryl Gard D: Debbie Allen	Dawn Lewis, Jasmine Guy, Kadeem Hardison, Charnele Brown, Cree Summer, Darryl Bell, Sinbad, Glynn Turman, Lou Myers	475,000
Alf	Mon.	8:00	30	Alien Prods.	EP: Tom Patchett, Bernie Brillstein SP: Lisa A. Bannick P: Paul Fusco CP: Steve Hollander	Max Wright, Anne Schedeen, Andrea Elson, Benji Gregory, Alf, JM J. Bullock, Liz Sheridan, John LaMotta	500,000
Amen	Sat.	8:30	30	Carson Prods.	EP: Ed. Weinberger, Artie Julian, Eric Cohen SP-D: Shelley Jensen P: Marty Nadler CP: Bill Daley, Ken Johnston, Reuben Cannon	Sherman Hemsley, Clifton Davis, Anna Maria Horsford, Barbara Montgomery, Roz Ryan, Jester Hairston	475,000
Baywatch	Fri.	8:00	60	GTG Entertainment	SEP: Robert Silberling EP: Douglas Schwarts, Michael Berk, Ernie Wallengren, Jill Donner P: Gregory Bonann, Jill Donner, Bill Schwartz	David Hasselhoff, Parker Stevenson, Shawn Weatherly, Billy Warlock, Erika Eleniak, Peter Phelps, Monte Markham, Brandon Call	900,000
Cheers	Thu.	9:00	30	Charles/Burrows/Charles Prods.-Paramount Network TV	EP: Glen & Les Charles, James Burrows P: Cheri Eichen, Bill Steinkellner, Phoef Sutton, Tim Berry	Ted Danson, Rhea Perlman, George Wendt, John Ratzenberger, Woody Harrelson, Kelsey Grammer, Kirstie Alley, Roger Rees	500,000
Cosby Show, The	Thu.	8:00	30		EP: Marcy Carsey, Tom Werner, John Martin P: Terry Guarnieri	Bill Cosby, Phylicia Rashad, Lisa Bonet, Sabrina LeBeauf, Malcolm-Jamal Warner, Tempestt Bledsoe, Keshia Knight Pulliam, Geoffrey Owens, Joseph C. Phillips, Raven-Symone	575,000

Series Title	Day	Hr.	Mins.	Supplier	Production principals	Cast regulars & semiregulars	Estimated prod. fee per episode
Dear John	Thu.	9:30	30	Ed. Weinberger Prods.-Paramount Network TV	EP: Ed. Weinberger, Hal Cooper, Rod Parker P: Bob Ellison, Mike Milligan, Jay Moriarty, Mark Reisman, Jeremy Stevens, Georg Sunga D: Hal Cooper	Judd Hirsch, Jane Carr, Jere Burns, Isabella Hofmann, Harry Groener, Billie Bird	475,000
Empty Nest	Sat.	9:30	30	Witt-Thomas-Harris Prods.-Touchstone TV	EP: Paul Junger Witt, Tony Thomas, Susan Harris, Gary Jacobs SP: Arnie Kogen, David Tyron King P: Susan Beavers, Gilbert Junger D: Steve Zuckerman	Richard Mulligan, Kristy McNichol, Dinah Manoff, David Leisure, Park Overall, Bear (dog)	475,000
Golden Girls, The	Sat.	9:00	30	Witt-Thomas-Harris Prods.-Touchstone TV	EP: Paul Junger Witt, Tony Thomas, Susan Harris, Marc Sotkin, Terry Hughes SP: Philip Lasker, Tom Whedon P: Gail Parent, Robert Bruce, Marty Weiss CP: Tracy Gamble, Richard Vaczy D: Terry Hughes	Bea Arthur, Betty White, Rue McClanahan, Estelle Getty	500,000
Hardball	Fri.	9:00	60	Columbia Pictures TV-NBC Prods.	EP: Frank Lupo, John Ashley, David Hemmings	John Ashton, Richard Tyson	900,000
Hogan Family, The	Mon.	8:30	30	Miller/Boyett Prods.-Lorimar TV	EP: Thomas L. Miller, Robert L. Boyett SP: Chip & Doug Keyes, Judy Pioli P: Bob Keyes, Deborah Oppenheimer CP: Harriette Ames-Regan	Sandy Duncan, Jason Bateman, Danny Ponce, Jeremy Licht, Edie McClurg, Josh Taylor	475,000
Hunter	Sat.	10:00	60	Stephen J. Cannell Prods.	EP: Fred Dryer, Larry Kubik, Marv Kupfer SP: Jo Swerling Jr., Paul Waigner P: Terry Nelson	Fred Dryer, Stepfanie Kramer, Charles Hallahan, Garrett Morris	950,000
In The Heat Of The Night	Tue.	9:00	60	Fred Silverman Co.-Jadda Prods.-MGM/UA TV	EP: Fred Silverman, Carroll O'Connor SP: Mark Rodgers, Ed DeBlasio P: Edward Ledding CP: Nancy Bond	Carroll O'Connor, Howard Rollins, Alan Autry, Anne-Marie Johnson, David Hart, Geoffrey Thorne, Hugh O'Connor	925,000
L.A. Law	Thu.	10:00	60	20th Fox TV	EP: David E. Kelley, Rick Wallace SP: William M. Finkelstein P: Elodie Keene, Michael M. Robin CP: Robert M. Breech	Harry Hamlin, Susan Dey, Jill Eikenberry, Corbin Bernsen, Michael Tucker, Michele Greene, Alan Rachins, Jimmy Smits, Susan Ruttan, Blair Underwood, Richard Dysart, Larry Drake	1-million
Magical World Of Disney, The	Sun.	7:00	60	Walt Disney TV		Michael D. Eisner	900,000
Brand New Life				NBC Prods.	EP: Chris Carter P: George W. Perkins	Barbara Eden, Don Murray, Shawnee Smith, David Tom, Jennie Garth, Alison Sweeney, Brian Patrick, Bryon Thames	
Parent Trap III				Walt Disney TV		Hayley Mills, Barry Bostwick, Creel Triplets	
Mancuso FBI	Fri.	10:00	60	Steve Sohmer Inc.-NBC Prods.	EP: Steve Sohmer, Jeff Bleckner SP: R.W. Goodwin P: Ken Solarz, Steve Bello	Robert Loggia, Lindsay Frost, Fredric Lehne, Randi Brazen, Charles Siebert	900,000
Matlock	Tue.	8:00	60	Fred Silverman Co. Dean Hargrove Prods.-Viacom Prods.	EP: Fred Silverman, Dean Hargrove, Joel Steiger SP: Jeff Peters P: Richard Collins	Andy Griffith, Nancy Stafford, Julie Sommars, Don Knotts, Clarence Gilyard	925,000
Midnight Caller	Tue.	10:00	60	December 3rd Prods.-Gangbuster Films-Lorimar TV	EP: Robert Singer SP: David Israel, Stephen Zito P: John F. Perry CP: Randy Zisk, John Schulian	Gary Cole, Wendy Kilbourne, Arthur Taxier, Dennis Dun	925,000
My Two Dads	Sun.	8:30	30	Columbia Pictures TV	EP: Michael Jacobs, Bob Myer SP: Chuck Lorre P: Mark Brull, Roger Garrett, David Steven Simon CP: Arlene Grayson	Paul Reiser, Greg Evigan, Staci Keanan, Florence Stanley, Vonni Ribisi, Amy Hathaway, Chad Allen	475,000
NBC Monday Night At The Movies	Mon.	9:00	120	Various			2.5-million

1989–90 Network Primetime At a Glance

Series Title	Day	Hr.	Mins.	Supplier	Production principals	Cast regulars & semiregulars	Estimated prod. fee per episode
NBC Sunday Night At The Movies	Sun.	9:00	120	Various			2.5-million
Night Court	Wed.	9:00	30	Starry Night Prods.-Warner Bros. TV	EP: Gary Murphy, Larry Strawther SP: Nancy Steen, Neil Thompson P: Fred Rubin, Bob Underwood, Tim Steele	Harry Anderson, John Larroquette, Markie Post, Charles Robinson, Richard Moll, Marsha Warfield	500,000
Nutt House, The	Wed.	9:30	30	Broksfilms TV-Alan Spencer Prods.-Touchstone TV	EP: Mel Brooks, Alan Spencer, Bob Brunner SP: Alan Mandel P: Ronald E. Frazier D: Gary Nelson	Cloris Leachman, Harvey Korman, Brian McNamara, Molly Hagan, Gregory Itzin, Mark Blakfield	450,000
Quantum Leap	Wed.	10:00	60	Belisarius Prods.-Universal TV	EP: Donald P. Bellisario SP: Deborah Pratt, Paul Belous, Robert Wolterstorff P: Parker Wade CP: Paul Brown, Jeff Gourson, Chris Ruppenthall	Scott Bakula, Dean Stockwell	900,000
Sister Kate	Sun.	8:00	30	Lazy B/FOB Prods.-Mea Culpa Prods.-20th Fox TV	EP: Frank Dungan & Jeff Stein, Tony Sheehan P: Patricia Rickey D: Jeff Melman	Stephanie Beacham, Harley Cross, Hannah Cutrona, Jason Priestley, Erin Reed, Joel Robinson, Penina Segall, Alexaundra Simmons	450,000
227	Sat.	8:00	30	Lorimar TV	EP: Irma Kalish SP: John Boni P: Roxie Wenk Evans, Larry Spencer D: Gerren Keith	Marla Gibbs, Jackee, Hal Williams, Alaine Reed-Hall, Paul Winfield, Regina King, Helen Martin, Curtis Baldwin, Barry Sobel	475,000
Unsolved Mysteries	Wed.	8:00	60	Cosgrove/Meurer Prods.	EP: John Cosgrove, Terry Dunn Meurer SP: Chris Pye, Edward Horwitz	Robert Stack	500,000

Abbreviations: SEP: supervising exec producer; EP: exec producer; SP: supervising producer; P: producer; CP: coproducer; D: director

Syndication At a Glance

VARIETY 9/13/89

Network programs' projected syndication dates

(Based on 4-year network play)

Program	Length	Distributor (Producer)	Network	Initial Air Date	Episodes To Date	Syndate
A Different World	30	(Carsey/Werner)	NBC	Sept. '87	30	Sept. '91
Alien Nation	60	Johnson Prods.	Fox	Sept. '89	n/a	Sept. '93
Anything But Love	30	20th Century Fox	ABC	Feb. '89	6	Sept. '92/'93
Baywatch	60	GTG Entertainment	NBC	Sept. '89	n/a	Sept. '93
Beauty And The Beast	60	(Witt/Thomas)	CBS	Sept. '87	30	Sept. '91
Booker	60	Stephen Cannell	Fox	Sept. '89	n/a	Sept. '93
Chicken Soup	30	Carsey/Werner	ABC	Sept. '89	n/a	Sept. '93
China Beach	60	Warner Bros.	ABC	April '88	12	Sept. '93
Coach	30	MCA	ABC	March '89	n/a	Sept. '93/'94
Dear John	30	Paramount	NBC	Oct. '88	n/a	Sept. '92/'93
Designing Women	30	Columbia	CBS	Sept. '86	52	Sept. '90
Doogie Howser, M.D.	30	Steven Bochco/20th Century Fox	ABC	Sept. '89	n/a	Sept. '93
Duet/Open House	30	Ubu Prods.	Fox	April, '87	36	Sept. '90
Empty Nest	30	(Witt-Thomas-Harris)	NBC	Oct. '88	n/a	Sept. '92/'93
Family Matters	30	Lorimar (WB)	ABC	Sept. '89	n/a	Sept. '93
Famous Teddy Z	30	Columbia	CBS	Sept. '89	n/a	Sept. '93
Free Spirit	30	Columbia	ABC	Sept. '89	n/a	Sept. '93
Full House	30	Lorimar/Telepictures	ABC	Sept. '87	28	Sept. '91
Hardball	60	Columbia	NBC	Sept. '89	n/a	Sept. '93
Hogan Family (ex Valerie's)	30	Lorimar/Telepictures	NBC	March '86	36	Sept. '90
Homeroom	30	Castle Rock Entertainment	ABC	Sept. '89	n/a	Sept. '93
In The Heat Of The Night	60	(Fred Silverman)	NBC	March '88	12	Sept. '92
Island Son	60	Lorimar (Time Warner)	CBS	Sept. '89	n/a	Sept. '93
It's Garry Shandling's Show	30	Viacom	Fox	March '88	30	Sept. '91/'92
Jake And The Fatman	60	Viacom	CBS	Sept. '87	48	Sept. '91
Just The 10 Of Us	30	Warner Bros.	ABC	April '88	12	Sept. '92
L.A. Law	60	20th Century Fox	NBC	Sept. '86	48	Sept. '90
Life Goes On	60	Warner Bros.	ABC	Sept. '89	n/a	Sept. '93
Living Dolls	30	Columbia	ABC	Sept. '89	n/a	Sept. '93
MacGyver	60	Paramount	ABC	Sept. '85	54	Sept. '89
Major Dad	30	MCA/Universal	CBS	Sept. '89	n/a	Sept. '93
Mancuso FBI	60	NBC Prods.	NBC	Sept. '89	n/a	Sept. '93

Program	Length	Distributor (Producer)	Network	Initial Air Date	Episodes To Date	Syndate
Married ... With Children	30	Columbia	Fox	April '87	36	Sept. '90/'91
Matlock	60	Viacom	NBC	Sept. '86	48	Sept. '90
Midnight Caller	60	Lorimar/Telepictures	NBC	Nov. '88	n/a	Sept. '92/'93
Mission: Impossible	60	Paramount	ABC	Oct. '88	12	Sept. '92/'93
Murphy Brown	30	Warner Bros.	CBS	Nov. '88	n/a	Sept. '92/'93
My Two Dads	30	Televentures	ABC	Sept. '87	24	Sept. '91
Nutt House	30	Buena Vista	NBC	Sept. '89	n/a	Sept. '93
Paradise	60	Lorimar/Telepictures	CBS	Jan. '89	n/a	Sept. '92/'93
Peaceable Kingdom	60	Columbia	CBS	Sept. '89	n/a	Sept. '93
People Next Door	30	Lorimar (WB)	CBS	Sept. '89	n/a	Sept. '93
Quantam Leap	60	MCA	NBC	April '89	n/a	Sept. '93/'94
Rescue 911	60	CBS Entertainment	CBS	Sept. '89	n/a	Sept. '93
Roseanne	30	(Carsey-Werner)	ABC	Oct. '88	n/a	Sept. '92/'93
Sister Kate	30	20th Century Fox	NBC	Sept. '89	n/a	Sept. '93
Snoops	60	Viacom	CBS	Sept. '89	n/a	Sept. '93
thirtysomething	60	MGM	ABC	Sept. '87	30	Sept. '91
Top Of The Hill	60	Stephen Cannell	CBS	Sept. '89	n/a	Sept. '93
Totally Hidden Video	30	Quantum Media	Fox	July '89	3	Sept. '93
Tour Of Duty	60	New World	CBS	Sept. '87	30	Sept. '91
Tracey Ullman	30	(Fox)	Fox	April '87	36	Sept. '90
Unsolved Mysteries	60	(Cosgrove-Meurer)	NBC	Oct. '88	n/a	Sept. '92/'93
Wiseguy	60	Televentures	CBS	Sept. '87	30	Sept. '91
Wolf	60	CBS Entertainment	CBS	Sept. '89	n/a	Sept. '93
Wonder Years	30	(New World)	ABC	Jan. '88	19	Sept. '93
Young Riders	60	MGM	CBS	Sept. '89	n/a	Sept. '93

Future off-network half-hours

(Note: C = cash; B = barter; each + is equivalent to one 30-second distributor commercial)

Program	Distributor	Episodes	Runs	Terms	Available	Comments
227	Columbia	116	7.8	C	Sept. '90	4 1/2-year deal — 910 telecasts
Alf	Warner Bros.	100	8	C	Sept. '90	delay to 91 OK w/10% price increase
Amen	MCA	110	8	C	Sept. '90/'91	
Dennis The Menace	Qintex	65	4	B (2/4)	Sept. '90	colorized/original; 1-year deal
Golden Girls	Buena Vista	130	6	C	Sept. '90/'91	
Head Of The Class	Warner Bros.	110	8	C	Sept. '90	6-year deal/episodes guaranteed
Hogan Family	Warner Bros.	97	6	C	Sept. '90/'91	3-year deal/syndex/91 opt./V. Harper-32 episodes
Leave It To Beaver	Qintex	130	8	C	Sept. '92	colorized original; 4-year deal
Perfect Strangers	Warner Bros.	100	8	C	Sept. '90	delay to 91 OK w/10% price increase

Future off-network hours

Program	Distributor	Episodes	Runs	Terms	Available	Comments
Highway To Heaven	Genesis Entertainment	87	3	B (5/7)	now	1-year only

Future firstrun half-hour strips

Program	Distributor	Episodes	Runs	Terms	Available	Comments
30 Minutes	King World	260	1	C++	Sept. '90	host, format TBA
Jokers Wild	Orbis	195	1+	C++	Sept. '90	host TBA
Monopoly	King World	195	1+	C++	Sept. '90	host, format TBA
Name That Tune	Orion	195	1+	C++	Sept. '90	host, format TBA
Out Of This World	MCA	72	6	C	Sept. '90	24 additional episodes available in '91
Scruples	Worldvision	196	1+	C++	Sept. '90	based on boardgame; host, format TBA
Tic Tac Dough	ITC	195	1+	C++	Sept. '90	host, format TBA
Wiseguys	Tribune Entertainment	196	1+	C++	Sept. '90	host, format TBA; produced by Jay Wolpert
And You Thought You Were So Smart	Warner Bros.	195	1+	C++	Sept. '90	terms/cond. tentative
Celebrity	GTG	195	1+	C++	Sept. '90	terms/cond. tentative
Concentration	Lexington	195	1+	C++	Sept. '90	terms/cond. tentative
House Party	Buena Vista	195	1+	C++	Sept. '90	terms/cond. tentative
Love Thy Neighbor	GTG	195	1+	C++	Sept. '90	terms/cond. tentative
Odd Man Out	Tribune Entertainment	195	1+	C++	Sept. '90	terms/cond. tentative
Today's People	Buena Vista	195	1+	C++	Sept. '90	terms/cond. tentative
Trump Card	Warner Bros.	195	1+	C++	Sept. '90	terms/cond. tentative
You Bet Your Life	Warner Bros.	195	1+	C++	Sept. '90	terms/cond. tentative

Future firstrun hour strips

Program	Distributor	Episodes	Runs	Terms	Available	Comments
21 Jump Street	Televentures	107	6	C	Sept. '91	for '90 start; 83 episodes available
Cristina Ferrare Show	MCA	195	1+	C++++	Sept. '90	daytime talk; title, format TBA
Entertainment Coast To Coast	Viacom	195	1+	C++++	Sept. '90	terms/cond. tentative
Open House	Group W (w/NBC Station Group)	195	1+	C++++	Sept. '90	terms/cond. tentative

Future firstrun weekly half-hours

Program	Distributor	Episodes	Runs	Terms	Available	Comments
Munsters	MCA	72	8	C	Sept. '91	
Water Sports World	Great Entertainment	26	2	B (3/3+)	Oct. '89	water events
Adam 12	MCA	26	2	C++	Sept. '90	terms/cond. tentative; remake
Dragnet	MCA	26	2	C++	Sept. '90	terms/cond. tentative; remake
Top Of The List	Viacom	26	2	C++	Sept. '90	terms/cond. tentative
What A Dummy	MCA	26	2	C++	Sept. '90	terms/cond. tentative; sitcom

Future firstrun weekly hour

Program	Distributor	Episodes	Runs	Terms	Available	Comments
Geraldo After Dark	Tribune Entertainment	26	2	C++++	Sept. '90	terms/cond. tentative

Future firstrun animated half-hours

Program	Distributor	Episodes	Runs	Terms	Available	Comments
Disney Afternoon — B	Buena Vista	65	8	B (3/3)	Sept. '90	"Tale Spin;" 4th-quarter B (2/4)
Disney Afternoon — C	Buena Vista	65	8	B (2+/3+)	Sept. '90	"Gummi Bears;" 4th-quarter B (2/4)
Disney Afternoon — D	Buena Vista	95	8	B (2+/3+)	Sept. '90	"Duck Tales;" 4th-quarter B (2/4)
Disney Afternoon — A	Buena Vista	65	8	B (3/3)	Sept. '90	"Chip 'N Dale;" 4th-quarter B (2/4)
G.I. Joe	Claster	106	2+	B (2+/3+	Sept. '90	2d-4th-quarter B (2/4); 24 new episodes; 1-year
Merry Melodies	Warner Bros.	65	8	B (2+/4)	Sept. '90	2-year deals
Police Academy	Lexington	65	8	B (2+/3+)	Oct. '89	4th-quarter B (2/4); 2-year deals
Tiny Toons Adventures	Warner Bros.	65	8	B (3/3+)	Sept. '90	Steven Spielberg Prod.
New Adventures Of He Man	Lexington	65	8	B (2+/3+)	Sept. '90	all new episodes

Current off-network half-hours

Program	Distributor	Episodes	Runs	Terms	Available	Comments
9 To 5	20th Century Fox	85	6	C	now	33 off net; 52 1st run
Alice	Warner Bros.	202	6	C	now	
All In The Family	Viacom	207	6	C	now	
Andy Griffith	Viacom	249	3	C	now	
Angie	Paramount	37	6	C	now	
Archie's Place	Columbia	97	6	C	now	
B.J./Lobo	MCA	86	6	C	now	also available as 1 hour
Barney Miller	Columbia	170	6	C	now	
Batman	20th Century Fox	120	6	C	now	
Benson	Columbia	158	6	C	now	
Best Of Groucho	W.W. Entertainment	130	6	C	now	
Beverly Hillbillies	Viacom	274	6	C	now	
Bewitched	DFS	252	2	B (2/4)	now	
Bob Newhart	Viacom	142	6	C	now	
Bosom Buddies	Paramount	37	6	C	now	
Brady Bunch	DFS	117	neg	B (2/4)	now	
Branded	King World	48	6	C	now	
Car 54, Where Are You?	Republic	60	6	C	now	
Carol Burnett	C.B. Distribution	150	6	C	now	
Carson Classics	Columbia	130	6	C	now	edited from net shows
Cheers	Paramount	112	7	C	now	
The Cosby Show	Viacom	150	8	C++	now	1 run daily; 4-year deals
Dick Van Dyke	Viacom	158	6	C	now	w/Mary Tyler Moore
Diff'rent Strokes	Columbia	189	6	C	now	
Facts Of Life	Columbia	209	6	C	now	
Fame, Fortune And Romance	TPE	115	4	C	now	off ABC-TV; 2-year deals
Family Affair	Viacom	138	6	C	now	
Family Ties	Paramount	98	7	C	now	
Fantasy Island	Columbia	200	6	C	now	also available as 1 hour
Flying Nun	Columbia	82	6	C	now	
Get Smart	Republic	138	5	C	now	
Gidget	Lexington	80	6	C	now	48 1st run; 32 off net
Gilligan's Island	Turner Program Sales	98	4	C	now	
Gimme A Break	MCA	85	6	C	now	
Gomer Pyle	Viacom	150	6	C	now	
Good Times	Columbia	133	6	C	now	
Growing Pains	Warner Bros.	110	8	C	now	
Guns Of Will Sonnett	King World	50	6	C	now	
Happy Days	Paramount	255	6	C	now	
Here's Lucy	Warner Bros.	144	6	C	now	
Hitchcock Presents	MCA	265	6	C	now	run negotiable
Hogan's Heroes	Viacom	168	2	C	now	
Honeymooners	Viacom	107	6	C	now	
I Dream Of Jeannie	DFS	139	6	B (2/4)	now	
I Love Lucy	Viacom	179	3	C	now	
I Married Joan	Weiss Global	98	6	C	now	b&w
Jeffersons	Columbia	253	6	C	now	
Kate & Allie	MCA	96	6	C	now	
Knight Rider	MCA	90	6	C	now	also available as 1 hour
Laugh-In	Warner Bros.	130	2	C+	now	
Laverne And Shirley	DFS	178	neg	B (2/4)	now	
Leave It To Beaver	Paramount	234	2	C	now	original series
Life Of Riley	New World	120	6	C	now	w/William Bendix (b&w)
Life Of Riley	New World	146	6	C	now	w/William Bendix (includes 26 w/Gleason) (b&w)

Syndication At a Glance

Program	Distributor	Episodes	Runs	Terms	Available	Comments
Life Of Riley	New World	26	6	C	now	w/Jackie Gleason (b&w)
Love Boat II	Worldvision	115	6	C	now	also available as 1 hour
Mash	20th Century Fox	255	6	C	now	
Make Room For Daddy	Weiss Global	161	4	C	now	
Maude	Columbia	141	6	C	now	
Mayberry, R.F.D.	Warner Bros.	78	6	C	now	
McHale's Navy	Qintex	130	8	C	now	colorized original episodes; 4-year deals
Monkees	Lexington	58	1+	B (2+/4)	now	original series
Mork & Mindy	DFS	95	neg	B (2/4)	now	
Mr. Belvedere	20th Century Fox	110	6	C	now	
My Favorite Martian	Warner Bros.	107	6	C	now	
My Little Margie	Weiss Global	126	4	C	now	
Newhart	MTM	134	6	C	now	
Night Court	Warner Bros.	101	8	C	now	5-year deals
Night Gallery	MCA	97	6	C	now	
Odd Couple	DFS	114	neg	B (2/4)	now	
One Day At A Time	Columbia	209	6	C	now	
Partridge Family	DFS	96	6	B (2/4)	now	
Soap	Columbia	93	3	C	now	
Square Pegs	Columbia	20	6	C	now	
Tales Of The Texas Rangers	Columbia	52	6	C	now	all b&w
Taxi	Paramount	93	6	C	now	
That Girl	Worldvision	136	6	C	now	
That's Incredible	MCA	165	6	C	now	
That's My Mama	Columbia	39	6	C	now	
The Ropers	Taffner	26	6	C	now	
Three's Company	Taffner	174	6	C	now	
Too Close For Comfort	Taffner	122	6	C	now	includes 22 "Ted Knight" episodes
Topper	King World	78	neg	C	now	
Twilight Zone	MGM/UA	94	2+	B (2+/4)	now	combined w/30 1st run episodes; 1 year only
Twilight Zone	Viacom	136	6	C	now	also available as 1 hour
We Love Lucy	Viacom	26	2	C	now	also available as 1 hour
Webster	Paramount	98	12	C	now	
What's Happening	Lexington	131	6	C	now	65 off net; 66 1st run
Who's The Boss?	Columbia	120	6	C	now	4 1/2-year deals; 910 plays
WKRP In Cincinnati	Victory	90	6	C	now	
Wyatt Earp	Columbia	130				

Current off-network hours

Program	Distributor	Episodes	Runs	Terms	Available	Comments
12 O'Clock High	20th Century Fox	78	6	C	now	
A-Team	MCA	128	6	C	now	
Airwolf	MCA	80	6	C	now	4-year deal; some episodes produced by USA Network
Avengers	Orion	83	5	C	now	
B.J./Lobo	MCA	86	6	C	now	also available as 1/2-hour
Barnaby Jones	Worldvision	177	6	C	now	
Black Sheep Squadron	MCA	35	4	C++	now	
Blue Knight	Warner Bros.	23	2	C	now	
Bonanza	Republic	268	5	C	now	
Buck Rogers	MCA	37	6	C	now	
CHiPs	MGM/UA	138	6	C	now	
Cagney & Lacey	Orion	125	6	C	now	on Lifetime cable
Cannon	Viacom	124	6	C	now	
Charlie's Angels	Columbia	115	4	C	now	
Crazy Like A Fox	Lexington	37	2	B (5/8)	now	
Dallas	Warner Bros.	161	6	C	now	
Dukes Of Hazzard	Warner Bros.	143	6	C	now	
Dynasty	20th Century Fox	178	6	C	now	could buy 1 run at reduced price
Eight Is Enough	Warner Bros.	112	6	C	now	
Falcon Crest	Warner Bros.	157	4	C	now	
Fall Guy	20th Century Fox	111	8	C	now	
Fantasy Island	Columbia	130	6	C	now	also available as 1/2-hour
Gunsmoke	Viacom	402	4	C	now	
Hardcastle And McCormick	Lexington	67	6	B (5/7)	now	
Hart To Hart	Columbia	112	6	C	now	
Hawaii Five-0	Viacom	200	6	C	now	
High Chaparral	Republic	98	5	C	now	
Hill Street Blues	Victory	146	6	C	now	
Hitchcock Hour	MCA	93	6	C	now	
Hunter	Televentures	107	6	C	now	
Incredible Hulk	MCA	85	6	C	now	
Jacques Cousteau	Turner Program Sales	36	neg	C	now	
Knight Rider	MCA	90	6	C	now	also available as 1/2-hour
Knots Landing	Warner Bros.	128	6	C	now	
Kojak	MCA	118	6	C	now	
Little House On The Prairie	Worldvision	216	6	C	now	
Lost In Space	20th Century Fox	83	6	C	now	
Love Boat I	Worldvision	140	6	C	now	
Love Boat II	Worldvision	115	6	C	now	also available as 1/2-hour
Magnum, P.I.	MCA	129	6	C	now	
Mannix	Paramount	130	6	C	now	
Matt Houston	Warner Bros.	68	6	C	now	
Mission: Impossible	Paramount	171	4	C	now	original series
Mystery Movies	MCA	124	8	C	now	"Columbo," etc.
Perry Mason	Viacom	271	4	C	now	
Police Story	Columbia	105	6	C	now	
Police Woman	Columbia	91	6	C	now	
Remington Steele	MTM	94	6	C	now	
St. Elsewhere	MTM	116	2	C	now	will sell 260 plays for 1-year deal

Program	Distributor	Episodes	Runs	Terms	Available	Comments
Star Trek	Paramount	79	4	C	now	original series
Streets Of San Francisco	Worldvision	119	6	C	now	
T.J. Hooker	Columbia	90	6	C	now	
That's Incredible	MCA	107	6	C	now	
The Man From U.N.C.L.E.	Turner Pgm. Sales	132	6	C	now	
The Prisoner	ITC	17	4	C	now	
Trapper John	20th Century Fox	132	6	C	now	
Twilight Zone	Viacom	18	6	C	now	
Vegas	20th Century Fox	68	8	C	now	
Voyage To The Bottom Of The Sea	20th Century Fox	110	6	C	now	
Waltons	Warner Bros.	221	6	C	now	
We Love Lucy	Viacom	13	2	C	now	also available as 1/2-hour
Wonder Woman	Warner Bros.	61	4	C	now	
Wonderful World Of Disney	Buena Vista	185	4	C	now	

Current firstrun half-hour strips

Program	Distributor	Episodes	Runs	Terms	Available	Comments
3rd Degree	Warner Bros.	195	1+	C++	now	host: Bert Convy
After Hours	Worldvision	130	2	B (2+/4)	now	hosts: John Major & Heidi Bohi; late fringe
Benny Hill	Taffner	100	8	C	now	
Brothers	Paramount	114	6	C	now	previously aired on Showtime cable
Bumper Stumpers	MG/Perin	260	1	C	now	host: Al Dubois; previously aired on USA cable
Business This Morning	Viacom	260	1	B (3/3+)	now	
CNN Headline News	Turner Pgm. Sales	260	1	C+	now	
Couch Potato	Group W	95/75	1/2	C++	now	beyond 4th quarter TBA
Crimewatch Tonight	Orion	185	1+	C++	now	host: Ike Pappas/15 weeks updated reports
Crook And Chase	Intermedia Mgmt.	260	1	B (3/3+)	now	
Current Affair	20th Century Fox	260	1	C	now	
Divorce Court	Blair Entertainment	160	1/2	C++	now	
Entertainment Tonight	Paramount	260	1	C+	now	
Everyday With Joan Lunden	Michael Krauss Prods.	195	1+	B (2/4+)	now	also available as 1 hour
Family Feud	Lexington	195	1/2	C++	now	
First Business	Biz Net	260	1	B (3/3)	now	host: Carl Grant; 5:30 & 6 a.m. sat feeds
Hard Copy	Paramount	195	1+	C++	now	host: Allan Frio
Independent Network News: USA Tonight	INN News	260	1	B (3+/3+)	now	daily newscast/satellite feed
Inside Edition	King World	260	1	C++	now	host: Bill O'Reilly
Inside Report	MCA	260	1	C++	now	host: Penny Daniels
It's A Living	Warner Bros.	120	6	C	now	5-year deals; 27 off-net episodes included
Jackpot	Palladium	175	1+	C++	now	host: Geoff Edwards
Jeopardy	King World	195	1/2	C++	now	host: Alex Trebek
Leave It To Beaver	Qintex	105	8	C	now	all new episodes
Littlest Hobo	Silverbach/Lazarus	130	6	C	now	
Lone Ranger	Palladium	221	6	C	now	39 color/182 b&w; all on tape
Love Connection	Warner Bros.	170	1/2	C+	now	
Mama's Family	Warner Bros.	160	6	C	now	includes 35 off-net episodes
Morning Stretch	PSS	130	2	B (2/4)	now	host: Joanie Greggains
People's Court	Warner Bros.	195	1/2	C++	now	
P.M. Magazine	Group W	195	1/2	C+	now	
Small Wonder	20th Century Fox	96	6	C	now	
Tales Of The Unexpected	Orbis	90	3	B (2+/4)	now	
Talkabout	Taffner	195	1+	C++	now	host: Wayne Cox; off Canadian tv
The Judge	Genesis Ent.	160	1/2	C++	now	
The Last Word	Turner Pgm. Sales	175	1+	B (2/4)	now	host: Wink Martindale
Trial By Jury	Viacom	160	1+	B (3/3+)	now	host: Raymond Burr
USA Today On TV	GTG	260	1	C++	now	
Bullwinkle	DFS	98	unl	B (2/4)	now	
C.O.P.S.	Claster	65	8	B (2/4)	now	
Care Bears	SFM	65	8	B (2+/4)	now	4-run, 1-year deal possible
Chip 'N Dale's Rescue Rangers	Buena Vista	65	4	B (3/3)	now	4th-quarter B (2/4)
Dennis The Menace	DFS	65	4	B (2/4)	now	
Denver The Last Dinosaur	World Events	52	5	B (2+/3+)	now	includes 13 episodes from '88/'89; 1-year deal
Duck Tales	Buena Vista	95	3	B (2+/3+)	now	4th-quarter B (2/4)
Dudley Do-Right	DFS	38	unl	B (2/4)	now	
Felix The Cat	Columbia	65	8	C	now	also available as segments
Flintstones	DFS	166	1/2	B (2/4)	now	
Funtastic World — Hanna/Barbera	Worldvision			B (6/12)	now	2-hour weekly
Gumby (new series)	Warner Bros.	65	8	B (2/4)	now	new episodes ('88/'89)
Gumby (original)	Ziv Intl.	130	unl	C	now	
Heathcliff	Lexington	86	4	C++	now	
Inch High Private Eye	DFS	13	unl	B (2/4)	now	
Jem	Claster	75	7	B (2/4)	now	
Jetsons	Worldvision	75	4	B (2/4)	now	
M.A.S.K.	Lexington	75	10	C	now	
Maxie's World	Claster	70	3+	B (2/4)	now	1st-, 3d-quarter B (2+/3+)
Mighty Mouse And Friends	Viacom	130	3	C	now	2-year deal/Alvin, Heckle
Muppet Babies	Claster	65	8	B (2+/3+)	now	4th-quarter B (2/4); 2-year deal
My Little Pony And Friends	Claster	65	8	B (2/4)	now	
New Archies	Claster	13	4	B (2/4)	now	
Real Ghostbusters	Lexington	99	8	B (2/4)	now	
Rocky And Friends	DFS	78	unl	B (2/4)	now	
Scooby Doo	DFS	110	2+	B (2/4)	now	
Smurfs	Worldvision	130	2	B (2/4)	now	
Snorks	Worldvision	65	8	B (2+/4)	now	
Space Kidettes	DFS	20	unl	B (2/4)	now	
Super Mario Brothers	Viacom	65	8	B (2+/4)	now	Nintendo characters; some live
Super Sunday	Claster	14	4	B (2/4)	now	
Superfriends	Lexington	110	2+	B (2/4)	now	
Teenage Mutant Ninja Turtles	Group W	65	8	B (2+/3+)	now	

Program	Distributor	Episodes	Runs	Terms	Available	Comments
Tennessee Tuxedo	DFS	140	unl	B (2/4)	now	
Thunderbirds	ITC	24	unl	C	now	
Thundersub	Lionheart (BBC)	27	unl	C	now	
Uncle Waldo	DFS	52	unl	B (2/4)	now	
Valley Of The Dinosaurs	DFS	16	unl	B (2/4)	now	
Visionaries	Claster	13	4	B (2/4)	now	
Wheelie And The Chopper Bunch	DFS	13	unl	B (2/4)	now	
Yogi Bear	Worldvision	65	8	B (2/4)	now	
Young Samson	DFS	20	unl	B (2/4)	now	

Current children's live action

Program	Distributor	Episodes	Runs	Terms	Available	Comments
Cisco Kid	Blair Entertainment	156	6	C	now	
Dr. Fad	Fox/Lorber	26	2	B (2+/4)	now	
Fun House	Warner Bros.	170	1+	B (2+/4)	now	
Littlest Hobo	Warner Bros.	96	3	C	now	
Muppets	ITC	120	neg	C	now	
Peppermint Place	Electra Pictures	52	1	B (2+/3)	now	
Super Sloppy Double Dare	Viacom	130	2	B (2+/4)	now	host: Mark Summer
Superman	Warner Bros.	104	4	C	now	
Young Universe	Behrens	26	2	C+	now	

Current short-length animation

Program	Distributor	Episodes	Runs	Terms	Available	Comments
Bugs Bunny/Porky Pig	Warner Bros.	256	unl	C	now	100 Bugs/156 Porkys
Casper The Friendly Ghost	Worldvision	244	unl	C	now	
Felix The Cat	Columbia	260	unl	C	now	
Hercules	Columbia	130	unl	C	now	
New Three Stooges	Muller Media	156	unl	C	now	
Tom And Jerry	Turner Program Sales	308	unl	C	now	

Current inserts

Program	Distributor	Episodes	Runs	Terms	Available	Comments
20th Century Woman	SFM/20th Century Fox	65	5	C+	now	
CNN News	Turner Program Sales	365	unl	C+	now	
Entertainment Report	Group W/All American	260	1	C	now	daily material
GVC Auto Tips	SPR News Service	15	unl	C	now	20-second length
GVC Health Tips	SPR News Service	15	unl	C	now	20-second length
Holiday Moments	Carter/Grant	36	4	C	now	50-second message
Mother's Minutes	Michael Krauss Prods.	285	3	C	now	w/Joan Lunden
Mr. Food	King World	260	unl	C	now	90-second length
N.I.W.S.	Warner Bros.	52	1	C	now	news/entertainment feature stories
News Travel Network	NTN	260	unl	C	now	
Newsfeed	Group W	130	1	C	now	news and entertainment features
Sylvia Porter's Money Tips	MG/Perin	156	unl	C	now	host: Carol Sinclair/3 per week/30 seconds
Your Pet And The Vet	World Events	52	unl	C	now	
Wheel Of Fortune	King World	195	1/2	C++	now	
Win, Lose Or Draw	Buena Vista	185	1/2	C++	now	host: Rob Weller

Current firstrun hour strip

Program	Distributor	Episodes	Runs	Terms	Available	Comments
Arsenio Hall	Paramount	200	1/2	B (7/7)	now	late fringe
Dr. Who	Lionheart (BBC)	260	1	C	now	normally airs on PBS
Everyday With Joan Lunden	Michael Krauss Prods.	195	1+	C++++	now	also available as 1/2-hour
Geraldo	Paramount/Tribune Ent.	240	1/2	C++++	now	host: Geraldo Rivera
Joan Rivers	Paramount	200	1+	C++++	now	
Live With Regis And Kathie Lee	Buena Vista	230	1/2	C++++	now	hosts: Regis Philbin/Kathy Lee Gifford
Oprah Winfrey Show	King World	240	1/2	C++++	now	host: Oprah Winfrey
Phil Donahue	Multimedia	230	1/2	C++++	now	host: Phil Donahue
Sally Jessy Raphael	Multimedia	230	1/2	C++++	now	host: Sally Jessy Raphael

Current firstrun weekly half-hours

Program	Distributor	Episodes	Runs	Terms	Available	Comments
9 To 5	20th Century Fox	26	2	C	now	
America's Top 10	All American	48/4	1/2	B (2+/3+)	now	
At The Movies	Tribune Entertainment	48/4	1/2	B (2+/4)	now	can air twice per week
Better Your Home With/BHG	Worldvision	26	2	B (2+/4)	now	host: Gerry Connell
Charles In Charge	MCA	26	2	B (3/4)	now	
College Madhouse	Warner Bros.	26	2	B (3/3+)	now	host: Greg Kinnear/fun house for college
Computer Show	Victory	39/13	1/2	B (3/3)	now	satellite feed Friday/air weekends
Crime Stopper 800	All American	39/13	1/2	B (3/3+)	now	
Ebony/Jet Showcase	Carl Meyers	26	2	B (3/3+)	now	

Program	Distributor	Episodes	Runs	Terms	Available	Comments
George And Mildred	Taffner	38	2	C	now	Britcom
Gidget	Lexington	44	1	B (3/3+)	now	
INN Magazine	INN News	52	1	B (3/3)	now	
In Sport	Select Media	50/2	1/2	B (3/3+)	now	hosts: Ahmad Rashad/Robin Swoboda
Inside Video This Week	MG/Perin	52	1	B (3/3+)	now	hosts: Eric Burns/Paula McClure
It's A Living	Warner Bros.	25	2	B (3/4)	now	
Jeopardy	King World	52	1	C	now	weekend version
Keep It In The Family	Taffner	31	2	C	now	Britcom
Lassie 4	MCA	26	2	B (3/4)	now	all new episodes
Mama's Family	Warner Bros.	68	2	B (3/4)	now	
Man About The House	Taffner	39	2	C	now	Britcom
Missing Reward	Group W	24	1/2	B (3/3+)	now	host: Stacy Keach
Monsters	Tribune Entertainment	26	2	B (3+/3+)	now	
Motorweek Illustrated	Orbis	52	1	B (2+/3+)	now	all about cars
Munsters (new)	MCA	24	2	B (3/4)	now	new cast
Music City, USA	Multimedia	26	2	B (2+/3+)	now	
My Secret Identity	MCA	26	2	B (3/4)	now	
Out Of This World	MCA	24	2	B (3/4)	now	
Punky Brewster	Columbia	22	2	B (3/4)	now	will be paired w/off-net episodes
Remote Control	Viacom	39/13	1/2	B (3/3+)	now	host: Ken Ober; new episodes; same at MTV
Runaway With/Rich And Famous	TPE	26	2	B (3+/3+)	now	
Secret World	Turner Pgm. Services	24	2	B (2/4)	now	nature/animal
Secret World	Turner Pgm. Services	24	2	C	now	
Secrets And Mysteries	ITC	26	2	B (3/3+)	now	host: Edward Mulhare/off cable
Siskel And Ebert	Buena Vista	46/6	1/2	B (2+/3+)	now	
Small Wonder	20th Century Fox	24	2	C++	now	
Superboy	Viacom	26	2	B (3/3+)	now	
T&T	Qintex	24	2	B (3/3+)	now	thru mid-Jan. '90
Tales From The Darkside	Tribune Entertainment	26	2	B (3/3+)	now	
That's My Mama	Columbia	22	2	C+++	now	combined w/39 off-net
The Making Of...	Muller Media	26	2	C	now	
Twlight Zone	MGM/UA	90	2	B (3/3+)	now	now inclues 60 off-net episodes
War Chronicles	Orbis	13	3	C	now	
Wheel Of Fortune	King World	52	1	C	now	weekend version
World Class Women	Select Media	13	2	B (2+/4)	now	host: Randi Hall
Youth Quake	J.M. Ent.	26	2	B (3/3+)	now	also 7 1-hour spex

Current firstrun weekly hours

Program	Distributor	Episodes	Runs	Terms	Available	Comments
American Gladiators	Samuel Goldwyn	26	2	B (6/6)	now	hosts: Joe Theismann/Mike Adamle
Blake's 7	Lionheart	52	2	C	now	also plays on PBS
Byron Allen	Genesis	26	2	B (6/7)	now	variety/entertainment
Classic Country	Genesis Entertainment	91	6	C	now	
Cop Talk — Behind The Shield	Tribune Entertainment	26	2	B (6/6)	now	host: Sonny Grosso; through Dec. '89
Entertainment This Week	Paramount	51	1	B (6/6)	now	6-year
Fairie Tale Theater	Silverbach/Lazarus	26	4	B	now	6-year deal/aired on Showtime cable
Freddie's Nightmare	Warner Bros.	22	2	B (6/6)	now	
Friday The 13th	Paramount	26	2	B (6/6)	now	
G.L.O.W.	MG/Perin	26	2	C	now	female wrestling
Hee Haw	Gaylord	26	2	B (5/5)	now	
Jacques Cousteau	Turner Pgm. Services	12	2	B (5/5)	now	play 1 per quarter/yearly
Lifestyles Of The Rich And Famous	TPE	26	2	B (6+/6)	now	
Michelob Presents Night Music	Fox/Lorber	24	2	B (5/8)	now	hosts: Jools Holland/David Sanborn
National Geographic On Assignment	Turner Pgm. Services	12	2	B (5/5)	now	monthly
National Geographic Explorer	Turner Pgm. Services	12	2	B (5/5)	now	
National Geographic Specials	Genesis Entertainment	96	6	C	now	
Robin Hood	Samuel Goldwyn	26	6	C	now	w/2-hour movie from Showtime cable
Roller Games	Qintex	13	2	B (6+/6)	now	roller derby meets "Starlight Express"
Showtime At The Apollo	Raymond Horn	26	2	B (6/6)	now	
Smithsonian Treasures	Lexington	6	2	B (12/12)	now	2-hour episodes/2-year deals
Soul Train	Tribune Entertainment	40/12	1/2	B (5+/8+)	now	
Space 1999	ITC	40	5	C	now	
Star Search	TPE	26	2	B (6+/6+)	now	
Star Trek: The Next Generation	Paramount	26	2	B (7/5)	now	
The Eyes Of War	Vestron	8	1	5/7	now	host: Robert Mitchum/or 2-hr. qtrly. (B11/13)
Tuff Trax	Qintex	52	1	B (6/8)	now	monster trucks/tractor pulls; all new
USA Today On TV	GTG	52	1	B (5/7)	now	all new material
War Of The Worlds	Paramount	24	2	B (6+/5+)	now	
World At War	Taffner	36	2	C	now	
Youth Quake	J.M. Entertainment	7	1	B (5/7)	now	1 per month/2 runs if needed

Current children's animated

Program	Distributor	Episodes	Runs	Terms	Available	Comments
Alvin And The Chipmunks	Warner Bros.	65	12	B (2+/3+)	now	
Animated Classics	Taffner	8	2	C	now	

By BRIAN LOWRY

Hollywood CBS had the numbers in terms of overall wins, but NBC stole the spotlight during the 41st annual Primetime Emmy Awards by winning best drama for "L.A. Law" and best comedy for "Cheers."

CBS finished with 27 Emmys, seven for miniseries "Lonesome Dove" and a quartet for freshman sitcom "Murphy Brown," including the best actress in a comedy series award to Candice Bergen, the first non-"Golden Girl" win since 1985.

The awards show proved to be a mixed bag of old and new, with few themes emerging, and the programs that took home the most statues were often deprived of the top honor in their category. For example, ABC's massive "War And Remembrance" shot down "Dove," the clear favorite and winner of six technical prizes during the non-televised awards Sept. 16, for outstanding miniseries.

NBC tallied 25 Emmys. During the Sept. 17 televised awards, NBC took 12, vs. six each for CBS and ABC. NBC finished third in terms of overall Emmys last year and watched ABC take home the top two prizes, for "The Wonder Years" and "thirtysomething."

ABC totaled 13 this year, followed by eight for PBS. After being shut out in its first two seasons despite nine nominations, "The Tracey Ullman Show" accounted for all of Fox Broadcasting Co.'s four Emmys, including the award for variety, music or comedy program. "Star Trek: The Next Generation" beamed up the only awards for syndicated fare, both tech awards for sound.

Cable, in its second year of eligibility, received a total of four trophies, three going to HBO (equaling its take from last year), while another went to superstation TBS for Hal Holbrook's performance in the informational program "Portrait Of America: Alaska." HBO was the only cabler to receive an Emmy last year.

The telecast was the shortest for an Emmy Awards since 1979. At two hours, 56 minutes, it ran 24 minutes shorter than last year's show and more than an hour shorter than the 4-hour marathon in 1987, the first year of Fox' Emmy contract.

"Murphy Brown," "Tracey Ullman" and ABC's "thirtysomething were the only series to receive four awards, while "Cheers" nabbed a trio, as did "War And Remembrance."

First-time winners included Bergen and Dana Delany, lead actress in a drama series for ABC's "China Beach," the program's sole Emmy.

By contrast, their male counterparts were both repeat winners: Richard Mulligan, in the comedy category for NBC's "Empty Nest," a previous honoree for "Soap," and Carroll O'Connor, a surprise victor for drama for "In The Heat Of The Night" and a 4-time Emmy recipient in the past for "All In The Family."

The controversial NBC telefilm "Roe Vs. Wade" tied CBS' "Day One" as outstanding drama/comedy special, while Holly Hunter was feted as lead actress in a miniseries or spec for the former, which dealt with the landmark 1973 abortion decision.

Hunter thanked the network for "getting this movie on the air" despite advertiser pull-outs and the real-life "Jane Roe," Norma McCorvey, for "continuing to fight to keep women from being second-class citizens."

James Woods picked up his second statuette for lead actor in a miniseries or special, for "My Name Is Bill W." — a "Hallmark Hall Of Fame' special that costarred James Garner, as did "Promise," for which he won in 1986.

"L.A. Law's" win was its second in three years for drama series. That show's Larry Drake repeated as supporting actor in a drama for his role as the mentally retarded Benny, while "thirtysomething's" Melanie Mayron — a first-time nominee — beat out a trio of "Law" nominees as supporting actress.

"Cheers" won for the first time since back-to-back wins in 1983 and 1984 and swept the awards for comedy supporting actors, as Rhea Perlman claimed her fourth nod for the show while Woody Harrelson was recognized for the first time. John Larroquette, who had a stranglehold on the category with his "Night Court" role, bowed out of this year's competition.

Colleen Dewhurst was the recipient of two supporting actress Emmys over the course of the weekend: one win in the miniseries/special category for NBC's "Those She Left Behind" and another nod as guest performer in a comedy series as Murphy Brown's mother.

"Brown's" wins also included a writing Emmy to Diane English for the series pilot, beating out four entries for last year's comedy series winner, "The Wonder Years." That ABC show was held to one award, for director Peter Baldwin, despite 13 nominations.

"War And Remembrance's" Dan Curtis was elated by the mini's victory, thanking former ABC Entertainment chief Brandon Stoddard and the network "for having the guts to pony up the dough" to fund the $110-million enterprise.

"Saturday Night Live" edged NBC brethren "The Tonight Show" and "Late Night With David Letterman" for writing on a variety or music program, the latenight show's first Emmy since 1983 and eighth in its 15-year history. Jim Henson claimed the directing award in that category for his short-lived "The Jim Henson Hour" on ABC.

The Emmy show had a nostalgic tint to it, with Milton Berle and Bob Hope receiving standing ovations, the latter as he presented the posthumous Governors Award to Lucille Ball. It was accepted by the late comedian's husband, Gary Morton.

Seven of the 29 winners were no-shows, though the producers used their absence to insert comedy vignettes.

When Robert Altman wasn't on hand as winner in directing for HBO's "Tanner '88," the clip showed an exclusive acceptance simulation from ABC News — alluding to the recent flap over its use of a re-creation on "World News Tonight."

As usual, Fox liberally used the telecast as a vehicle to showcase and promote its primetime lineup.

Among other notable footnotes, Dwight Hemion, who has more Emmys on his mantel than anyone, else, was given his 16th for directing "The Kennedy Center Honors."

A complete list of winners follows:

COMEDY SERIES

"Cheers" — James Burrows, Glen Charles, Les Charles, executive producers; Cheri Eichen, Bill Steinkellner, producers; Tim Berry, Phoef Sutton, coproducers (NBC).

DRAMA SERIES

"L.A. Law" — Steven Bochco, executive producer; Rick Wallace, coexecutive producer; David E. Kelley, supervising producer; Scott Goldstein, Michele Gallery, producers; William M. Finkelstein, Judith Parker, coproducers; Philip M. Goldfarb, Alice West, coordinating producers (NBC).

MINISERIES

"War And Remembrance" — Dan Curtis, executive producer; Barbara Steele, producer (ABC).

VARIETY, MUSIC OR COMEDY PROGRAM

"The Tracey Ullman Show" — James L. Brooks, Heidi Perlman, Jerry Belson, Ken Estin, Sam Simon, executive producers; Richard Sakai, Ted Bessell, producers; Marc Flanagan, coproducer; Tracey Ullman, host (Fox).

DRAMA/COMEDY SPECIAL

"Day One" ("AT&T Presents") — Aaron Spelling, E. Duke Vincent, executive producers; David W. Rintels, producer (CBS).

"Roe Vs. Wade" — Michael Manheim, executive producer; Gregory Hoblit, producer; Alison Cross, coproducer.

CLASSICAL PROGRAM IN THE PERFORMING ARTS

"Bernstein At 70!" ("Great Performances") — Harry Kraut, Klaus Hallig, executive producers; Michael Bronson, Thomas P. Skinner, producers (PBS).

INFORMATIONAL SERIES

"Nature" — David Heeley, executive producer; Fred Kaufman, series producer (PBS).

INFORMATIONAL SPECIAL

"Lillian Gish: The Actor's Life For Me" ("American Masters") — Freida Lee Mock, Susan Lacy, executive producers; Terry Sanders, producer; William T. Cartwright, coproducer (PBS).

ANIMATED PROGRAM (ONE HOUR OR LESS)

"Garfield: Babes And Bullets" — Phil Roman, producer; Jim Davis, writer; Phil Roman, director; John Sparey, Bob Nesler, codirectors (CBS).

CHILDREN'S PROGRAM

"Free To Be ... A Family" — Marlo Thomas, Christopher Cerf, executive producers, U.S.; Robert Dalrymple, producer, U.S.; Leonid Zolotarevsky, executive producer, USSR; Igor Menzelintsev, producer, USSR; Vern T. Calhoun, coproducer (ABC).

SPECIAL EVENTS

"Cirque Du Soleil" ("The Magic Circus") — Helene Dufresne, producer (HBO).

"The 11th Annual Kennedy Center Honors: A Celebration Of The Performing Arts" — George Stevens Jr., Nick Vanoff, producers (CBS).

"The 42d Annual Tony Awards" — Don Mischer, executive producer; David J. Goldberg, producer; Jeffrey Lane, coproducer (CBS).

"The 17th Annual American Film Institute Life Achievement Award: A Salute To Gregory Peck" — George Stevens Jr., producer; Jeffrey Lane, coproducer (NBC).

ACTOR IN COMEDY SERIES

Richard Mulligan — "Empty Nest" (NBC).

ACTOR IN DRAMA SERIES

Carroll O'Connor — "In The Heat Of The Night" (NBC).

ACTOR IN MINISERIES OR SPECIAL

James Woods — "My Name Is Bill W." ("Hallmark Hall Of Fame") (ABC).

ACTRESS IN COMEDY SERIES

Candice Bergen — "Murphy Brown" (CBS).

ACTRESS IN DRAMA SERIES

Dana Delany — "China Beach" (ABC).

ACTRESS IN MINISERIES OR SPECIAL

Holly Hunter — "Roe Vs. Wade" (NBC).

SUPPORTING ACTOR IN COMEDY SERIES

Woody Harrelson — "Cheers" (NBC).

SUPPORTING ACTOR IN DRAMA SERIES
Larry Drake — "L.A. Law" (NBC).

SUPPORTING ACTOR IN MINISERIES OR SPECIAL
Derek Jacobi — "The Tenth Man" ("Hallmark Hall Of Fame") (CBS).

SUPPORTING ACTRESS IN COMEDY SERIES
Rhea Perlman — "Cheers" (NBC).

SUPPORTING ACTRESS IN DRAMA SERIES
Melanie Mayron — "thirtysomething" (ABC).

SUPPORTING ACTRESS IN MINISERIES OR SPECIAL
Colleen Dewhurst — "Those She Left Behind" (NBC).

GUEST ACTOR IN DRAMA SERIES
Cleavon Little — "Dear John" ("Stand By Your Man") (NBC).

GUEST ACTOR IN DRAMA SERIES
Joe Spano — "Midnight Caller" ("The Execution Of John Saringo") (NBC).

GUEST ACTRESS IN COMEDY SERIES
Colleen Dewhurst — "Murphy Brown" ("Mama Said") (CBS).

GUEST ACTRESS IN DRAMA SERIES
Kay Lenz — "Midnight Caller" ("After It Happened...") (NBC).

INDIVIDUAL PERFORMANCE IN VARIETY OR MUSIC PROGRAM
Linda Ronstadt — "Canciones De Mi Padre" ("Great Performances") (PBS).

WRITING IN A COMEDY SERIES
Diane English — "Murphy Brown" ("Respect") (Pilot) (CBS).

WRITING IN A DRAMA SERIES
Joseph Dougherty — "thirtysomething" ("First Day/Last Day") (ABC).

WRITING IN VARIETY OR MUSIC PROGRAM
Jim Downey, head writer; **John Bowman, A. Whitney Brown, Gregory Daniels, Tom Davis, Al Franken, Shannon Gaughan, Jack Handey, Phil Hartman, Lorne Michaels, Mike Myers, Conan O'Brien, Bob Odenkirk, Herb Sargent, Tom Schiller, Robert Smigel, Bonnie Turner, Terry Turner, Christine Zander**, writers; **George Meyer**, additional sketches — "Saturday Night Live" (NBC).

WRITING IN MINISERIES OR SPECIAL
Abby Mann, Rubin Vote, Ron Hutchinson — "Murderers Among Us: The Simon Wiesenthal Story" (HBO).

DIRECTING IN COMEDY SERIES
Peter Baldwin — "The Wonder Years" ("Our Miss White") (ABC).

DIRECTING IN DRAMA SERIES
Robert Altman — "Tanner '88" ("The Boiler Room") (HBO).

DIRECTING IN VARIETY OR MUSIC PROGRAM
Jim Henson — "The Jim Henson Hour" ("Dog City") (NBC).

DIRECTING IN MINSERIES OR SPECIAL
Simon Wincer — "Lonesome Dove" (CBS).

ACHIEVEMENT IN CHOREOGRPAHY
Walter Painter — "Disney/MGM Studios Theme Park Grand Opening" (NBC).
Paula Abdul — "The Tracey Ullman Show" ("The Wave Girls," "D.U.I.," "The Cure," "Maggie In Peril, Pt. 1") (Fox).

SOUND EDITING FOR SERIES
William Wistrom, supervising sound and ADR editor; **James Wolvington, Mace Matiosian**, sound editors; **Wilson Dyer, Guy Tsujimoto; Gerry Sackman**, supervising music editor — "Star Trek: The Next Generation" ("Q Who") (Syn).

SOUND EDITING FOR MINISERIES OR SPECIAL
David McMoyler, supervising sound editor; **Joseph Melody**, cosupervising sound editor; **Mark Steele, Rick Steele, Michael J. Wright, Gary Macheel, Stephen Grubbs, Mark Friedgen, Charles R. Beith, Scot A. Tinsley, Karla Caldwell, G. Michael Graham, George B. Bell**, sound editors; **Kristi Johns**, supervising ADR editor; **Tom Villano, Jamie Gelb**, supervising music editors — "Lonesome Dove" (Part 3 — "The Plains") (CBS).

CINEMATOGRAPHY FOR SERIES
Roy H. Wagner — "Quantum Leap" (Pilot) (NBC).

CINEMATOGRAPHY FOR MINISERIES OR SPECIAL
Gayne Rescher — "Shooter" (NBC).

SOUND MIXING FOR COMEDY SERIES OR SPECIAL
Klaus Landsberg, production mixer; **Craig Porter, Alan Patapoff**, re-recording mixers — "Night Court" ("The Last Temptation Of Mac") (NBC).

SOUND MIXING FOR VARIETY OR MUSIC SERIES OR SPECIAL
Robert Douglass, SFX mixer; **David E. Fluhr**, re-recording mixer; **Ed Greene**, production mixer, **Larry Brown**, music mixer — "Kenny, Dolly And Willie: Something Inside So Strong" (NBC).

SOUND MIXING FOR DRAMA SERIES
Chris Haire, re-recording mixer, dialog; **Doug Davey**, re-recording mixer, effects; **Richard Morrison**, re-recording mixer, music and Foley; **Alan Bernard**, sound mixer, production — "Star Trek: The Next Generation" ("Q Who") (Syn).

SOUND MIXING FOR DRAMA MINISERIES OR SPECIAL
Don Johnson, sound mixer; **James L. Aicholtz**, dialog mixer; **Michael Herbick**, music mixer; **Kevin O'Connell**, sound effects mixer — "Lonesome Dove" (Part 4) — "The Return") (CBS).

LIGHTING DIRECTION (ELECTRONIC) FOR COMEDY SERIES
Mark Levin, lighting director — "Who's The Boss?" ("A Spirited Christmas") (ABC).

LIGHTING DIRECTION (ELECTRONIC) FOR DRAMA SERIES, VARIETY SERIES, MINISERIES OR SPECIAL
Robert Andrew Dickinson, lighting director — "The Magic Of David Copperfield XI: The Explosive Encounter" (CBS).

EDITING FOR SERIES (SINGLE CAMERA PRODUCTION)
Steven J. Rosenblum — "thirtysomething" ("First Day/Last Day") (ABC).

EDITING FOR MINISERIES OR SPECIAL (SINGLE CAMERA PRODUCTION)
Peter Zinner, John F. Burnett — "War And Remembrance" (Part 10) (ABC).

EDITING FOR SERIES (MULTICAMERA PRODUCTION)
Tucker Wiard — "Murphy Brown" ("Respect" pilot) (CBS).

EDITING FOR MINISERIES OR SPECIAL (MULTICAMERA PRODUCTION)
Mark D. West — "Dance In America: Gregory Hines Tap Dance In America" ("Great Performances") (PBS).

TECHNICAL DIRECTION/ CAMERA/VIDEO FOR SERIES
Robert G. Holmes, technical director; **Leigh V. Nicholson, John Repczynski, Jeffrey Wheat, Rocky Danielson**, camera operators; **Thomas G. Teimpidis**, senior video control — "Night Court" ("Yet Another Day In The Life") (NBC).

TECHNICAL DIRECTION/ CAMERA/VIDEO FOR MINISERIES OR SPECIAL
Ron Graft, technical director; **Richard G. Price, Kenneth Patterson, Greg Harms**, camera operators; **Mark Sanford**, senior video control — "The Meeting" ("American Playhouse") (PBS).

MAKEUP FOR SERIES
Thomas R. Burman, Bari Dreiband-Burman, special makeup; **Carol Schwartz, Robin Lavigne**, makeup artists — "The Tracey Ullman Show" ("The Subway") (Fox).

MAKEUP FOR MINISERIES OR SPECIAL
Manlio Rocchetti, makeup supervisor; **Carla Palmer, Jean Black**, makeup artists — "Lonesome Dove" (Part 4 — "The Return") (CBS).

HAIRSTYLING FOR A SERIES
Virginia Kearns" — "Quantum Leap" ("Double Identity") (NBC).

HAIRSTYLING FOR MINISERIES OR SPECIAL
Betty Glasow, chief hairstylist; **Stevie Hall, Elaine Bowerbank**, hairstylists — "Jack The Ripper" (Part 1) (CBS).

COSTUME DESIGN FOR SERIES
Judy Evans — "Beauty And The Beast" ("The Outsiders") (CBS).

COSTUME DESIGN FOR MINISERIES OR SPECIAL
Van Broughton Ramsey — "Lonesome Dove" (Part 2 — "On The Trail") (CBS).

COSTUME DESIGN FOR VARIETY OR MUSIC PROGRAM
Daniel Orlandi — "The Magic Of David Copperfield XI: The Explosive Encounter" (CBS).

COSTUMING FOR SERIES
Patrick R. Norris, men's costumer; **Julie Glick**, women's costumer — "thirtysomething" ("We'll Meet Again") (ABC).

COSTUMING FOR MINISERIES OR SPECIAL
Paula Kaatz, costume supervisor; **Andrea Weaver**, women's costumer, Los Angeles; **Janet Lawler**, women's costumer, Dallas; **Stephen Chudej**, men's costumer — "Pancho Barnes" (CBS).

ART DIRECTION FOR SERIES
James J. Agazzi, production designer; **Bill Harp**, set decorator — "Moonlighting" ("A Womb With A View") (ABC).

ART DIRECTION FOR VARIETY OR MUSIC PROGRAM
Bernie Yeszin, art director; **Portia Iversen**, set decorator — "The Tracey Ullman Show" ("All About Tammy Lee; Maggie In Peril — Part 2") (Fox).

ART DIRECTION FOR MINISERIES OR SPECIAL
Jan Scott, production designer; **Jack Taylor**, art director; **Edward J. McDonald**, set decorator — "I'll Be Home For Christmas" (NBC).

MUSIC COMPOSITION FOR SERIES (DRAMATIC UNDERSCORE)
Joel Rosenbaum — "Falcon Crest" ("Dust To Dust") (CBS).

MUSIC COMPOSITION FOR MINISERIES OR SPECIAL (DRAMATIC UDNERSCORE)
Basil Poledouris — "Lonesome Dove" ("Part 4 — The Return") (CBS).

MUSIC DIRECTION
Ian Fraser, music director, **Chris Boardman, J. Hill**, principal arrangers — "Christmas In Washington" (NBC).

MUSIC, LYRICS
Lee Holdridge, composer; **Melanie**, lyricist — "Beauty And The Beast" ("A Distant Shore"); song; "The First Time I Love Forever" (CBS).

INDIVIDUAL ACHIEVEMENT INFORMATIONAL PROGRAMMING
Performance
Hal Holbrook — "Portrait Of America: Alaska" (TBS).
Writing
John Heminway — "The Mind" ("Search For Mind") (PBS).

INDIVIDUAL ACHIEVEMENT SPECIAL EVENTS
Directing
Dwight Hemion — "Th 11th Annual Kennedy Center Honors: A Celebration Of The Performing Arts" (CBS).
Performance
Billy Crystal — "The 31st Annual Grammy Awards" (CBS).
Sound Mixing
Ed Greene, music production mixer; **Don Worsham**, dialog production mixer; **Carroll Pratt**, production mixer; audience reaction; **Paul Sandweiss**, production mixer, nomination categories — "The 31st Annual Grammy Awards" (CBS).
Writing
Jeffrey Lane — "The 42d Annual Tony Awards" (CBS).

INDIVIDUAL ACHIEVEMENT CLASSICAL MUSIC/ DANCE PROGRAMMING
Performance
Mikhail Baryshnikov — "Dance In America: Baryshnikov Dances Balanchine" ("Great Performances") (PBS).

SPECIAL VISUAL EFFECTS
Charles Stauffel, Martin Guttridge, Bill Schirmer, effects supervisors; **Bill Cruse, Simon Smith, Steve Anderson, Ed Williams**, miniature designers; **Egil Woxholt, Godfrey Godar**, directors of photography — "War And Remembrance" (ABC).

CASTING FOR MINISERIES OR SPECIAL
Lynn Kressel, "Lonesome Dove" (CBS).

GOVERNORS AWARD
Lucille Ball

(Through Dec. 31, 1989)

Calendar year 1989 saw three additions to this chart — all of them made-for-tv features. The chart, originally compiled by Carol Levine Sussman (and updated by VARIETY staffers since her death), first appeared in this paper in September 1974 and lists all primetime network movies of both types aired since September 1961 which received a Nielsen rating of 24.0 or better.

Selection of the 24.0 rating figure was arbitrary, based on the assumption that such a rating result represented an audience of major proportions, and the movie justifiably could be termed a hit. Since 1961, the tv household count has practically doubled, so the more recent the rating, the larger the actual audience.

The holdover champions for repeat appearances on the chart remain "The Wizard Of Oz" for theatrical films and "The Homecoming" for made-for-tv features. "Oz," with its 15 listed telecasts, is the one exception of the time limitation, having first been shown in 1956. "Homecoming" has five listings on the chart.

*Movie Made For TV (R)-Repeat

Rank	Title	Web	Day	Date	Rtg.	Share
1.	Gone With The Wind-Pt. 1	NBC	Sun	11/ 7/76	47.7	65
2.	Gone With The Wind-Pt. 2	NBC	Mon	11/ 8/76	47.4	64
3.	The Day After*	ABC	Sun	11/20/83	46.0	62
4.	Airport	ABC	Sun	11/11/73	42.3	63
	Love Story	ABC	Sun	10/ 1/72	42.3	62
6.	The Godfather-Pt. 2	NBC	Mon	11/18/74	39.4	57
7.	Jaws	ABC	Sun	11/ 4/79	39.1	57
8.	Poseidon Adventure	ABC	Sun	10/27/74	39.0	62
9.	True Grit	ABC	Sun	11/12/72	38.9	63
	The Birds	NBC	Sat	1/ 6/68	38.9	59
11.	Patton	ABC	Sun	11/19/72	38.5	65
12.	Bridge On The River Kwai	ABC	Sun	9/25/66	38.3	61
13.	Helter Skelter-Pt. 2*	CBS	Fri	4/ 2/76	37.5	60
	Jeremiah Johnson	ABC	Sun	1/18/76	37.5	56
15.	Ben-Hur	CBS	Sun	2/14/71	37.1	56
	Rocky	CBS	Sun	2/ 4/79	37.1	53
17.	The Godfather-Pt. 1	NBC	Sat	11/16/74	37.0	61
18.	Little Ladies Of The Night*	ABC	Sun	1/16/77	36.9	53
19.	Wizard Of Oz (R)	CBS	Sun	12/13/59	36.5	58
20.	The Burning Bed*	NBC	Mon	10/ 8/84	36.2	52
21.	Wizard Of Oz (R)	CBS	Sun	1/26/64	35.9	59
22.	Planet Of The Apes	CBS	Fri.	9/14/73	35.2	60
	Helter Skelter-Pt. 1*	CBS	Thu	4/ 1/76	35.2	57
24.	Wizard Of Oz (R)	CBS	Sun	1/17/65	34.7	49
25.	Born Free	CBS	Sun	2/22/70	34.2	53
26.	Wizard Of Oz	CBS	Sat	11/ 3/56	33.9	53
27.	Sound Of Music	ABC	Sun	2/29/76	33.6	49
28.	The Waltons' Thanksgiving Story*	CBS	Thu	11/15/73	33.5	51
29.	Bonnie & Clyde	CBS	Thu	9/20/73	33.4	38
30.	Ten Commandments	ABC	Sun	2/18/73	33.2	54
	Night Stalker*	ABC	Tue	1/11/72	33.2	48
32.	The Longest Yard	ABC	Sun	9/25/77	33.1	53
	A Case Of Rape*	NBC	Wed	2/20/74	33.1	49
34.	Wizard Of Oz (R)	CBS	Sun	12/ 9/62	33.0	55
	Return To Mayberry*	NBC	Sun	4/13/86	33.0	49
	Dallas Cowboys Cheerleaders*	ABC	Sun	1/14/79	33.0	48
37.	Brian's Song*	ABC	Tue	11/30/71	32.9	48
38.	Wizard Of Oz (R)	CBS	Sun	12/11/60	32.7	52
	Fatal Vision-Pt. 2*	NBC	Mon	11/19/84	32.7	49
40.	Beneath The Planet Of The Apes	CBS	Fri	10/26/73	32.6	54
41.	Wizard Of Oz (R)	CBS	Sun	12/10/61	32.5	53
42.	Women In Chains*	ABC	Tue	1/24/72	32.3	48
	Cat On A Hot Tin Roof	CBS	Thu	9/28/67	32.3	50
44.	Jesus Of Nazareth-Pt. 1*	NBC	Sun	4/ 3/77	32.3	50
45.	Sky Terror	ABC	Sun	9/19/76	32.0	51
	Apple Dumpling Gang	NBC	Sun	11/14/76	32.0	47
	Magnum, P.I. Finale	CBS	Sun	5/ 1/88	32.0	48
48.	Butch Cassidy & The Sundance Kid	ABC	Sun	9/26/76	31.9	51
	The Sting	ABC	Sun	11/ 5/78	31.9	48
	Something About Amelia*	ABC	Mon	1/9/84	31.9	46
51.	Heidi*	NBC	Sun	11/17/68	31.8	47
	Smokey & The Bandit	NBC	Sun	11/25/79	31.8	44
53.	Oh, God!	CBS	Sun	11/25/79	31.7	45
	Guyana Tragedy: The Story Of Jim Jones-Pt. 2*	CBS	Wed	4/16/80	31.7	50
	My Sweet Charlie*	NBC	Tue	1/20/70	31.7	48
56.	Airport 1975	NBC	Mon	9/20/76	31.6	46
	Feminist And The Fuzz*	ABC	Tue	1/26/71	31.6	46
58.	Something For Joey*	CBS	Wed	4/ 6/77	31.5	51
	Dawn: Portrait Of A Teenage Runaway*	NBC	Mon	9/27/76	31.5	46
60.	Great Escape-Pt. 2	CBS	Fri	9/15/67	31.3	55
	Kenny Rogers As The Gambler*	CBS	Tue	4/ 8/80	31.3	50
62.	McLintock!	CBS	Fri	11/ 3/67	31.2	54
63.	Ballad Of Josie	NBC	Tue	9/16/69	31.1	56
	Great Escape-Pt. 1	CBS	Thu	9/14/67	31.1	51
	Wizard Of Oz (R)	CBS	Sun	1/ 9/66	31.1	49
	Goldfinger	ABC	Sun	9/17/72	31.1	49
	Coward Of The County*	CBS	Wed	10/ 7/81	31.1	48
68.	Amazing Howard Hughes-Pt. 2*	CBS	Thu	4/14/77	31.0	53
	The Robe	ABC	Sun	3/26/67	31.0	53
	Sarah T.-Portrait Of A Teenage Alcoholic*	NBC	Tue	2/11/75	31.0	44
71.	Call Her Mom*	ABC	Tue	2/15/72	30.9	46
72.	A Death Of Innocence*	CBS	Fri	11/26/71	30.8	55
	Ten Commandments-Pt. 2 (R)	ABC	Mon	2/18/74	30.8	48
	Autobiography Of Miss Jane Pitman*	CBS	Thu	1/31/74	30.8	47
75.	Charlie's Angels*	ABC	Sun	3/21/76	30.7	49
76.	Three Days Of The Condor	CBS	Sun	11/27/77	30.5	47
	The Graduate	CBS	Thu	11/ 8/73	30.5	48
78.	Rescue From Gilligan's Island-Pt. 1*	NBC	Sat	10/14/78	30.4	52
	The Dirty Dozen	CBS	Thu	9/24/70	30.4	53
	Tribes*	ABC	Tue	11/10/70	30.4	45
	Yuma*	ABC	Tue	3/ 2/71	30.4	44
	Brian's Song (R)*	ABC	Tue	11/21/72	30.4	43
83.	Mr. & Mrs. Bo-Jo Jones*	ABC	Tue	11/16/71	30.2	45
	Superman-Pt. 2	ABC	Mon	2/ 8/82	30.2	42
85.	Earthquake-Pt. 2	NBC	Sun	10/ 3/76	30.1	46
86.	The War Wagon	NBC	Sat	10/31/70	30.0	53
	Lilies Of The Field	CBS	Fri	3/24/67	30.0	50
	Airport (R)	ABC	Sun	2/ 9/75	30.0	42
89.	Melvin Purvis, G-Man*	ABC	Tue	4/ 9/74	29.8	49
90.	Your Cheatin' Heart	CBS	Fri	4/ 5/68	29.7	50
91.	Gidget Goes Hawaiian	CBS	Thu	3/31/66	29.6	49
	Superman-Pt. 1	ABC	Sun	2/ 7/82	29.6	42
	The Gambler-The Adventure Continues-Pt. 2*	CBS	Tue	11/29/83	29.6	45
94.	The Gambler-The Adventure Continues-Pt. 1*	CBS	Mon	11/28/83	29.5	42
	Mrs. Sundance*	ABC	Tue	4/ 9/74	29.5	43
	Fatal Vision-Pt. 1*	NBC	Sun	11/18/84	29.5	44
97.	The Waltons' Easter Story*	CBS	Thu	4/19/73	29.4	48
	Charlie's Angels (R)*	ABC	Tue	9/14/76	29.4	47
	Maybe I'll Come Home In The Spring*	ABC	Tue	2/16/71	29.4	42
	Five Branded Women	CBS	Fri	1/ 6/67	29.4	42
101.	Jesus Of Nazareth-Pt. 2*	NBC	Sun	4/10/77	29.3	48
	Alias Smith And Jones*	ABC	Tue	1/ 5/71	29.3	44
103.	Hooper	CBS	Sun	2/ 8/80	29.2	41
	Every Which Way But Loose	CBS	Sun	11/ 1/81	29.2	42
105.	P.T. 109	CBS	Fri	1/13/67	29.1	50
	Escape From Planet Of The Apes	CBS	Fri	11/ 6/73	29.1	50
	Roustabout	ABC	Wed	1/ 3/68	29.1	48

Rank	Title	Web	Day	Date	Rtg.	Share
	Hombre	ABC	Sun	1/25/70	29.1	45
	Harper Valley PTA	NBC	Sun	2/24/80	29.1	42
110.	Guyana Tragedy: The Story Of Jim Jones-Pt. 1*	CBS	Tue	4/15/80	28.9	46
	Green Berets	NBC	Sat	11/18/72	28.9	45
	West Side Story-Pt. 1	NBC	Tue	3/14/72	28.9	41
113.	Cat Ballou	ABC	Wed	10/ 2/68	28.8	48
	Raid On Entebbe*	NBC	Sun	1/ 9/77	28.8	41
	Gone With The Wind (R)-Pt. 2	CBS	Mon	2/12/79	28.8	40
	Help Wanted: Male*	CBS	Sat	1/16/82	28.8	47
117.	Valley Of The Dolls	CBS	Fri	9/22/72	28.7	50
	The Homecoming (R)*	CBS	Fri	12/ 7/73	28.7	49
	Sybil-Pt. 2*	NBC	Mon	11/15/76	28.7	43
120.	Wizard Of Oz (R)	CBS	Sun	2/12/67	28.6	50
	Splendor In The Grass	CBS	Thu	10/12/67	28.6	47
	Walking Tall (R)	ABC	Sun	11/ 9/75	28.6	46
	Survive!	ABC	Sun	2/27/77	28.6	44
	The Hospital	ABC	Sun	11/18/73	28.6	44
	Incredible Journey Of Dr. Meg Laurel*	CBS	Tue	1/ 2/79	28.6	42
	In Harm's Way-Pt. 2	ABC	Mon	1/25/71	28.6	42
	9 To 5	CBS	Sun	2/27/83	28.6	42
128.	To Kill A Mockingbird	NBC	Sat	11/ 9/68	28.5	49
	Mario Puzo's The Godfather-Pt. 4	NBC	Tue	11/15/77	28.5	43
	Gidget Gets Married*	ABC	Tue	1/14/72	28.5	40
131.	The Jericho Mile*	ABC	Sun	3/18/79	28.4	46
	The Runaways*	CBS	Tue	4/ 1/75	28.4	44
	Dr. Cook's Garden*	ABC	Tue	1/19/71	28.4	41
	Fallen Angel*	CBS	Tue	2/24/81	28.4	42
135.	W.W. & The Dixie Dancekings	ABC	Sun	1/ 2/77	28.3	43
136.	Flying High*	CBS	Mon	8/28/78	28.2	46
	That Touch Of Mink	NBC	Tue	1/ 9/68	28.2	43
	The Women's Room*	ABC	Sun	9/14/80	28.2	45
139.	Cactus Flower	NBC	Sat	9/30/72	28.1	46
	The Last Child*	ABC	Tue	10/15/71	28.1	44
	Battlestar Galactica*	ABC	Sun	9/17/78	28.1	43
142.	SST-Death Flight*	ABC	Fri	2/25/77	28.0	47
	Buster And Billie	ABC	Mon	3/22/76	28.0	44
	Mario Puzo's The Godfather-Pt. 3	NBC	Mon	11/14/77	28.0	42
145.	Madame X	NBC	Mon	9/16/68	27.9	47
	Oklahoma	CBS	Thu	11/26/70	27.9	47
	The Outlaw Josey Wales (R)	NBC	Sun	10/19/80	27.9	44
148.	North By Northwest	CBS	Fri	9/29/67	27.8	50
	Tora! Tora! Tora!	CBS	Fri	9/21/73	27.8	47
	Serpico	ABC	Sun	9/21/75	27.8	47
	Girl Most Likely To*	ABC	Tue	11/ 6/73	27.8	42
	The Cowboys	ABC	Tue	11/13/73	27.8	42
	Earthquake-Pt. 1	NBC	Sun	9/26/76	27.8	41
	The End	NBC	Tue	2/19/80	27.8	41
155.	Wizard Of Oz (R)	NBC	Sun	3/15/70	27.7	50
	Second Time Around	NBC	Tue	10/ 3/67	27.7	48
	Sons Of Katie Elder	ABC	Sun	11/17/68	27.7	46
	In Search Of America*	ABC	Tue	3/23/71	27.7	42
159.	What A Way To Go!	NBC	Sat	9/16/67	27.6	50
	The Carpetbaggers	ABC	Sun	2/16/69	27.6	48
	Cry Rape!*	CBS	Tue	11/27/73	27.6	43
	Wild Women*	ABC	Tue	10/20/70	27.6	41
163.	Doomsday Flight*	NBC	Tue	12/13/66	27.5	48
	Billy Jack	NBC	Sat	11/20/76	27.5	46
	What's Up, Doc?	ABC	Fri	1/23/76	27.5	44
	Longest Hundred Miles*	NBC	Sat	1/21/67	27.5	43
	Run, Simon, Run*	ABC	Tue	12/ 1/70	27.5	43
	The Red Pony*	NBC	Sun	3/18/73	27.5	42
169.	Ten Commandments (R)	ABC	Sun	3/25/79	27.4	48
	Send Me No Flowers	NBC	Tue	9/19/67	27.4	47
	I Want To Live!	CBS	Thu	2/15/68	27.4	43
	It Happened One Christmas*	ABC	Sun	12/11/77	27.4	42
	Elvis*	ABC	Sun	2/11/79	27.4	40
	Second Chance*	ABC	Tue	2/ 8/72	27.4	40
	Jacqueline Bouvier Kennedy*	ABC	Wed	10/14/81	27.4	42
176.	Man From Atlantis*	NBC	Fri	3/ 4/77	27.3	46
	Smash-Up On Interstate 5*	ABC	Fri	12/ 3/76	27.3	45
	Blue Hawaii	NBC	Tue	11/29/66	27.3	45
	Jane Eyre*	NBC	Wed	3/24/71	27.3	43
	That's Entertainment	CBS	Tue	11/18/75	27.3	41
181.	Spencer's Mountain	CBS	Fri	10/13/67	27.2	49
	Battle Of The Bulge-Pt. 2	CBS	Fri	2/19/71	27.2	43
	Victims*	ABC	Mon	1/11/82	27.2	40
	Hawaii	CBS	Fri	1/11/74	27.2	42
	A Taste Of Evil*	ABC	Tue	10/12/71	27.2	41
	Hardcase*	ABC	Tue	2/ 1/72	27.2	40
	The Victim*	ABC	Tue	11/14/72	27.2	40
	Perry Mason Returns*	NBC	Sun	12/ 1/85	27.2	39
189.	McLintock! (R)	NBC	Sat	2/27/71	27.1	44
	Stepford Wives	ABC	Sun	10/24/76	27.1	43
	Loneliest Runner, The*	NBC	Mon	12/20/76	27.1	42
	West Side Story-Pt. 2	NBC	Wed	3/15/72	27.1	42
	Diary Of A Mad Housewife	NBC	Mon	1/24/73	27.1	42
	The Mating Game	NBC	Mon	10/21/63	27.1	41
	In Harm's Way-Pt. 1	ABC	Sun	1/24/71	27.1	41
	V-Pt. 2*	NBC	Mon	5/ 2/83	27.1	40
197.	Walking Tall	ABC	Sat	3/ 1/75	27.0	45
	Girl Who Came Gift-Wrapped*	ABC	Tue	1/29/74	27.0	40
199.	Hot Spell	NBC	Wed	3/17/65	26.9	44
	She Waits*	CBS	Fri	1/28/72	26.9	44
	African Queen	CBS	Thu	3/ 5/70	26.9	43
	Gator	CBS	Sun	2/12/78	26.9	41
	Savage Bees*	NBC	Mon	11/22/76	26.9	41
	Sybil-Pt. 1*	NBC	Sun	11/14/76	26.9	40
	Crowhaven Farm*	ABC	Tue	11/24/70	26.9	40
206.	Diamonds Are Forever (R)	ABC	Sun	12/26/76	26.8	47
	I'll Take Sweden	NBC	Tue	9/17/68	26.8	47
	Guess Who's Coming To Dinner?	CBS	Mon	9/19/71	26.8	44
	Five Easy Pieces	ABC	Mon	4/ 5/76	26.8	44
	The Last Voyage	NBC	Wed	3/24/65	26.8	43
	Over-The-Hill Gang*	ABC	Tue	10/ 7/69	26.8	42
	Wizard Of Oz (R)	CBS	Sun	3/14/76	26.8	42
	Love Hate Love*	ABC	Tue	2/ 9/71	26.8	38
	Murder In Texas-Pt. 2*	NBC	Mon	5/ 4/81	26.8	41
215.	Gone With The Wind-Pt. 1 (R)	CBS	Sun	3/22/81	26.7	40
	Saturday Night Fever	ABC	Sun	11/16/80	26.7	38
	Thrill Of It All	NBC	Sat	11/25/67	26.7	46
	Fame Is The Name Of The Game*	NBC	Sat	11/26/66	26.7	44
	Father Goose	NBC	Sat	1/ 4/69	26.7	42
	The Big Country-Pt. 2 (R)	NBC	Tue	3/30/71	26.7	42
	Hardcastle & McCormick*	ABC	Sun	9/18/83	26.7	43
	Stir Crazy	ABC	Sun	11/6/83	26.7	41
223.	Killer Who Wouldn't Die*	ABC	Sun	4/ 4/76	26.6	45
	Tickle Me	CBS	Fri	12/ 8/67	26.6	44
	The Homecoming (R)*	CBS	Fri	12/ 8/72	26.6	43
	Life & Times Of Grizzly Adams	NBC	Mon	5/17/76	26.6	43
227.	Wizard Of Oz (R)	CBS	Sun	3/20/76	26.5	41
	Man In The Iron Mask*	NBC	Mon	1/17/77	26.5	39
	If Tomorrow Comes*	ABC	Tue	12/ 7/71	26.5	38
230.	Marnie	NBC	Sat	11/ 4/67	26.4	47
	Shadow Over Elveron*	NBC	Tue	3/ 5/68	26.4	43
	Wizard Of Oz (R)	NBC	Sun	4/ 8/73	26.4	43
	Fantasy Island*	ABC	Fri	1/14/77	26.4	42
	House On Greenapple Road*	ABC	Sun	1/11/70	26.4	40
	A Cry In The Wilderness*	ABC	Tue	3/26/74	26.4	39
	Intimate Strangers	CBS	Wed	1/ 1/86	26.4	38
	The Karen Carpenter Story	CBS	Sun	1/ 1/89	26.4	41

Rank	Title	Web	Day	Date	Rtg.	Share
238.	Harlow	ABC	Sun	10/15/67	26.3	43
	Trapped*	ABC	Wed	11/ 4/73	26.3	43
	From Russia With Love	ABC	Mon	1/14/74	26.3	42
	Green Eyes*	ABC	Mon	1/ 3/77	26.3	40
	Texas Across The River*	NBC	Mon	1/19/70	26.3	39
	Cagney & Lacey*	CBS	Thu	10/ 8/81	26.3	42
244.	Strange Bedfellows	NBC	Sat	2/24/68	26.2	45
	Shenandoah	NBC	Sat	9/27/69	26.2	45
	The Big Country-Pt. 1 (R)	NBC	Mon	3/29/71	26.2	43
	Naked Jungle	NBC	Wed	2/10/65	26.2	42
	Horror At 37,000 Ft.*	CBS	Tue	2/13/73	26.2	41
	Wizard Of Oz (R)	NBC	Sun	3/ 9/69	26.2	40
	Firecreek	NBC	Mon	10/ 9/72	26.2	40
	Yours, Mine & Ours	NBC	Mon	11/ 3/73	26.2	40
	Dark Secret Of Harvest Home-Pt. 2*	NBC	Tue	1/24/78	26.2	39
	Playing For Time*	CBS	Tue	9/30/80	26.2	41
254.	Mysterious Island	CBS	Thu	11/25/65	26.1	50
	Man's Favorite Sport	NBC	Sat	10/ 7/67	26.1	46
	Little House On The Prairie*	NBC	Sat	3/30/74	26.1	44
	The Lively Set	NBC	Tue	1/23/68	26.1	43
	Angel In My Pocket	NBC	Sat	2/14/70	26.1	43
	B.J. And The Bear*	NBC	Wed	10/ 4/78	26.1	41
	To Sir, With Love	CBS	Sun	10/ 3/71	26.1	41
	Dark Secret Of Harvest Home-Pt. 2*	NBC	Mon	1/23/78	26.1	40
	Playmates*	ABC	Tue	10/ 3/72	26.1	40
	Like Mom, Like Me*	CBS	Sun	10/22/78	26.1	39
	Mario Puzo's The Godfather-Pt. 2	NBC	Sun	11/13/77	26.1	39
	Follow That Dream	ABC	Sun	1/24/65	26.1	38
	Smokey & The Bandit (R)	NBC	Sun	3/23/80	26.1	40
267.	Carnival Of Thrills*	CBS	Sun	4/20/80	26.0	39
	How The West Was Won	ABC	Sun	10/24/71	26.0	46
	Delicate Delinquent	CBS	Fri	1/20/67	26.0	45
	Vegas*	ABC	Tue	4/25/78	26.0	44
	Hush, Hush Sweet Charlotte	ABC	Sun	1/21/68	26.0	43
	A Summer Place	CBS	Thu	1/12/67	26.0	41
	Three Faces Of Eve	ABC	Sun	2/ 6/66	26.0	40
	Strongest Man In The World	NBC	Sun	2/27/77	26.0	39
275.	Coming Home	NBC	Mon	9/17/79	25.9	42
	The Apartment	CBS	Thu	9/21/67	25.9	45
	Andromeda Strain	NBC	Sat	2/10/73	25.9	45
	River Of No Return	NBC	Sat	10/26/62	25.9	44
	Chisum	NBC	Sat	10/27/73	25.9	43
	Bullitt	CBS	Thu	11/ 1/73	25.9	43
	The Longest Day	ABC	Sun	11/14/71	25.9	42
	All My Darling Daughters*	ABC	Wed	11/22/72	25.9	42
	Along Came A Spider*	ABC	Tue	2/ 3/70	25.9	38
	Giant-Pt. 2	NBC	Mon	11/13/72	25.9	38
285.	Sergeants Three	NBC	Sat	11/ 1/69	25.8	46
	Parent Trap (R)	NBC	Sun	5/15/76	25.8	44
	Move Over Darling	ABC	Sun	10/ 2/66	25.8	43
	The Stripper	CBS	Thu	3/13/69	25.8	42
	Patch Of Blue	NBC	Sat	2/13/71	25.8	41
	House Without A Christmas Tree*	CBS	Thu	12/ 6/73	25.8	40
	Amelia Earhart*	NBC	Mon	10/25/76	25.8	38
292.	Man Who Shot Liberty Valance	CBS	Fri	9/22/67	25.7	46
	Live And Let Die (R)	ABC	Sun	2/26/78	25.7	43
	Coffee, Tea Or Me?*	CBS	Tue	9/11/73	25.7	43
	Don't Give Up The Ship	NBC	Sat	2/ 6/65	25.7	41
	Guess Who's Coming To Dinner? (R)	CBS	Thu	10/26/72	25.7	41
	Wilma*	NBC	Mon	12/19/77	25.7	40
	The Sheepman	ABC	Sun	2/12/67	25.7	40
	Incredible Mr. Limpet	CBS	Fri	1/ 3/69	25.7	40
	Secret Ceremony	NBC	Tue	1/12/71	25.7	40
	Execution Of Private Slovik*	NBC	Wed	3/13/74	25.7	40
	Heaven Knows, Mr. Allison	NBC	Mon	2/18/63	25.7	38
	Over-The-Hill Gang Rides Again*	ABC	Tue	11/17/70	25.7	38
	The California Kid*	ABC	Wed	9/25/74	25.7	38
	Suddenly Single*	ABC	Tue	10/19/71	25.7	37
	The Best Little Girl In The World*	ABC	Mon	5/11/81	25.7	39
	The Outlaw Josey Wales	NBC	Sun	9/23/79	25.7	41
	Nobody's Child	CBS	Sun	4/ 6/86	25.7	39
309.	Whatever Happened To Baby Jane?	ABC	Sun	10/ 1/67	25.6	45
	Rear Window	NBC	Sat	10/ 1/66	25.6	44
	Story Of Pretty Boy Floyd*	ABC	Tue	5/ 7/74	25.6	41
	Now You See Him, Now You Don't-Pt. 2 (R)	NBC	Sun	10/15/78	25.6	40
	Day The Earth Stood Still	NBC	Sat	3/ 3/62	25.6	39
	Goodbye, Columbus	ABC	Sun	2/ 4/73	25.6	39
	The Night They Took Miss Beautiful	NBC	Mon	10/24/77	25.6	38
	And I Alone Survived*	NBC	Mon	11/27/78	25.6	37
317.	Chitty Chitty Bang Bang	CBS	Thu	11/23/72	25.5	45
	The Brass Bottle	NBC	Sat	12/ 2/67	25.5	44
	Fun In Acapulco	NBC	Tue	9/12/67	25.5	43
	The V.I.P.'s	ABC	Sun	1/15/67	25.5	41
	Five Desperate Women*	ABC	Tue	9/28/71	25.5	40
	You're Never Too Young	NBC	Sat	1/22/66	25.5	39
	House That Wouldn't Die*	ABC	Tue	10/27/70	25.5	38
	A Short Walk To Daylight*	ABC	Tue	10/24/72	25.5	37
	Young Pioneers*	ABC	Mon	3/ 1/76	25.5	37
326.	The Dirty Dozen-Pt. 2 (R)	CBS	Fri	10/15/71	25.4	44
	Days Of Wine & Roses	CBS	Thu	11/ 2/67	25.4	43
	Nevada Smith	ABC	Sun	9/28/69	25.4	43
	Banacek*	NBC	Mon	3/20/72	25.4	43
	Hud	ABC	Sun	10/29/67	25.4	42
	Sons Of Katie Elder (R)	ABC	Sun	3/ 1/70	25.4	42
	A Shot In The Dark	CBS	Fri	1/12/68	25.4	40
	Houdini	NBC	Sat	1/30/65	25.4	39
	Carrie	CBS	Tue	10/ 3/78	25.4	38
	Tenspeed & Brown Shoe*	ABC	Sun	1/27/80	25.4	36
	V-Pt. 1*	NBC	Sun	5/ 1/83	25.4	40
337.	The Amityville Horror	CBS	Sun	3/ 1/81	25.3	37
	Ironside (R)*	NBC	Sat	9/ 2/67	25.3	47
	Nightmare In Badham County*	ABC	Fri	11/ 5/76	25.3	42
	Pleasure Of His Company	CBS	Thu	2/ 2/67	25.3	42
	Under The Yum Yum Tree	ABC	Sun	10/ 8/67	25.3	41
	Cat On A Hot Tin Roof (R)	CBS	Thu	1/30/69	25.3	41
	The Homecoming*	CBS	Sun	12/19/71	25.3	39
	Rockabye	CBS	Sun	1/12/86	25.3	38
	Superdome*	ABC	Mon	1/ 9/78	25.3	38
	Grease	ABC	Sun	11/ 8/81	25.3	38
	Rags To Riches	NBC	Sun	3/ 9/87	25.3	36
348.	Amazing Howard Hughes-Pt. 1*	CBS	Wed	4/13/77	25.2	43
	White Christmas (R)	NBC	Sat	12/16/67	25.2	43
	The French Connection	CBS	Thu	10/30/75	25.2	43
	Last Dinosaur, The*	ABC	Fri	2/11/77	25.2	41
	Death Wish	CBS	Wed	11/10/76	25.2	39
	Born Innocent*	NBC	Tue	9/10/74	25.2	39
	Hello, Dolly!	CBS	Thu	2/28/74	25.2	38
	Midway-Pt. 2	NBC	Mon	2/ 6/78	25.2	36
	Aunt Mary*	CBS	Wed	12/ 5/79	25.2	40
	On Golden Pond	NBC	Sun	2/ 5/84	25.2	37
358.	Star Wars	CBS	Sun	2/26/84	25.1	35
	My Mother's Secret Life*	ABC	Sun	2/ 5/84	25.1	36
	Silver Streak	CBS	Thu	11/15/79	25.1	40
	The Thrill Of It All (R)	NBC	Sat	4/ 6/68	25.1	43
	Girls! Girls! Girls!	NBC	Tue	10/10/67	25.1	42
	Girl Happy	CBS	Fri	1/24/69	25.1	41

Rank	Title	Web	Day	Date	Rtg.	Share
	The Glass House*	CBS	Fri	2/ 4/72	25.1	41
	Story Of Jacob & Joseph*	ABC	Sun	4/ 7/74	25.1	40
	High Plains Drifter (R)	ABC	Sun	3/14/76	25.1	40
	But I Don't Want To Get Married*	ABC	Tue	10/ 6/70	25.1	38
	Death Stalk*	NBC	Tue	1/21/75	25.1	37
	The Gauntlet	NBC	Sun	11/ 2/80	25.1	39
	A Very Brady Christmas	CBS	Sun	12/18/88	25.1	39
	Elvis & Me-Pt. 2	ABC	Mon	2/ 8/88	25.1	36
	Those She Left Behind	NBC	Mon	3/ 6/89	25.1	38
374.	The Dirty Dozen-Pt. 1 (R)	CBS	Thu	10/14/71	25.0	42
	Lords Of Flatbush	ABC	Fri	3/26/76	25.0	42
	Secrets*	ABC	Sun	2/20/77	25.0	40
	Ten Commandments-Pt. 2 (R)	ABC	Sun	3/30/75	25.0	40
	Daughters Of Joshua Cabe*	ABC	Wed	9/31/72	25.0	38
	The Old Man Who Cried Wolf*	ABC	Tue	10/13/70	25.0	37
380.	I'd Rather Be Rich	NBC	Sat	3/ 2/68	24.9	42
	Stranded*	NBC	Mon	9/22/86	24.9	38
	Magnum Force (R)	CBS	Tue	2/28/78	24.9	41
	Murder At The World Series*	ABC	Sun	3/20/77	24.9	40
	Wonder Woman*	ABC	Tue	3/12/74	24.9	38
	Killer Bees*	ABC	Tue	2/26/74	24.9	36
	Guilty Or Innocent: The Sam Sheppard Murder Case*	NBC	Mon	11/17/75	24.9	36
	The Sting (R)	ABC	Sun	4/20/80	24.9	38
	The Ewok Adventure*	ABC	Sun	11/25/84	24.9	36
389.	Palm Springs Weekend	CBS	Fri	11/10/67	24.8	44
	The Ghost And Mr. Chicken (R)	NBC	Tue	9/ 9/69	24.8	44
	Elephant Walk	NBC	Wed	12/30/64	24.8	42
	Hang 'Em High (R)	ABC	Sun	5/23/76	24.8	41
	Sabrina	NBC	Sat	1/14/67	24.8	40
	Then Came Bronson	NBC	Mon	3/24/69	24.8	40
	P.J.	NBC	Sat	2/28/70	24.8	40
	The Scalphunters	NBC	Mon	2/25/74	24.8	39
	Home From The Hill	NBC	Tue	3/ 1/66	24.8	38
	Car Wash	NBC	Mon	1/ 9/78	24.8	37
	The Homecoming (R)*	CBS	Sun	12/ 8/74	24.8	37
	Having Babies*	ABC	Sun	10/17/76	24.8	36
	Death Of Richie*	ABC	Mon	1/10/77	24.8	35
	Call To Glory	ABC	Mon	8/13/84	24.8	44
	Blood Vows: The Story Of A Mafia Wife	NBC	Sun	1/18/87	24.8	37
404.	In The Heat Of The Night	NBC	Sat	9/16/72	24.7	45
	Stalag 17	NBC	Sat	10/23/65	24.7	44
	Lover Come Back	NBC	Tue	10/ 4/66	24.7	43
	Sunshine*	CBS	Fri	11/ 9/73	24.7	41
	Murder On Flight 502*	ABC	Fri	11/21/75	24.7	41
	Bermuda Depths	ABC	Fri	1/27/78	24.7	40
	Shane	ABC	Sun	2/18/68	24.7	40
	A Family Upside Down*	NBC	Sun	4/ 9/78	24.7	39
	The King And I	ABC	Wed	10/25/67	24.7	33
	Gone With The Wind-Pt. 2 (R)	CBS	Tue	3/24/81	24.7	39
414.	Fighting Back*	ABC	Sun	12/ 7/80	24.6	37
	O'Hara, U.S. Treasury*	CBS	Fri	4/ 2/71	24.6	43
	The Mating Game (R)	NBC	Sat	9/19/64	24.6	41
	The Hustler	ABC	Sun	1/23/66	24.6	41
	Love With The Proper Stranger	ABC	Wed	10/11/67	24.6	41
	The Way We Were	ABC	Sun	10/ 3/76	24.6	40
	Rafferty & The Highway Hustlers	NBC	Mon	3/14/77	24.6	40
	The Long Hot Summer	NBC	Sat	2/16/63	24.6	40
	The Caretakers	CBS	Thu	2/ 9/67	24.6	40
	Man Who Shot Liberty Valance (R)	CBS	Thu	2/ 3/68	24.6	40
	Glass Bottom Boat (R)	NBC	Sat	3/ 6/71	24.6	40
	It's A Mad, Mad, Mad Mad World	NBC	Sat	10/28/72	24.6	40
	The Longest Yard (R)	ABC	Sun	1/21/79	24.6	39
	The Ghost And Mr. Chicken	NBC	Mon	2/24/69	24.6	39
	Congratulations, It's A Boy!*	ABC	Tue	9/21/71	24.6	39
	Evel Knievel	ABC	Thu	9/13/72	24.6	39
	Get Christie Love*	ABC	Tue	1/22/74	24.6	35
	The Deep-Pt. 1	ABC	Sun	2/10/80	24.6	36
432.	Cleopatra-Pt. 1	ABC	Sun	2/13/72	24.5	44
	McLintock! (R)	CBS	Fri	3/15/68	24.5	43
	Intimate Strangers*	ABC	Fri	11/11/77	24.5	42
	Good Neighbor Sam (R)	CBS	Fri	2/10/67	24.5	41
	Hustling*	ABC	Sat	2/22/75	24.5	41
	Vacation In Hell*	ABC	Mon	5/21/79	24.5	39
	Cleopatra-Pt. 2	ABC	Mon	2/14/72	24.5	39
	Man With The Golden Gun (R)	ABC	Sun	1/22/78	24.5	38
	Amos	CBS	Sun	9/29/85	24.5	37
	Kansas City Bomber	CBS	Thu	2/ 7/74	24.5	36
	Man With The Golden Arm	ABC	Mon	1/10/77	24.5	34
	Boy In The Plastic Bubble*	ABC	Fri	11/12/76	24.5	31
444.	The Russians Are Coming, The Russians Are Coming	NBC	Sat	10/17/70	24.4	43
	Buona Sera, Mrs. Campbell	NBC	Mon	8/15/77	24.4	42
	No Man Is An Island	NBC	Sat	1/27/68	24.4	41
	Cool Hand Luke	NBC	Sat	9/22/73	24.4	41
	A Catered Affair	NBC	Wed	3/10/65	24.4	40
	Fantastic Voyage	ABC	Sun	10/12/69	24.4	40
	Duel At Diablo (R)	NBC	Sat	1/16/71	24.4	40
	The Five-Hundred Pound Jerk*	CBS	Tue	1/ 2/73	24.4	39
	The Letters*	ABC	Tue	3/ 6/73	24.4	38
	Where Eagles Dare-Pt. 2	ABC	Mon	1/31/72	24.4	37
	The Rookies*	ABC	Tue	3/ 7/72	24.4	35
	Mafia Princess	NBC	Sun	1/19/86	24.4	37
456.	Joe Kidd (R)	ABC	Sun	8/20/78	24.3	42
	Babe*	CBS	Thu	10/23/75	24.3	42
	Homecoming (R)*	CBS	Fri	12/12/75	24.3	42
	Ride To Hangman's Tree	NBC	Sat	3/29/69	24.3	42
	Hatari-Pt. 2 (R)	CBS	Fri	2/13/70	24.3	41
	Smile	CBS	Wed	12/29/76	24.3	40
	Teacher's Pet	NBC	Sat	1/28/67	24.3	40
	Battle Of The Bulge-Pt. 1	CBS	Thu	2/18/71	24.3	39
	The Forgotten Man*	ABC	Tue	9/14/71	24.3	39
	20,000 Leagues Under The Sea/Beaver Valley	NBC	Sat	2/23/74	24.3	39
	Ghost Of Cypress Swamp*	NBC	Sun	3/13/77	24.3	38
	Do Not Fold, Spindle Or Mutilate*	ABC	Tue	11/ 9/71	24.3	38
	Reluctant Heroes*	ABC	Tue	11/22/71	24.3	37
	Adventures Of Nick Carter*	ABC	Sun	2/20/71	24.3	37
	Gone With The Wind-Pt. 1 (R)	CBS	Sun	2/11/79	24.3	36
	Longstreet*	ABC	Tue	2/23/71	24.3	35
	Nurse*	CBS	Wed	4/ 9/80	24.3	39
474.	The Pink Panther	NBC	Sat	9/23/67	24.2	45
	Viva Las Vegas	CBS	Fri	10/ 6/67	24.2	43
	Dirty Mary, Crazy Larry	ABC	Fri	2/18/77	24.2	42
	The Nanny	CBS	Thu	10/31/68	24.2	42
	Prescription: Murder*	NBC	Tue	2/20/68	24.2	40
	Doctor Zhivago-Pt. 1	NBC	Sat	11/22/75	24.2	40
	In Search Of Noah's Ark	NBC	Mon	5/ 2/77	24.2	39
	Texas Across The River (R)	NBC	Sat	12/ 5/70	24.2	39
	The Bravos*	ABC	Sun	1/ 9/72	24.2	38
	Murderer's Row	ABC	Mon	1/10/72	24.2	38
	The Screaming Woman*	ABC	Sat	1/29/72	24.2	38
	The Sex Symbol*	ABC	Tue	9/17/74	24.2	37
	Girl Most Likely To (R)*	ABC	Wed	3/26/75	24.2	37
	The Couple Takes A Wife*	ABC	Tue	12/ 5/72	24.2	37
	Play Misty For Me	NBC	Mon	9/17/73	24.2	37
	Angel Dusted*	NBC	Mon	2/16/81	24.2	36
	Bigfoot, The Mysterious Monsters	NBC	Mon	2/28/77	24.2	36

Hit Movies on U.S. TV Since 1961

Rank	Title	Web	Day	Date	Rtg.	Share
	Nowhere To Run*	NBC	Mon	1/16/78	24.2	35
	Eleanor and Franklin-Pt. 2*	ABC	Mon	1/12/76	24.2	35
	Moonraker	ABC	Sun	11/22/81	24.2	39
	Fantasies*	ABC	Mon	1/18/82	24.2	37
494.	The Kid With The Broken Halo*	NBC	Mon	4/ 5/82	24.1	36
	Four For Texas	CBS	Fri	1/17/69	24.1	42
	El Dorado	ABC	Sun	9/19/71	24.1	40
	Bridge On The River Kwai-Pt. 2 (R)	CBS	Fri	1/ 8/71	24.1	39
	The Gun & The Pulpit*	ABC	Wed	4/ 3/74	24.1	39
	Hombre (R)	ABC	Sun	1/31/71	24.1	38
	Battleground	NBC	Sat	3/ 6/65	24.1	36
	The Other Man*	NBC	Mon	10/19/70	24.1	36
	True Grit (R)	ABC	Sun	1/13/74	24.1	36
	Ice Station Zebra-Pt. 2	ABC	Mon	2/ 7/72	24.1	35
	The Apple Dumpling Gang (R)	NBC	Sun	2/24/80	24.1	36
	Kate's Secret*	NBC	Mon	11/17/86	23.1	36
506.	The Wizard Of Oz (R)	CBS	Fri	3/ 7/80	24.0	40
	Stagecoach (R)	ABC	Sun	5/ 2/71	24.0	42
	Wild Women (R)*	ABC	Tue	6/ 1/71	24.0	41
	Dirty Harry	NBC	Sat	2/21/76	24.0	41
	Live & Let Die	ABC	Sun	10/31/76	24.0	40
	The Birds (R)	NBC	Sat	3/ 8/69	24.0	40
	Back Street	NBC	Sat	2/ 4/67	24.0	39
	Stripes	ABC	Sun	11/27/83	24.0	36
	Rough Night In Jericho	ABC	Sun	11/22/70	24.0	39
	Cheyenne Social Club	CBS	Thu	10/24/74	24.0	39
	White Lightning	NBC	Mon	9/ 8/75	24.0	38
	Hatari-Pt. 1 (R)	CBS	Thu	2/12/70	24.0	38
	With Six You Get Eggroll	NBC	Mon	4/18/72	24.0	37
	The Great Escape-Pt. 1 (R)	NBC	Mon	2/11/74	24.0	37
	The Devil's Daughter*	ABC	Tue	1/ 9/73	24.0	36
	The Six Million Dollar Man*	ABC	Tue	3/ 7/73	24.0	36
	The Royal Romance Of Charles & Diana*	CBS	Mon	9/20/82	24.0	37
	The Executioner's Song-Pt. 2*	NBC	Mon	11/29/82	24.0	36

*Movie Made For TV (R)-Repeat

3-Network Primetime Share

9/13/89

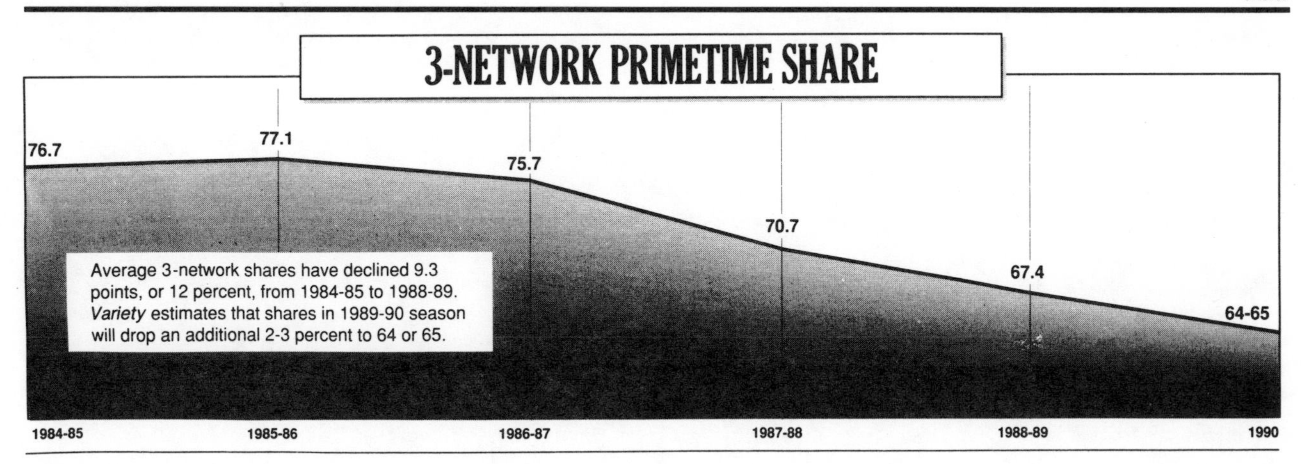

1988–1990 Kidvid Sked—1st Draft

5/3/89

Effective Sept. 9 (New shows in caps)

SATURDAY MORNING

Time	ABC	CBS	NBC
8:00	A Pup Named Scooby Doo	DINK, THE LITTLE DINOSAUR	Kissyfur
8:30	Disney's Gummi Bears/ Winnie The Pooh Hour	Jim Henson's Muppet Babies	CAMP CANDY
9:00			CAPTAIN N: THE GAMEMASTER
9:30	Slimer & The Real Ghostbusters	Pee-wee's Playhouse	THE KARATE KID
10:00		THE CALIFORNIA RAISINS	Smurfs
10:30	BEETLEJUICE	Garfield & Friends	
11:00	Bugs Bunny & Tweety Show		The Chipmunks
11:30		RUDE DOG & THE DWEEBS	SAVED BY THE BELL
12:00	Animal Crack-Ups	The Adventures Of Raggedy Ann & Andy	Alf
12:30	ABC Weekend Specials	CBS Storybreak	Alf-Tales

Radio

In 1989, several new formats (like the "Elvis" sound and "gameshow" radio) were introduced, enjoying a certain amount of popularity. However, traditional mass apppeal programming still seemed to attract the largest listener numbers in America.

Similarly, the demographic numbers reported for radio over the previous years were most impressive. Revenues had increased by 8.2%, and further growth seemed probable, especially with the development of more sophisticated research methods to track radio audiences. Also, preliminary talks with CATV executives about future cooperative ventures led most media critics to believe that radio was stronger than ever.

Controversy abounded as well. "Shock radio," like "trash television," became a very risky proposition for many program executives. NBC left the radio airwaves after over six decades of service and, on May 26, 8000 AM and FM stations observed 30 seconds of silence to make the public more aware of radio's place in the world of media.

Elvis Presley 'Alive and Kicking' at WCVG-AM in Cincinnati

VARIETY 3/1/89

By BRUCE INGRAM

Chicago When 500-watt WCVG-AM Cincinnati dropped its country music/classified ads programming last August and became the world's first all-Elvis Presley station, it received an extraordinary amount of media attention — and numerous predictions for failure.

Now, more than six months later, all-Elvis is alive and well and is engendering advertising rates three times higher than the station's previous format. John Stolz, WCVG g.m., explains why.

VARIETY: **How did the all-Elvis format come to exist?**

Stolz: Actually it was Steve Parton's idea — our former program director. He got a promotion after all of this and became the general manager of one of our other stations.

Back in July of 1988, the station was still floundering and going nowhere fast. We were having a meeting with the programming and sales departments and I said, "I don't have any idea what we can do with this station with the limitations of our half-kilowatt." Steve said, "Why don't we do all-Elvis?" We all just started laughing and all of a sudden he said, "I'm serious." A couple of days later he came back with some good ideas. So I said, okay, let's put it in motion.

I talked the ownership into doing it, basically as a publicity gimmick. But boy, it just took off, and it's still doing well today.

VARIETY: **Were you surprised by the amount of attention WCVG received when it went all-Elvis?**

Stolz: Amazed. I was not an Elvis Presley fan, but I've found that the King still lives, today, in the minds of an awful lot of people.

VARIETY: **Did the national press attention surprise you?**

Stolz: Oh yeah. On Aug. 3d, a friend of mine called me from a cruise ship off of Anchorage, Alaska. He said, "You aren't going to believe this. I'm sitting here reading the newspaper and you're on the front page." In the Star newspaper in South Africa there was a full page on the Elvis station. The only network that did not cover us was NBC. It was just amazing. Every time we turned around we wound up with more articles and more footage.

VARIETY: **Despite the hubbub, WCVG still hasn't made it onto the Arbitron survey of Cincinnati. Does that bother you?**

Stolz: No, not a bit, simply because it's the nature of the animal. This is a half-kilowatt radio station with a good, listenable range of about 30 miles. It does not cover the entire Cincinnati marketplace.

There are successes with ratings and there are successes without ratings. The bottom line is what makes a station a success or a failure. AM stations, in the years to come, are going to have to diversify in order to survive.

VARIETY: **How has the new format affected advertising rates and revenues?**

Stolz: Our revenues were up 700% for the first three months of the format. The revenues have leveled off, but they're in an area we can live with. In our previous format we were charging $5 for a 1-minute spot, then it went up to about $25 a holler. Now it's leveled off to about $15-17.

VARIETY: **What sort of demographic category does the diehard Elvis fan fit into?**

Stolz: We did our first Elvis cruise back in September: 375 people on a boat that holds 350 people with Elvis impersonators and lookalikes onboard. And the demographic ranged from 12-15 years all the way up to about 55-60. It was 75% female, but there were an awful lot of guys who are closet listeners.

VARIETY: **Is there a danger of burnout among your listeners?**

Stolz: Yes, there is a very definite problem with burnout, but our program director Rudy Owen is playing only about 350 songs at a time from the Elvis catalog of 652 songs. We keep those 350 songs in the studio for three weeks, then we rotate 50 of those out and bring in 50-60 more. But yes, that's a bridge that we're going to have to cross someday.

The Elvis format was discontinued in August, 1989. A business format now is in existence.

Ratings, Money Prizes for AMers Who Gamble on Gameshows

VARIETY 8/9/89

By BRUCE INGRAM

Chicago After seven years of successfully hosting and producing his original gameshow package at various radio stations, Mark Richards has moved into syndication with "The Radio Game Show," targeting struggling AMers. First to pay the $6,000 license fee was WNLK-AM New Canaan, which signed on January and has reported positive results.

Richards, who currently presides over the games at KENO-AM Las Vegas and previously popped the questions on short-lived cable tv shows for Turner Broadcasting and The Nashville Network, explains why the time is right for gameshows to return to radio — where they came from in the first place.

VARIETY: **Why did you decide to offer a gameshow on radio?**

Richards: I've always wanted to be a gameshow host on television. In fact, I've been on five gameshows as a contestant. I figured if I can't get on tv as a gameshow host, I might as well do it on radio. So I approached the program/operations director at KOGO-AM San Diego back in the spring of '82 with a concept. He said, "Let's try it out on a Saturday night 7-9." We tried it and the phone lines went wild and crazy from the first week. After four weeks, the show went to six nights a week, Monday-Saturday, 7-9, then it went 6-8 Monday-Saturday. So it went from 2-12-18 hours a week.

The ratings were just fantastic. We were beating KFMB, which was the main AM station in San Diego. Even during the Padres baseball season, we were either tying or beating KFMB during that time period.

VARIETY: **How do you describe the show?**

Richards: "The Game Show" is basically three hours of music, fun and games. We play two songs back-to-back and after every two songs, we do a game. There are 12 different games we play during a 3-hour period. Each game is only three minutes in length and we have two contestants calling in and competing against each other for a game win. We have bells, buzzers, applause, sound effects, just like a regular television gameshow. The winner of each game stays on the line and takes on a new challenger for the very next game.

VARIETY: **Do you think of your show as a throwback to gameshows from the early days of radio?**

Richards: Yes. But the gameshows back then were all half-hour programs with one game and there was no calling in. My show enables gameshow fans to pick up the phone, be the right caller and, for three minutes, to be a star on the radio.

VARIETY: **How has "The Game Show" fared, typically, in terms of ratings?**

Richards: The ratings have done very well. When I was in San Diego for almost two years I was either No. 1 or No. 2 in my time period. The only reason I left is KOGO sold the station and went to a news/talk format. That's when I joined KFI-AM Los Angeles and the show ran there for a year, 6-9 Monday-Saturday nights — another 18 hours a week in primetime. It went well up there until they dropped my show and joined Westwood One. My show was doing well, if not better, than Lohman and Barkley's morning program at KFI.

In Portland (KKSN-AM), I went on the air without any hoopla or promotion in August '87 and earned a 2.7 in adults 25-49 in the fall book. The morning team had a 1.4. Middays got a 1.3, afternoon drive got a 2.2 and I got a 2.7 from 7-midnight. In men 35-54 I pulled in 6.8, the morning show pulled a 2.0, middays got a 1.7 and drive time got a 5.7.

In other words, in almost every demographic breakdown I either doubled or tripled the rest of the dayparts. Now here's something unbelievable. I left KKSN at the beginning of the fall book in '88. In that book, every daypart did something except for 7-midnight, my old shift. It was all zeros. The person who showed me the fall book at Arbitron couldn't believe his eyes. He'd never seen a drop where nobody listened unless the station went dark or completely changed formats.

VARIETY: **Have you found that program directors are hesitant to accept the notion of a radio gameshow?**

Richards: Absolutely. When I was in New York I called several stations. I said: "My name is Mark Richards, I'm from Las Vegas and I have an idea for you. Would you be interested in a gameshow on the radio?" They said: "Nope, thanks for calling, good-bye." That happens all the time. Program directors and general managers, instead of asking how the show works, say no thanks.

VARIETY: **How do you sell the program?**

Richards: Every week we get 12 non-competitive advertisers. They all pay $600 and get $600 worth of spots a week on the station. That brings in the station $7,200. During that week, 60 of my listeners are going to win $30 gift certificates: a total of $1,800 in gift certificates. Those 60 people have 14 days to go to any of the 12 sponsors and redeem their certificates. At the end of the 2-week spending period we contact every advertiser and buy back every gift certificate they've received dollar-for-dollar.

In other words, they're spending $600 for advertising and if they get 20 $30 certificates out of the 60 we've mailed out, they've got their money back. And the station makes $5,400 a week. You get advertisers to buy into this program because they have a chance to get some of their money back and some new customers. Every store is hoping that they'll get 40-50 of those certificates back.

VARIETY: **Do you think it would be feasible for a radio station to offer a format made up entirely of gameshows?**

Richards: Sure, if it's the way I'm doing it. You can't do a 24-hour block just playing games. You've got to have music in between. In a 3-hour program like I'm doing in Las Vegas, I play 12 games and I play 30 songs. You could do that on a 24-hour basis and it would work. In fact, that's my main goal, to take a station that's about ready to go into the dumper and say "I've got an idea; let's try it."

Radio's Most Programmed Format: Adult Contemporary/Soft Rock

VARIETY 9/27/89

Chicago Adult Contemporary/-Soft Rock and Oldies are respectively the most-programmed and fastest-growing formats in radio today, per the 1989 Radio Format Trends report published by McGavren Guild Marketing Research.

Of the 16 formats tracked in the report, Adult Contemporary was the most widely used, programmed on 670, or 18.3%, of the stations surveyed. However, AC did show its first usage decrease in five years, dropping from 19.2% in 1988.

McGavren Guild suggests that the 1.1% decrease could be the start of a downward trend rather than a natural fluctuation.

Country music was the second most widespread format, accounting for 16.7% of the surveyed stations, continuing a downward trend that began in 1987 when it peaked at 17.9%. Contemporary Hit Radio showed with 12.6% of the surveyed stations, down slightly from 12.8% last year.

Fastest-growing honors went to the Oldies format, which has increased consistently over the past five years, per McGavren, from 2.9% to 7.7%, or 281 of the total surveyed. Report predicts continued growth, at a slower pace, into the early 1990s, but cautions that overexposure of oldies music on radio, tv and feature films could dilute the format's appeal.

New Age/Jazz was the next most-rapidly proliferating format, riding an upward trend from 0.3% in 1987 to 1.2% this year. McGavren reports 43 stations programming the format in the U.S.

All Talk came in third with 1.0% of the surveyed stations compared with .9% in 1988. Talk radio was the least-used of the 16 formats studied, but enjoyed a 43% increase over last year's usage level, causing McGavren to predict the format is primed for growth.

Primary resource for the Radio Format Trends report is the Interep Market Profile System, a database with info on 3,600 stations nationwide.

Good Times Roll for Radio; RAB Sez Revenues Up 8.2%

3/22/89

Chicago Total radio industry revenues rose to roughly $7.9-billion in 1988, per estimates from a Radio Advertising Bureau study of 97 markets nationwide.

Figure which represents an 8.2% increase from '87, combines increased revenues from network, national spot and local ads. Data for the study was provided by participating stations.

Cost-efficient medium

"Radio is going through a good time, basically," said RAB communications v.p. Joan Voukides. "Probably because the other media have become so expensive, advertisers are looking for more efficient ways to get the word out and they're using radio to do that."

Network advertising improved 3.2% over last year for a total of $382-million.

National spot ads moved up 6.6% over 1987 to $1.402-billion. Biggest regional winner in that category was the western U.S. (representing 25 markets including Denver, Honolulu, San Francisco, Los Angeles and Seattle), which increased its take 13.4%

Local billings showed the most growth in 1988, gaining 9.0% over last year to $6.109-billion. Biggest winner again was the western region, jumping 13.6% over 1987 earnings.

Only region to show a drop in business during 1988, per the RAB study, was the southwestern U.S. (represented by ten markets including Austin, Little Rock, Houston and Oklahoma City), which decreased 1.8%. *—Bruce Ingram*

RIC Unveils On-Line Radio Database

VARIETY 5/3/89

Chicago New York-based Radio Information Center has unveiled an online computer service designed to provide radio networks, program syndicators, advertising agencies, media brokers and investors with easy access to key information about virtually every radio station in the U.S.

The system, called Radio/Link, expands on the RIC's present database of basic facts about the country's approximately 11,000 radio stations to provide a variety of info on any station, including format, present and past call letters, network affiliations, market rank, audience share and, for qualified subscribers, detailed Arbitron data.

Maurie Webster, 30-year veteran of CBS Radio who founded the RIC in 1978 with his son, Scott, said the purpose of the organization had always been to provide a single, comprehensive source of information about the radio industry, which is often perceived as a complex medium.

Radio/Link, per Webster, represents the first time a reference of basic information about stations has been matched with extensive audience analysis and the ability to instantly present the data in a variety of combinations via computer.

Fee for use of the information system, which is expected to debut June 1, will range in price from approximately $50-300 per month for radio stations, depending on how much use is made of the system. For radio networks and ad agencies, Webster said the price would probably be substantially higher, ranging from $500-1,000 per month.

Radio, Cable Made for Each Other

VARIETY 6/28/89

Detroit Radio and cable tv have a lot to offer each other in the way of promotion and marketing, according to representatives of both businesses at the Broadcast Promotion and Marketing Executives confab.

"You can't hook up a cable system to your car and you can't watch a radio, so there are obvious ways we can help each other," said Lloyd P. Trufelman, director of communications for the Cabletelevision Advertising Bureau. He was speaking at a June 22 meet to discuss radio-cable synergies.

Trufelman was referring to cable's ability to provide affordable tv advertising to radio stations and radio's ability to target potential subscribers for cable systems — a theme that was reiterated throughout the session.

Jim Barrett, promotion director of CapCities' full-service market leader WJR-AM Detroit, said cable is an important means of maximizing a station's tv advertising budget.

Barrett said tv advertising is particularly important for AM stations like WJR, which must attract new listeners from a generation that has grown up listening mostly to FM radio.

Unfortunately, he said, radio stations' tv ad budgets are relatively small. Barrett gave an example of a recent campaign he executed for WJR on a budget of $100,000. Producing eight, 60 second spots ate up $40,000 of his funds, he said, and $60,000 doesn't go very far toward buying an effective flight of commercial tv time.

For that reason, Barrett limited his commercial tv buys to essential placements in late news slots and used the remainder of his budget for spots on Continental Cablevision.

Barrett said he got a lot of mileage out of his cable buys because the spots cost as little as $15 and the cable company was more than willing to barter for time on WJR. "It's a whole heck of a lot cheaper than commercial tv," he said.

Walter Maude, area manager for Continental Cablevision, acknowledged there are limitations in advertising on cable. For one thing, cable systems cannot respond as quickly to requests for specific time placements as commercial broadcasters can because cablers make up reels of spots only once or twice a week. Maude pointed out that subscriptions, not commercials, are cable's most important revenue source.

Maude said radio can be used effectively as a promotion device to increase subscribers by targeting potential viewers, informing them about cable programming and trying to counteract image problems resulting from dissatisfied consumers in the early days of the business.

Maude added that radio also appeals to cable because the medium is relatively easy on the wallet. "We have limited promotion budgets too," he said.

—Bruce Ingram

Shockjocks, Trash TV Take Beating; Scare Sponsors, Lawyers, NAB Told

VARIETY 5/3/89

By BRIAN LOWRY

Las Vegas Trash tv and shock radio morning-drive teams may be the trend in broadcasting, but the antics that are attracting large audiences while scaring off some advertisers pose additional hazards: legal actions against stations.

A panel of lawyers conversant with liability rules took on the issue at the National Assn. of Broadcasters convention, and NAB counsel and panel moderator Steve Bookchester noted afterward that there has been an increase in suits against radio stations — coinciding, perhaps, with widening standards and an increase in the number of outlandish morning-drive programs.

So-called shock jocks and morning zoos have to be wary of actionable issues such as invasion of privacy and libel. "Just because somebody else called in and said it doesn't mean the radio station is off the hook," said attorney Daniel M. Waggoner.

There are also potential dangers in calling someone and putting him on the air without permission, as well as what attorney Frederick A. Polner described as "a dangerous and troubling trend" of female personalities on morning zoos seeking damage for sexual harassment if they are the constant target of jokes.

While acknowledging that the station doesn't want to inhibit spontaneity, the panelists pointed out that First Amendment rights belong to the station, not the talent. The station can establish whatever guidelines it sees fit. That requires reigning in "sophomoric" talent, said attorney Harry Cole, who could be willing "to flush the station's license down the toilet.

Compensation from deejay

Polner said some stations now include clauses in radio talent contracts that allow stations to seek compensation from the deejay if reckless antics put the station at risk. He advised a careful screening process that will make certain that on-air personalities are responsive to station concerns.

Bookchester said lawsuits over "shock" antics are rare, accounting for about one-third of suits brought against stations (the majority are inspired by the news department). They also noted that "Donahue" is being sued by a Galveston woman whose mother revealed information of incest on the program.

NBC Bids a Fond Farewell

Variety 4/5/89

Chicago NBC, the company that originated network broadcasting in 1926, prepared to slip quietly out of the radio business March 16 by agreeing to sell its last radio property, KNBR-AM San Francisco, to Susquehanna Broadcasting for about $21-million.

Sale, subject to Federal Communications Commission approval, will be the last step in a dismantling process that has taken place over the past 20 months, garnering roughly $253-million for parent company General Electric.

NBC's exit from the radio biz has attracted considerably less attention than its pioneering debut Nov. 15, 1926, when the company broadcast entertainment from 8 p.m.-midnight from the Waldorf-Astoria Hotel in New York.

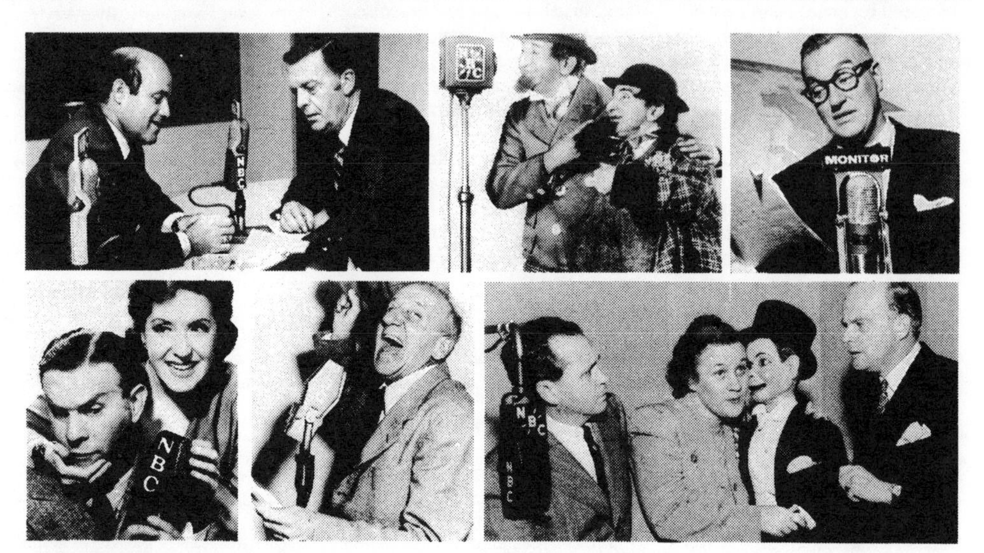

NBC Radio's galaxy (clockwise, from top left): Joe Garagiola and Chet Huntley; vaude's Weber & Fields; Dave Garroway; Burns & Allen; Jimmy Durante; Fibber McGee & Molly with Edgar Bergen & Charlie McCarthy

NBC was the brainchild of David Sarnoff, its founder and chairman. As early as 1916, when he was an assistant traffic manager at the Marconi Wireless Telegraph Co. of America, Sarnoff had envisioned "a radio music box" that "would make radio a household utility," suggesting that such a device could bring entertainment and information programs into every home.

On the rise

By 1922, Sarnoff had risen to general manager of the Radio Corp. of America, company formed shortly after World War I by several electrical equipment manufacturers, including GE, Westinghouse and American Telephone & Telegraph. He wrote the chairman of General Electric that what the newly born radio industry needed was a programming service, a national broadcasting company providing first-rate material to the country's radio stations, which often supplied amateurish and sporadic programs to their listeners.

By 1926, Sarnoff's vision had become a reality after years of corporate wrangling. The National Broadcasting Co. was incorporated in Dover, Del., Sept. 9, 1926, with ownership divided among GE (50%), Westinghouse (20%) and RCA. RCA became sole owner in 1930.

NBC began inventing the art of radio news by covering events such as Charles Lindbergh's return to Washington, D.C., after his solo transatlantic flight in 1927. NBC placed four reporters on the story so the audience could hear the huge throngs that gathered throughout the city, one at the Navy Yard where Lindbergh's ship docked, one in the Capitol dome to give an account of the official reception, one in the Treasury Building to describe the parade down Pennsylvania Avenue and one on top of the Washington Monument to wrap the story up.

NBC began coverage of national campaigns with the 1928 presidential election, and brought millions more voters to the polls. Some analysts even attributed Al Smith's defeat by Herbert Hoover to listeners' dislike of his heavy New York accent. Likewise, pundits ascribed Hoover's defeat in 1932 to Franklin Delano Roosevelt's charismatic radio presence.

Entertainment programming on NBC was a success literally from day one. Immediately after the big broadcast Nov. 15, 1926, web unveiled its first drama series, "Great Moments In History," and soon after, "Biblical Dramas," both employing name talent and orchestral backgrounds.

NBC quickly branched out with a smorgasbord of programming, adding education, religious, children's and women's programming.

This was a far cry from the type of programming that many local radio stations offered their listeners. Historic KKDA-AM, for example, one of the first radio stations in the country, broadcast only one hour per day from the coatroom of Westinghouse electric meter factory in Pittsburgh — usually offering selections played by an employee band.

Theater, music — both classical and popular — and sports all were popular features of NBC's programming virtually from the start.

For many listeners in small towns and rural areas throughout the country, radio was the only contact with such fare available in the large cities.

Along with motion pictures, radio succeeded in putting the kibosh on vaudeville permanently. NBC, however, made restitution by providing an alternative venue for many displaced performers who, by way of mass exposure, became legendary entertainment figures. To name a few: Eddie Cantor, Al Jolson, Rudy Vallee, Fred Allen, Jack Benny, Groucho Marx, Bob Hope, Jimmy Durante, Bing Crosby, Red Skelton, George Burns and Gracie Allen, and Edgar Bergen and Charlie McCarthy.

The development of news programming, which helped to sustain NBC in the face of competition from television, was an outgrowth of the web's coverage of World War II, the start of which the web marked by adding a fourth chime to its 3-chime signature. The fourth chime summoned personnel to their posts Dec. 7, 1941, moments before the first reports of the Japanese attack of Pearl Harbor.

NBC news covered the war from beginning to end. It was present at the signing of the Munich Pact and the invasions of Austria, Czechoslovakia, Poland, France and Russia. NBC reporters flew on nighttime bombing missions over Berlin, accompanied troops on D-Day landings in Normandy and witnessed both the German and Japanese surrenders.

All of NBC's established programs were put at the service of the war effort, becoming vehicles for selling War Bonds, increasing recruiting and educating the public on such subjects as conservation of supplies. Most programs also went remote to provide entertainment to troops in training in the U.S. and later in service around the world.

The arrival of tv meant a drastic change in radio's role as a national medium. As tv began to catch on in 1948-51, it reduced audience and ad revenues for NBC Radio. For several years, all radio networks, including NBC, operated at a loss — and the future was uncertain.

NBC countered the downward trend by switching from program sponsorships to spot ads and shifting its programming emphasis from general entertainment shows to a reduced number of specialized programs. The network also relied heavily on its respected news' division in the mid-'60s, developing the innovative "Emphasis" series, 5-minute news and commentary features on worldwide issues offered 40 times per week.

NBC became the first radio web to shake its tv-induced slump and get back into the black. In 1964, NBC Radio Network garnered 36% of the billing total earned by all national webs and registered its largest time and program billings since 1958, when its programming inventory was twice as large.

In 1975, the web made a major effort to attract unclaimed affils by launching its News and Information Service. Hoping to appeal to all-news stadium plagued by high overhead, NIS offered 45 minutes of news per hour at a rate that was less expensive than hiring a staff of reporters. However, number of affils needed to make the venture profitable did not materialize.

Since that time, NBC has kept pace with the rest of the radio industry by targeting specific demographic groups with specialized material — particularly with the development of youth-oriented web The Source in the late '70s and all-talk Talknet in the early '80s.

Even so, NBC's network radio operations were functioning in straitened circumstances compared with the web's heyday in the '30s and '40s — a situation that no doubt influenced GE's decision to divest the radio operation that launched the broadcast industry.

But Westwood One topper Norm Pattiz, who brought the NBC Radio Networks from GE, has stated that he does not consider network radio a disappearing medium, or even a medium with limited horizons. In an article in the Jan. 20, 1988, VARIETY, Pattiz wrote that on the contrary, he believes that glory years for network radio, once epitomized by the NBC Radio Networks, lie ahead.

"Thirty years ago it appeared that network radio was dead. Today, we are experiencing a rebirth in the medium of such magnitude that if one refers to the 1930s and '40s as radio's 'Golden Age,' then the 1980s and '90s are surely platinum!"

"The Golden Age Of Broadcasting" by Robert Campbell was used in researching this article.

80% of Radio Stations Go Silent in Support of NAB/RAB Campaign

VARIETY 6/17/89

Chicago Working on the principle that you don't know what you've got 'til it's gone, approximately 8,000 radio stations nationwide went silent May 26 for 30 seconds starting 7:42 a.m. local time. Widespread dead air was part of a 90-second spot extolling the virtues of radio, first in an ongoing campaign sponsored by the National Assn. Broadcasters/Radio Advertising Bureau Radio Futures Committee.

Jerry R. Lyman, co-chairman of the committee and president of Radio Ventures L.P. a recently formed station acquisition group, explains why he believes the radio awareness campaign got off to a good start.

VARIETY: **Are you pleased with the results of the campaign kickoff?**

Lyman: Yes, I'd say it was quite a success. Going into something like this you're not quite sure what to expect, particularly when it comes to trying to unify an industry like radio. We are extremely pleased with the estimate that some 80% of all radio broadcasters aired the spot or did something appropriate at 7:42 a.m.

One of the things that the Radio Futures Committee was concerned about was the feeling that radio was not getting major media coverage. It was interesting to me that this campaign kickoff did in fact get tremendous national coverage by all media.

VARIETY: **Is it true that some air talents treated the spot less than seriously?**

Lyman: You have to realize that these personalities are going to put their shtick into it. Don Imus (of Emmis Broadcasting's WFAN-AM) was paying homage to (Emmis prexy) Jeff Smulyan during the silence, mentioning that he's just signed a new contract with the boss and he thought he ought to go ahead and do the spot. But, to me, that just draws more attention to the campaign. We encouraged stations to have fun with this.

VARIETY: **Was there any resistance to this campaign among stations? Any dissension at all?**

Lyman: One trade publication (newsletter Inside Radio) really took off after us, but in spite of that there was not much resistance. There were, I think, a number of program directors across the country who were nervous about going silent for 30 seconds. I think the facts have proven that those worries were unfounded.

VARIETY: **Were most stations that took part in the campaign group- or individually owned?**

Lyman: There were a lot of individual stations. The large group operators came together in Indianapolis back in January and we presented the project to them. Bill Stakelin (formerly RAB prexy) gave a magnificent presentation of what we were trying to accomplish. It was at that meeting, with approximately 30-40 large-group operators who basically dominate radio in the top 30 markets, where we gained our pledges from them to contribute money and time for promotion to the campaign. At that time, we raised almost $700,000. We have since pushed that figure to almost $1-million.

Jerry Lyman, prez of Radio Ventures

VARIETY: **How did you get around competitive worries among individual stations?**

Lyman: We just hoped that this would be a radio-pride event and that everyone would feel that if we were all going silent simultaneously, nobody would get hurt. When stations didn't air the spot it was probably because of some paranoid program director who thought: "My God, if I do this, I'm going to lose 25% of my cume and I can't afford to do that." But I don't think that happened.

In this (Washington, D.C.) market, for example, both album rock stations did it. The one country station didn't run the spot, but they paid some sort of tribute at 7:42 to the radio industry doing this campaign. One urban station did it and the other didn't. So, we had some different responses among stations but I think that depended on how each program director felt about the spot.

VARIETY: **Is the campaign aimed primarily at advertisers or consumers?**

Lyman: The Radio Futures Committee has put fourth four goals for this ongoing project. First, to raise the public's awareness of how important radio is in their lives. Second, to heighten the awareness of our community leaders regarding radio's importance in community life. Third, to raise the pride of those employed in the industry. We're an industry that at times tends to get down on itself because we appear to be dominated by the other broadcast medium. There tends to be more so-called glamour attached to television. Finally, to strengthen the bottom line, to have all the different aspects of this public awareness campaign domino into the advertising community — all of whom are part of the public. We hope they will start saying maybe we're not looking at this medium the way we should be.

Marconi Winners Announced at NAB

Variety 9/20/89

New Orleans Radio '89 closed on a glitzy note Sept. 16 with the presentation of the first annual Marconi Radio Awards.

Twenty-three awards were handed out in four major categories: legendary station, stations of the year, personalities of the year and stations of the year by format.

Show, produced by Tony Quin of Los Angeles-based Film House, was emceed by Dick Clark and featured performances by John Candy, Donna Summer and Paul Shaffer & the World's Most Dangerous Band.

List of winners follows:

LEGENDARY STATION

WLS-AM Chicago

PERSONALITIES OF THE YEAR

Network/syndicated: Paul Harvey, ABC
Major market: Ron Chapman, KVIL-AM/FM Dallas
Large market: Bob Steele, WTIC-AM Hartford, Conn.
Medium market: Mark Summers, WBBQ-AM/FM Augusta, Ga.
Small market: Billie Oakley, KMA-AM Shenandoah, Iowa

STATIONS OF THE YEAR BY FORMATS

Country: KNIX-AM/FM Phoenix
AC/soft rock/oldies: KVIL-AM/FM Dallas
News/talk: KMOX-AM St. Louis
CHR/top 40: KPWR-FM Los Angeles
Jazz/new age: KTWV-FM Los Angeles
Spanish language: WQBA-AM/FM Miami
Black/urban: WVAZ-FM Chicago
Classical: WQXR-AM/FM New York
Big band/nostalgia: KMPC-AM Los Angeles
Religious/gospel: KLTY-FM Dallas
EZ listening/beautiful music: KABL-AM/FM San Francisco
MOR/variety: WGN-AM Chicago
AOR/classic rock: WMMR-FM, Philadelphia

STATIONS OF THE YEAR

Major market: KNIX-AM/FM Phoenix
Large market: WIVK-AM/FM Knoxville, Tenn.
Medium market: WBBQ-AM/FM Augusta, Ga.
Small market: KBOZ-FM Bozeman, Mont.

Net Ad Revenues Still Soaring

Variety 4/26/89

Chicago Advertiser's continued 10-month spree of increased spending on network radio in March when earnings increased 12.4% over the same time last year to $32,604,578, per the Radio Network Assn.

Upward trend culminated in a first-quarter 1989 increase of 12.4%, from $80,604,213 to $90,632,235.

Revenues from all territories posted double-digit gains, even the New York market, which overcame traditional losses brought on by post-election budget jitters to score its largest gain in the last 10 months — a 16.7% increase of $2.6-million over March 1988's $15,699,294.

Chicago enjoyed the largest increase of any market last month, picking up 27.7% to $10,072,622.

RNA President Peter Moora reports that a good portion of this across-the-board increase in ad spending is coming from new accounts and products. "They have discovered network radio while seeking relief from decreasing ad dollar efficiencies, fueled by fragmentation of their other media."

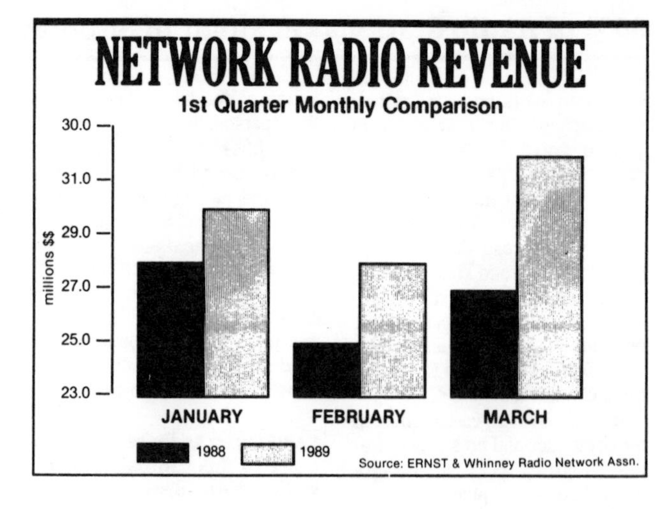

CATV

Cable television entered into 1989 with heated discussions over "syndex" (syndicated exclusivity), possible telephone company interference, and skyrocketing programming costs. However, despite these negative issues, CATV enjoyed a year of stability, prestige, and profitability. NBC, for example, divested itself from network radio interests, and moved directly into cable by launching a business channel. Pay-per-view (PPV) TV also became very profitable. And as a result of numerous popular PPV shows, as well as other original programming, many advertisers diverted their monies from broadcasting and put them into cable.

Technologically, new developments like fiber optics insured even more flexibility for future endeavors in cable, leading broadcasters to continue to worry about CATV's growing competitive edge.

Syndex, Must-Carry, Show Costs are Issues Confronting INTV

VARIETY 1/4/89

By JOHN DEMPSEY

New York Syndicated exclusivity, must-carry, the telephone company getting into cable, attempts to control the spiralling cost of programming: every year it seems as if these same issues keep cropping up as independent stations prepare for their annual INTV convention in Los Angeles.

Bob Kreek, president of the Fox-owned tv stations, calls it "the standard laundry list," problems that never seem to go away — they're so persistent they've almost become boring.

Syndex 'high-profile'

But their persistence doesn't make them any less vexing. Syndex (the shorthand term for syndicated exclusivity) "will be a high-profile item at the INTV convention — it'll get lots of talk among the stations, the syndicators and the regulatory community," says Randy Smith, president/g.m. of indie WPHL-TV Philadelphia. "Getting exclusive rights to a program he buys is the most important issue facing a station in today's marketplace. It's too easy for a show on my station to get lost" if it's competing with the same show being brought in on independent stations outside Philadelphia by the local cable systems.

Lev Pope, president of indie WPIX-TV New York, disagrees with Smith, saying syndex "won't be such a big issue" at INTV because "lots of people are going to go to court, no matter how the FCC decides the matter. This thing could drag on for months, or even years before anything gets resolved."

WGN's plight

Because WPIX is a Tribune Broadcasting station, Pope is against the reimposition of syndicated exclusivity. Tribune's WGN-TV Chicago harvests 60% of its viewing audience from outside the Chicago designated market area, via cable systems all over the country that pick up WGN's signal. Return of syndex could cost WGN a huge chunk of those outside viewers because cable systems would drop the station rather than go to the trouble of blacking out signals of individual shows scattered throughout WGN's various dayparts.

Dirk Brinkerhoff, v.p./g.m. of indie KTXH-TV Dallas, probably puts it best when he says, "Syndex has been plaguing us for years and I m afraid it's going to continue to plague us."

Syndex, must-carry and the phone company as a possible competitor to existing cable systems are all part of what Marty Colby, president/g.m. of indie XETV San Diego, calls "the reason that cable systems have had a government license to be a monopoly. Allowing the telephone company to provide cable signals would at least increase competition."

Preston Padden, president of INTV, says the association hasn't taken a position yet on whether the government should allow telco into the cable-tv business because there's a downside that has to be taken into account. "The last thing we want to do," he says, "is to exchange a 500-pound gorilla for a 1,000-pound gorilla."

But must-carry wouldn't be at issue if the phone company were providing cable-tv service to the home because, says Marty Brantley, president/g.m. of indie KPTV Portland (Ore.), "fiber optics would allow plenty of room to carry all of the local stations without any problem."

'Evil Empire'

WPIX' Pope is convinced that "the pendulum in Washington is swinging against cable," but INTV as a group, and most indie g.m.s as individuals, continue to regard cable as the tv equivalent of Reagan's "evil empire."

In programming, the independent stations are entering the era of the cheapskate. "My watchword is fiscal responsibility," says Dan Berkery, v.p./g.m. of indie WSBK-TV Boston. "In all of my programming purchases, I'm looking to save money."

Because of the circus-type atmosphere, Berkery continues, "I've found that the worst places to buy programs are at INTV and NATPE."

"I'm shying away from buying any program that's not going to make me money," adds Dallas' Brinkerhoff. "Stations are taking a more disciplined look at what shows they can and can't afford. I regard myself as a businessman first and a broadcaster second."

'Below top 50' appeal

Gerry Walsh, president/g.m. of WLVI-TV Boston, says, "I have all the product I need so I'm not going to INTV this year. All of the syndicators have been through Boston with their new product. As a place to buy product, INTV is important mainly to stations in markets below the top 50."

In terms of program genre, Walsh echoes many of the other indie g.m.s when he says he's most interested in the new major-studio movie packages that are either out in the market or about to be released. Industry observers say the package with the most "A" titles will be the new Paramount bundle, which may include blockbusters "Crocodile Dundee," "Beverly Hills Cop II," "Star Trek IV" and "Top Gun."

Portland's Brantley says the only majors that don't have a recently released package or a new one they're about to release are Warner Bros. and Disney's Touchstone.

But even though movies are a vital part of an indie station's schedule, particularly an indie that runs a primetime movie strip, the g.m.s say they're not going to hand over blank checks to the major studios. KTXA's Brinkerhoff says that in Dallas, one of the most competitive markets in the country, with five independent stations grappling with each other for increasingly thinner shares of the market, no station has yet bought four movie packages that their distributors have already pitched: Columbia, Orion, MGM/UA and Viacom.

"I'm not going to buy a package that has one or two 'A' titles and 19 or 20 mediocre to poor movies," Brinkerhoff continues. "I can't afford to buy a lot of pictures that I can't play in primetime."

The issue of unwired networks, which Berkery says will be "a big topic of discussion at INTV," is getting mixed reviews from indie g.m.s. "Any scheme that gives us less control over our commercial inventory is not necessarily a blessing," says Bob Fulstone, v.p./g.m. of KPLR-TV St. Louis.

Portland's Brantley, who's a member of the INTV committee formed to study the feasibility of an indie-station-controlled unwired network, says, "There's an $11-billion network pie out there that independent stations are not sharing in. My main concern is that an indie unwired network must not cannibalize the spot business."

"Independent stations are getting ratings that are competitive with the networks' but we haven't found a convenient way to get those network dollars," says WPHL Philadelphia's Smith. "There ought to be a mechanism that would get us involved directly in an unwired network rather than relying on the third parties that are doing it now."

The general consensus is that an indie-station-generated unwired network will see the light of day, probably sometime between now and early spring, because of the sluggish spot marketplace. But if advertisers should develop a new love affair for national and local spot, unwired networks could end up falling by the wayside.

Syndex Set for 1990, Cablers See Chaos

VARIETY 3/1/89

By PAUL HARRIS

Washington The FCC last week reaffirmed its decision to reimpose syndicated exclusivity, but granted a 4-month postponement of the rule's effective date. The regs now kick in Jan. 1, 1990, instead of next Aug. 18.

The agency nixed requests to scuttle entirely or make wholesale changes to its syndex rule. It also spurned pleas from some tv stations to impose exclusivity on existing contracts.

The brief delay in enactment was applauded by broadcasters, but cable industry reps again promised wholesale blackouts and anticipated mass complaints from viewers demanding return of popular programs. Factions of the cable industry will appeal the decision.

One key date that did not change under the revision is the June 19, 1989 deadline by which tv stations must alert their local cable operators about existing exclusivity contracts. It means that's the date cable operators will get their first real indication as to how burdensome the new syndex obligations will be.

Mooney's complaint

For the record, they are fearing the worst: "The impact of this on consumers will be that if you're expecting a popular show like "Star Trek" to be on an out-of-town channel at 7 p.m., it won't be there," said Jim Mooney, prez of the National Cable Television Assn. "Some filler program — or a blank screen — will be there instead, and if the cable operator finds it too impractical to make all the deletions and additions, he may be forced to drop the channel in its entirety."

All of this is being done "so Hollywood people can get more money from local broadcast stations, and so the local broadcast stations can cut down on competition for viewers. This represents a very odd notion of 'government for the people'," Mooney fumed.

The FCC made one clarification in response to complaints about a provision that permits "cherrypicking" by cable systems — the substitution of one distant signal for another. The film industry and sports interests feared systems could create a composite channel consisting of programs plucked from several distant signals, and would be excused from paying additional copyright fees.

Under the revision, a cable operator may run a substituted program to completion without incurring additional copyright obligations, but must then return to the regularly carried signal even if a program is in progress on that signal.

The commission also advised that its syndex regs will apply to bartered programming just as it will for cash transactions. It said in both cases, the rules "will maximize incentives for program production and will result in a more accurate market response to viewer preferences."

The commission also clarified how notice requirements would operate. It said that to be entitled to protection, requests to blackout a program must be furnished to cable systems within 60 days of the signing of the contract that grants those syndex rights. Cable operators also have a 60-day period before they must provide protection.

Commissioners disagree

The decision to postpone enactment of the syndex rule until yearend — where it will conform to the start of a copyright accounting period — was an obvious compromise among the three commission members. Commissioner Patricia Dennis wanted a longer delay, Commissioner James Quello wanted none, and Chairman Dennis Patrick admitted to being in between.

"The only certainty about syndex is that it will cause dislocation in the cable industry," said Dennis. "Cable systems will have to install new equipment and deal with the administration burden of compiling program information and arranging for deletions — signal by signal, day by day."

Only after June 19, the date stations must give syndex notification, will systems learn how many programs will be affected by the regs, she said.

Dennis said she was also concerned about how the broadcast and cable industries will cope with the flood of paperwork that could result from syndex. For example, the rules require each cable system to collect the syndex notices and then match the requests, program by program, with all the distant signals the cable system carries. (In the pre-1980 syndex rules, that burden was placed on stations).

Quello repeated his beef that the syndex rules should have granted exclusivity to existing contracts, and to do otherwise guts the underlying premise of the decision. Quello also argued that the 4-month delay is not fair to stations that purchased exclusive rights to programs based on the August 1989 enactment date.

'Real world' situation

NCTA followed last week's decision with a press briefing concerning how the rules might play in the "real world." Exec veepee Decker Anstrom cited Continental Cablevision's Holyoke, Mass. cable systems to illustrate the potential harm that the syndex rules might have on systems and their viewers.

Considering the system's eight local stations and five distant signals for the week of Feb. 18-24, a total of 219.75 hours of programming are duplicated and could conceivably be affected under the rules. That's 28% of programming on the five distant stations during the week, said Anstrom. But he conceded it is unlikely that stations would pay for exclusivity for many of the programs in question.

An official of Turner Broadcasting said WTBS-TV Atlanta can live with the latest rule revision. "From our point of view, we've been operating under them since last May," he said.

One area that remains uncertain is the effect that any court stay of the rules might have on the marketplace. For example, cable operators banking on a successful appeal of the regs might postpone the purchase of expensive equipment needed to blackout signals.

But Clay Felker, chief of the FCC's Mass Media Bureau, discounted such concerns. "I'm not convinced there will be a stay. We have a great set of rules here and I believe the court will uphold them," he said.

High Budget Original Programming Earning Big Payoff for Cable

VARIETY 5/3/89

By RICHARD HUFF

New York The big bucks being spent on original programming for cable is paying off at the nation's basic networks by adding both viewers and advertising revenues.

Original programming has long been looked upon as the driving force in cable's effort to attract viewers. And now that cable has reached penetration levels above 50%, advertisers are flocking to cable and its made-for-cable movies.

In fact, advertisers are taking an advertising spot on some made-fors and paying 50% more than they would pay buying the same spot during a run-of-schedule placement.

Wholesome stuff

"Advertisers are attracted to original programming like moths to a flame," said Farrell Reynolds, president of broadcast sales at Turner Broadcasting. "They want to be part of wholesome programs that are going to reach the right kinds of people."

Turner Network Television, one of four TBS networks, has been airing one original a month and will increase to two a month in October. Those productions have scored some of the highest ratings the network has seen. TNT's April 24 telecast of "Margaret Bourke-White" scored a 5.1 in its first airing at 8 p.m. (TNT repeats the originals at 11 p.m. ET.)

As a result, Turner has been charging more for ad time.

There is a catch. TNT does not allow advertisers to "cherry pick" the originals: they buy spots in the made-fors as part of a larger plan that includes other TNT programs. "We don't allow them to (cherry pick) or the originals would sell out immediately," explained Reynolds.

Unlike TNT, the USA Network does allow advertisers to buy its made-fors only, per John Silvestri, senior v.p. of ad sales at the network. "Original programming is definitely an anchor in our sales," said the executive. USA charges about $4.70 a CPM, up about 30% from its normal primetime fee of $3.70 a CPM. USA's "The Forgotten" racked up a 3.8 rating for its April 26 airing, overdelivering the network's advertising guarantee of a 3.5.

The cable networks have realized the benefits of made-for programming and have invested heavily in original production. TNT budgets more than $2-million for each of its made-fors; USA has committed more than $250-million to the cause; and CBN-The Family Channel spends $1.6-2-million for its originals.

"We just signed Kodak for a third quarter package which includes original productions only," said CBN's Dick Hammer, v.p. of advertising sales. "Advertiser interest in original productions is very high." CBN-The Family Channel is charging around $4.50 a CPM for its original works while its regular primetime rate is more like $3.25. "We have to get a higher rate to make it worthwhile," he said.

Hammer said the Kodak deal was the first time an advertiser has purchased a made-for package at the network, and he's not sure if he would sell it that way again. "It may have been a mistake.

Farrah Fawcett stars in 'Margaret Bourke-White,' which scored a 5.1 rating for Turner Network Television

Cabler Exulting as They Poise for Hefty Upfront Market

VARIETY 4/19/89

By RICHARD HUFF

New York Five years ago, when some cable networks were on shakey ground, naysayers said it was all over for the medium. Today, the nation's cable networks are readying themselves for the $300-400-million worth of advertising sales expected during the upfront buying season set to break during the next few weeks.

Unlike broadcast networks, cable has been without a focused upfront buying period. Cable advertising has been bought as an adjunct to the network buys well after the broadcast upfront season has ended. However, that system began to change in 1987, when cable penetration broke the 50% level, and advertisers now look at cable advertising in a different light.

50% penetration level

"When we hit the 50% penetration level it was like a light switch went on," says Doug Greenlaw, senior sales v.p. for MTV Networks. This season, for the first time, MTV has gone on the offensive in attracting upfront sales, taking its message to the advertisers through a traveling roadshow. "We will get our top 10 clients in upfront deals, or multiyear deals," explains Greenlaw.

According to Robert Alter, president of the Cabletelevision Advertising Bureau, somewhere from $300-400-million will be spent on cable advertising during the upfront period, expected to extend from late April to early July. The CAB forecasts national cable ad sales to reach nearly $1.3-billion by year's end, with the total cable picture, including regional and local, nearing $1.8-billion.

"Cable is now being considered significant and being bought in tandem with mainstream television," says Alter. He says about 40% of all national cable sales deals may be done in the upfront period. "Advertisers are now thinking of cable as part of their plans to reach national tv audiences."

Among the cable networks, ESPN, CNN, TBS and the USA Network grab most of the upfront advertising dollars because of their penetration and demographics. The Nashville Network, Lifetime, the MTV Networks and others follow.

"We would like to do more than 35% of our sales upfront," says John Barbera, senior sales v.p. at CNN. Barbera says that because his network has been selling out, sometimes as much as two months in advance, more advertisers are coming on upfront.

ESPN's director of eastern region sales, Michael Gannon, says the all-sports network also has avail problems, and on some programs can't accommodate advertisers for a few months. "We expect a lot of the larger advertising categories, such as automobiles, packaged goods, insurance, fast foods and sneakers, to come on early," says Gannon.

Because of the sellouts, advertisers are now positioning themselves for the market. According to John Silvestri, senior v.p. of sales for the USA Network, USA has been sold out for the second quarter since early March. "People are moving more quickly because they are finding in some cases the cable networks have sold out."

The growth of cable advertising is also being driven by the erosion of the broadcast networks' audiences. As the networks' share goes down, the networks are charging more for their ad time and reaching fewer people.

For advertisers the picture is not all that rosy. During the last season, basic cable networks recorded their highest ratings ever. Those audience increases will be costing advertisers some 30% more at the bargaining table, according to those queried for this report.

NBC Bets Big Bucks on Biz Channel; CNBC to Take Off Next Week

VARIETY 4/12/89

By RICHARD HUFF

Fort Lee, N.J. "This way there will be a new growth in the spring," says Michael Eskridge, president of CNBC, NBC's consumer-oriented business news channel, as he applies his pruning shears to an office tree.

Eskridge hopes more than just his tree will be growing this spring. In a matter of days, the studio lights at CNBC's 40,000-sq.-ft. facility will be on and its maiden signal beamed to nearly 13-million homes nationwide.

The service will officially launch at 6 a.m. April 17 with its first broadcast of "World Business," an hour-and-a-half program that will also serve as the start of CNBC's broadcast day. The cable industry will be watching closely to see what happens in Fort Lee.

"There will be no hard general news," says Eskridge. "We're sensitive about that point because some of our affiliates are." But CNBC will occasionally put a business spin on a national news event, such as the recent Exxon oil tanker accident in Alaska. Moreover, CNBC added to each of its affiliate contracts a declaration that it will not turn into an all-news service.

Competition?

"The biggest misconception about CNBC," says Eskridge, "is that we're getting into this business to compete with CNN. What we're doing has got nothing to do with what they do." He says the Financial News Network is not in direct competition with CNBC either. However, he admits there will be some overlap in the topics covered by the two services.

CNBC was first announced in May 1988 as part of NBC's $21-million purchase of Tempo Television from Tele-Communications Inc. A month later, citing tax implications, NBC reconfigured the deal from a straight buy of Tempo to a longterm lease. The pricetag remained the same. Since then, Eskridge and Tom Rogers, president of NBC Cable and Business Development, have put together 275 employees and set up a cable network in Fort Lee, N.J., across the Hudson River from NBC's New York City corporate headquarters.

"Our on-air people are not just talking heads. They all have solid business-news backgrounds," says Eskridge. In fact, CNBC has raided its "noncompetitors" for on-and-off-air talent. Sue Herera, a principal anchor on CNBC daytime programming, comes from FNN. Mary Alice Williams, now with NBC News and anchor of "Media Beat" for CNBC, spent nine years with CNN.

When searching for a cable niche, NBC put its in-house research department on the job. The results: Americans, according to NBC's calculations, are seeking information about personal finances. However, FNN has been targeting the business consumer for eight years, and CNN has had cablecast business programs for just as long. "Our information indicated there was an appetite for increased quality in business news," says Rogers.

17 hours of new shows

CNBC will program each weekday with 17 hours of newly produced shows and seven hours of repeats of the day's programming. A stock market ticker will occupy the bottom portion of the screen throughout the business day. On weekends, CNBC will use about 3½ hours of new programming, the balance consisting of wrap-up programs and compilations of the week's news. In addition to reports from CNBC talent, it will rely on stories from consumer correspondents at NBC's o&o's bureau reports and newsfeeds gathered from NBC's satellite news service, Skycom (see schedule).

At launch, the service's signal will be microwaved from a truck parked outside the Fort Lee studios to a location in Jersey City. It will then be transmitted via fiber optic cable to Cablevision System's Long Island, N.Y., uplink facility. Cablevision will then send the signal to the satellite. To date, the uplink is the only involvement of NBC cable partner Cablevision.

According to Eskridge, about $10-million has been spent on equipment. Approximately $2-million more in equipment used at the Summer Olympics has been incorporated into CNBC's facilities. The service will spend $5-million over five years to lease its headquarters. Eskridge won't say how much will be lost before the service makes a profit, but he calls FNN president David Meister's estimates that CNBC will lose $200-million before making a dime "absurd."

5-year plan

He says if CNBC stays near its 5-year business plan, it should earn money in either its third or fourth year. Eskridge estimates that profits in the fourth and fifth years should offset the start-up losses. In the early years 50% of the service's revenue will come from ad sales; the other 50% will come from license fees. After year five, according to the business plan, ad revenue will carry the service at about a 2-to-1 advertising-to-license-fee ratio.

Charter affiliates had to commit 25% of their total reach to the service in the first year, 75% in the second and 100% in the third year. Operators will pay 7¢ a subscriber in the first year. The cost per sub will increase a penny a year from there. Affiliates that are on at launch become eligible for an ad rebate program that kicks in during the third and fifth years. Under that plan, CNBC will set aside 20% of its net ad revenue in the third through fifth years for distribution to charter affiliates. By the end of the third year, CNBC officials estimate the subscriber base will surpass 20-million.

In preparation for the launch, CNBC staffer began real-time rehearsals on April 10. Those practice runs will last until the 16th.

Eskridge expects program changes to be made as the service evolves and as research indicates. NBC's cable research teams will go into action the day CNBC launches.

"This is going to last forever," says Eskridge. "We're not going to go on the air, declare victory and then go home."

Operators and Programmers Alike Are Pinning Their Hopes on P-P-V

VARIETY 1/11/89

New York Is it a squalling, lusty infant, or still little more than a gleam in cable's eye?

Whatever its present state, operators and programmers agree that pay-per-view is (or will be) a Blessed Event.

Thus far, feature films have been the chief grist for such money mills as the 10-million (and counting) addressable homes may provide. But p-p-v has not replaced homevideo as a major source of revenue for film companies and, in fact is far behind.

The very presence of homevideo is a serious impediment to the growth of p-p-v. As long as the former's window remains ahead of cable's, operators are reluctant to push up the timetable for installation of the costly addressable equipment.

So, the "event" comes front and center.

Revenue possibilities for events exclusive to p-p-v have become evident with World Wrestling Federation's Wrestlemania, and the Tyson-Spinks and Leonard-Lamonde fights, among others. To some extent, such programs are still carried on closed-circuit in non-cable or non-addressable areas, but p-p-v has eclipsed closed-circuit as a revenue source.

When NBC formed a partnership with Rainbow Programming Services to provide non-networked events including some portions of the 1992 Summer Olympics to a dedicated two or three p-p-v cable channels, both parties suggested that the games' presence on p-p-v would spark the growth of the medium by encouraging more addressable installation.

Not waiting

Others aren't waiting, though. Viacom Enterprises sold off its Viewer's Choice p-p-v distribution service Home Premiere Network in which a number of large MSO's are partnered. It had long been Viacom's position that because only about 10% of money generated by p-p-v went to the distributor, the present number of addressable homes grossed too little for the medium to be attractive. Viewer's Choice proved to be an attractive name though and Home Premiere took on that name for its surviving service.

Shut off Viewer's Choice, Viacom then established Showtime Event Television under the direction of Scott Kurnit, who had headed Viewer's Choice. That company's first in-house event will be a daredevil motorcycle jump in Las Vegas by the son of Evel Knievel.

Showtime Event also gave impetus to Jim Spence's Sports Television Intl.'s efforts to become a p-p-v supplier. Spence was senior v.p. of ABC Sports when Knievel's efforts were televised and he produced the upcoming "son of" special.

It is not required that a special company be set up to produce events. This fall, Request Television, with Viewer's Choice, put on its own 3-fight card and sold it to other distributors.

Other players

The Cablevision-NBC combine isn't the only provider of basic cable to get into the p-p-v game. Turner Broadcasting is about to distribute its own version of wrestling through a recent TBS buyout of a promotion company which has been mainly active in the south and has sold to tv stations there.

It has been estimated that there may be 25-million addressable homes by the time the 1992 Olympics roll around, and some say that is a conservative estimate. Such a number will certainly add to revenues from movie runs, but is still not seen as a strong enough base for the distribution end to be wildly profitable without special programming.

There are only so many championship fights available in any given year and wrestling may be a fad which could run its course almost any time, so it will be up to producers to come forward with programs so appealing they will convince an appreciable number of John Q. Publics to order the product from their local operator.

Cable Nibbling at TV's Ad Pie; Looking at $472-Mil in '89, Taking it From Stations

VARIETY 6/28/89

By RICHARD HUFF

New York While the broadcast networks are cashing in on some of the largest upfront advertising deals ever, cable systems that sell time on the spot market are quietly preparing to take their share of the national advertising pie.

In fact, more than $472-million is expected to be spent on local and national cable spot advertising this year alone, and the figure is expected to soar to over $599-million in 1990. The cable systems' gain is coming at the expense of some local affiliates and independents in the larger markets that are facing competition from a medium that sells demographics, not ratings.

"We are seeing cable take some broadcast dollars in the major markets," said Bill Lemanski, president of Cable Media Inc., a Chicago-based cable system rep firm. "Any market that has 50%-plus cable homes is going to see an erosion in local (broadcast) spot sales."

John Weidman, v.p. and director of advertising at Metrobase Cable Advertising, agrees: "As the cable channels take viewers away from broadcast, we'll take the dollars away."

High-level meet

Cable spot sales will be the topic of a high-level planning session of the Cabletelevision Advertising Bureau June 29 in Chicago. The meeting will bring together 70 top executives from the nation's largest multiple-systems operators and advertising agencies to figure out what the industry can do to increase its share of the national spot market.

Although estimates vary, an informal poll of cable reps indicates about 2,500 of the nation's 9,600 cable systems are inserting advertising into the time provided by the national cable services. (The basic cable networks give cable operators two to four minutes each hour for local sales.) While the number is small, those 2,500 systems reach approximately 90% of all cable homes. The relatively small number of systems inserting ads also stems from the fact that advertising makes up only 5% of the total revenue being generated by the MSOs. On a national level, consumers spend nearly $14.2-billion yearly on subscribers' fees, compared with the $1.7-billion that comes from all forms of advertising.

Cable advertising is not without its problems. Attempts to develop a traditional spot market have been hampered by the amount of paperwork and invoicing an advertiser would have to go through to place a spot. For example, an advertiser wanting to hit the Philadelphia marketplace could place spots on any of the independents or affiliates with one or two calls. In the past to do the same for cable would have meant dealing with as many as 43 cable systems to cover the same area. Some of the problems have been alleviated through the use of interconnects, or groups of cable systems being sold by one organization.

"The mechanics of planning, placing and paying invoices in the past has kept some agencies with national advertisers away," said Robert Fennimore, president and chief operating officer of Cable Networks Inc. "What we have done is put all of the information on one data base so we can place ads on 50 systems and the client sends one tape and gets one bill."

Local spots still tops

The cable spot marketplace is broken down into three categories: local, regional and national spot. For most reps queried, 90% of the business comes from the local and regional areas, with the balance coming from the national marketplace.

"We're seeing big incremental growth in regional and national spot," said Dave McGlade, president of Cable Adnet Inc. "However, local advertising is still the driving force. It's going to take an educational process and a rethinking on the part of the agencies."

Because cable is pitched as a frequency medium and not a reach vehicle, advertising agencies have had a hard time coming around with national advertisers. Cable executives say that while they may deliver only 1,000 homes, they deliver a quality 1,000 homes. "We've already deleted the old, poor and underemployed from the viewing mix," said Fennimore, referring to the fact that people in those demographics tend not to subscribe to cable. However, advertisers tend to approach the field looking for the medium with the largest reach unless the product is limited to a certain demographic.

"This is one of those businesses where you have to write the rules as you go," said Jeff Williams, president of Brannon Cable TV Advertising. "It's simply a matter of educating the agencies."

Cable Showing Up in Ratings; Makes Dent in May Sweeps, but Claims Under-reporting

VARIETY 7/5/89

By RICHARD HUFF

New York Cable services are not commanding the high ratings of their over-the-air counterparts; however, they are working their way into the A.C. Nielsen local market reports — reports the cable networks maintain vastly under-report cable viewing.

During the May sweeps, a number of the basic cable networks appeared in the local market reports, giving broadcast stations in those DMAs more competition in the advertising sales arena. The Assn. of Independent Television Stations plays down the results, saying cable is still too small to cut into their revenues.

Turner's Superstation TBS appeared in 186 of the 210 DMAs in the country during the May sweeps, according to the Nielsen Station Index reports. More important, the USA Network, which offers its affiliates local advertising time, appeared in 137 of those books. Chicago's superstation WGN was recorded in 74 books and ESPN showed up in 30. TBS and WGN do not offer their affiliates advertising avails.

Cable network local appearances are being driven partly by a change in Nielsen's reporting standards for cable, which took effect at the start of the 1988-89 broadcast year. Simply, 19.5% of the homes in a given market have to view the cable service once in any average week during the measurement period. The old standard was 20%.

On the whole, the number of appearances were down from those recorded during the February sweeps period, when HUT levels are higher. However, the figures are still significant on the local market level.

"This has got to be nettlesome for salesmen at the local stations," said USA research v.p. David Bender. "All of a sudden cable networks are popping up into the local market books."

"As the cable services get listed," said Turner research v.p. Robert Sieber, "those selling local avails will get some credence."

However, INTV's director of marketing, Rob Friedman, does not believe that cable's increase will hurt the independents in markets where cable is appearing in reports. "Fragmentation makes these services unacceptable as reach vehicles," explained Friedman. "Their percentages are big only because they're reflecting small bases." Friedman said the cable services won't take money away from the independent stations. "We're (independents) the principal beneficiary of network erosion," he said.

Friedman cited a recent INTV report indicating that nationally, on a 24-hour basis, indies command a 24 share; the three broadcast networks earn a 19, and the ad-supported cable networks record a 15 share of the viewing audiences.

Research executives attribute the high number of NSI appearances to the lower requirement and to the decent performances of several made-for-cable projects. Even so, all agree that the diary method of viewership reporting still underreports cable and independent viewers. "The diaries never report people who tune in for a little while and tune out, while the meters always do," said USA's Bender.

Cablers Move to Fiber Optics as Telco Competition Looms

VARIETY 2/8/89

By KATE BULKLEY

Denver Spurred by declining construction costs and the promise of enhanced capabilities, several of the largest cable-television companies in the nation have committed millions of dollars to putting fiber-optic technology into their cable-tv systems.

The decision to add fiber optics comes at a time when regional telephone companies are increasing their efforts to break the federal lock on the door to providing video services and restrictions that are the result of several federal actions, including the court-ordered breakup of the American Telephone & Telegraph Co., Federal Communications Commission rulings and the 1984 Cable Communications and Policy Act.

Fiber advantages

Advantages of fiber technology include expanded band width, which means larger channel capacities, and clearer pictures because fiber does not have to be boosted with amplifiers, as does coaxial cable, the traditional wire used to send cable signals to subscribers' homes.

Cable companies are putting fiber optics into their systems "because of technological advances, but they have been accelerated in response to the threat of telephone company entry," said Ken Goldman, cable-tv analyst at Denver brokerage Hanifen Imhoff Inc. "The cable companies are looking at ways to counter that threat."

In the last four months Denver's Jones Intl. Ltd. and TeleCommunications Inc., which own and manage systems serving nearly 12-million cable subscribers across the nation, have announced plans to put millions of dollars worth of fiber-optic technology in several of their systems. American Television & Communications Corp. is testing fiber-optic technology in several of their systems. ATC is the second-largest cable-tv operator in the country, serving about 3.7-million subscribers. At least two other cable companies, including Pennsylvania-based Comcast Corp. and New York's Cablevision Systems Corp., are also putting fiber into their systems.

Hardware scramble

For many years, ATC was a lone wolf in fiber optics research and testing, but now it appears that the time for fiber optic technology has arrived, as witnessed by the fact that at least 12 major hardware suppliers, including such giants as AT&T and the Jerrold division of Hatburo, Pa.-based General Instrument Corp., are scrambling to develop equipment compatible with fiber optic technology.

One hardware supplier, Skokie, Ill.-based Anixter Cable TV, will open a fiber optic product development laboratory in Denver this month to research and develop fiber optic hardware for the cable business. The field of research and development is still very much at the forefront of fiber technology for cable-tv systems and is something the industry is addressing through a recently formed R&D consortium of cable companies called Cable Labs (VARIETY, Feb. 1-7). Fiber optics and research into high-definition tv are at the top of the Denver-based Cable Labs' $8.5-million research agenda.

Part of the reason for the hurry to develop fiber optics on the cable tv side may be the much-publicized building of fiber optic cable-tv system in Cerritos, Calif., by Stamford, Conn.-based phone company GTE Corp. GTE received a special business waiver from federal authorities to build the system.

The 5-house system was called the "video world of the future" in a recent Wall Street Journal article, which said it will provide customers with video-on-demand services from movies to home shopping. But many in the cable industry said that publicity about the Cerritos system is giving people the perception that only a telephone company can provide such enhanced video services.

"That's the telephone company myth," said Jim Carlson, v.p. of corporate communications for Jones Intl.

Dissension in ranks

Despite the move toward fiber, there are differences of opinion in the industry about how fast and how fully fiber optics should be deployed.

"In our opinion there is absolutely no reason to wait (to deploy fiber)," said Robert Luff, Jones Group v.p. of technology at Jones Intl.

Jones officials said any renovation the company does will likely include fiber because the cost of putting fiber in a system is comparable to that of a traditional cable upgrade.

A more cautious assessment was provided by ATC. "At some point in the future we believe that it will be economical to replace the whole (cable) plant with fiber, but that's 10-15 years out," said Dave Pangrac, director of engineering and technology at ATC.

Top 10 Cable Advertisers, 1st Qtr. '89

VARIETY 6/14/89

New York Time Inc. was the top advertiser on six major cable networks monitored by Arbitron's Broadcast Advertisers Reports in the first quarter of 1989.

BAR measures advertising expenditures on CBN Cable Network, Cable News Network (CNN), Entertainment and Sports Programming Network (ESPN), MTV: Music Television, TBS Superstation in Atlanta and USA Network.

Total advertising expenditures for these networks for the first quarter of 1989 was $216,116,628, compared with $153,849,616 for the first quarter of 1988. This reflects a 40.5% increase.

The top 10 advertisers on six major cable networks for the first quarter of 1989, along with their rankings for the same period in 1988, according to Arbitron's BAR are:

Company	1st Qtr. '89	Rank 1st Qtr. '88	1st Qtr. '88	% Change From 1988
1. Time Inc.	$8,293,725	2	$5,848,177	+41.8
2. Procter & Gamble	8,073,544	1	$6,765,584	+19.3
3. Kohlberg Kravis Roberts & Co.*	5,372,542	—	310,459	+1,630.5
4. Philip Morris	5,204,679	4	4,351,793	+19.6
5. General Mills	5,088,123	3	4,688,460	+8.5
6. Anheuser-Busch	4,901,738	5	3,992,308	+22.8
7. Mars Inc.	4,554,168	8	2,792,684	+63.1
8. General Motors	4,363,245	—	1,927,881	+126.3
9. Sears Roebuck	2,840,434	—	384,736	+638.3
10. Pepsico	2,838,068	—	1,697,898	+67.2

*Includes RJR/Nabisco

Not appearing on the first quarter 1989 list are Eastman Kodak (first quarter 1988 expenditures: $2,991,572/first-quarter 1989 expenditures: $2,592,672), Chrysler Corp. ($2,660,645/$2,719,143) and American Home Products ($2,125,712/$2,660,569), which ranked 7, 9 and 10, respectively, on BAR's first quarter 1988 list.

Detailed logs and commercial activity for cable networks are published in BAR weekly reports. Estimated expenditures by parent company, brand/product and product class are included in cumulative monthly and quarterly summaries.

International

On the international front, import programming quotas as well as the usual technological restrictions dominated the year's activities. Satellite dish purchases ran much more sluggishly than predicted, causing concerns to be raised about the types of promotional tease programs which might be shown to increase sales. As for program distribution, the European Community's "Television without Frontiers" directive proposed that non-European shows be restricted by a quota system. Needless to say, American producers and syndicators were quite unhappy about the latter proposition, and attempted to counter the quota system with a multi-continental clearinghouse approach.

On a more positive note, coproductions, especially between Great Britain and the United States, seemed alive and well in 1989. Their future growth seemed to be equally optimistic.

Sky Has Limits, Media Analysts Believe, With Dishes Selling at 3,000 per Week

VARIETY 5/10/89

London On a hunch that its most likely audience will be young blue-collar males, Rupert Murdoch's fledgling direct-to-home satellite film channel, Sky Movies, is launching a summer schedule top-heavy with action-adventure and R-rated fare. Idea is to drum up a modest following before advertisers lose interst.

Sky TV has been on the air for three months and dish sales remain dismal — about 3,000 a week. At that rate only 150,000 British households will be equipped to receive Sky's satellite-delivered channels by the end of the first operating year, whereas Murdoch had pledged advertisers he'd be delivering 1.15-million (including cable homes) within 12 months.

Sources close to Sky, however, say morale hasn't had time to sink. Staffers remain gung-ho, and many are working overtime, they say.

Latest research, say Sky honchos, shows that among those who receive Sky and terrestrial channels, 51% of total viewing is for Sky's four channels. Of these, Sky Movies is by far the most popular, with 22% of total viewing.

"We've got to do better, but we're not disappointed," says Jonathan Miller, Sky TV spokesperson. "We're running a marathon, not a sprint."

Miller dismisses any suggestion that the promotional strategy of the operation will change saying Sky's £13-million ($22-million) advertising budget is being spent as originally planned. Much of that coin is going to hype Sky Movies, considered the service most likely to pull in viewers dissatisfied with the four off-air channels.

Most of the new pics that will be offered on Sky Movies over the next three months were acquired from the Hollywood majors over the past six months. They will be available free to the estimated 40,000 viewers who have so far purchased rooftop dishes as well as the 53,000 homes already wired to broadband cable systems in the U.K.

The approximately 150 films scheduled for May, June and July include "Aliens," "Terminator," "Big Trouble In Little China," "Once Upon A Time In America," "Bloody Mary," "Crimes Of Passion," "The Big Easy" and "Breathless."

'Modern, glossy, upbeat

According to Sky's head of acquisitions, Stewart Till, the channel is aiming for a "modern, commercial, glossy and upbeat" profile. There are no plans to dip into soft porn or excessively violent fare, as some had predicted.

So far Sky Movies has acquired pay-tv rights to some 2,500 feature films in more than 40 deals with distributors in both the U.S. and U.K. Those include major output deals with Warner Bros., Touchstone (through Murdoch's joint venture with the Disney Channel), 20th Century Fox (owned by Murdoch), the Rank Organization, Orion, Handmade and New World.

One-billion dollars has been spent in Hollywood over the last eight months by Sky Movies and its soon-to-be-launched rival — British Satellite Broadcasting's Movie Channel.

Many media and financial analysts here think only one of the two channels will eventually thrive. (For the next several years, Sky Movies and BSB's Movie Channel will be available on separate, incompatible dishes, broadcast in different transmission standards and encrypted differently.)

By the end of the year Sky Movies and the Disney Channel will be encrypted and sold as a pay package for £12 ($20) a month. Also at that time, BSB will be offering its Movie Channel for about £10 ($17.50).

—Elizabeth Guider

European TV Quotas: Angry Yanks and a Community Divided

VARIETY 5/24/89

By FRED HIFT

Paris The issue of quotas, ostensibly aimed at stimulating the production and exchange of television programming within the European Community and at stemming the tide of American shows, is rapidly becoming the hot potato of European tv broadcasting.

It is creating lively disputes among the Europeans themselves, not all of whom are convinced that a quota system is applicable or would stimulate local European tv production.

And it is generating an angry response from the Americans who consider quotas unfair, unrealistic and an artificial barrier designed to prevent them from reaching their maximum commercial potential on the Continent.

Anti-quota lobbying in

Washington is paying off. The U.S. State Dept. is understood to be readying an early demarche over the current French quota (60% of French programming, including primetime, must either be French or Common Market originated).

Valenti holds a meet

Jack Valenti, chairman of the Motion Picture Export Assn. of America, held a meeting with Jack Lang, the French culture minister, during the Cannes Festival to emphasize American industry objections to the ongoing quota discussions, which appear largely French-driven.

At the recent MIP, Lang advocated tough quota rules by way of promoting European production and said the quotas were not meant to be restrictive but were aimed at putting the Europeans into a better competitive position.

It all seems a muscle-flexing prelude to December 1992, when the new European Common Market is supposed to drop tariff barriers and — theoretically — turn into a single economic unit capable of competing with the United States, Japan and others, both as a producer and as a consumer.

A recent visit to MIP in Cannes and to Paris made it clear that the quota issue is not as clear-cut as some would like to present it, nor is it the kind of villainous anti-American plot that some U.S. executives envision.

Idealism not issue

The quota campaign, both for and against, has in fact little to do with any "idealism," cultural or otherwise. It seems firmly rooted in cold-blooded business considerations.

It seems doubtful that an eventual European quota of about 50% (if it is ever actually mandated) would seriously effect the sale of Yank programs in a rapidly expanding market in which U.S. features and series rarely occupy more than 45-50% of the networks' schedules. In fact, in most countries, local shows grab the top ratings.

There is a group, led by the French, which argues passionately that the large amount of American programming on European tv is "drowning" European culture and that, on top of that, it represents unfair competition on two grounds.

One is in the words of Jacques Dercourt, partner in France's Telecip production outfit, that the Americans are "dumping" largely amortized programs in Europe, selling for 400,000 francs an hour of entertainment that would cost the French 4-7-million francs

The second is that, in the view of the Europeans, there is no reciprocity in terms of the export of European programs to the U.S., where neither dubbed or subtitled European material is acceptable.

"The Americans are splendidly and safely isolated by their language," says Dercourt who, along with other European producers, sees an answer in inter-European coproduction.

The quota concept, while vocally opposed by many in European tv broadcasting and particularly by the growing number of privately owned networks, much to the annoyance of French officials, is advocated in many ways, though not always with complete candor.

"If France needs a quota it is because we have a need for a preference for European programs on our channels" says Michel Berthod, in charge of quota matters for the Ministry of Communication in Paris. "American programs are good and bad, like everywhere else. Quota has nothing to do with a value judgment. We simply want to protect our national cultural and artistic heritage."

Berthod and others, like Jean Rouilly, the powerful deputy chief at Antenne 2, in charge of programming and acquisition, feel that the American programs tend to undermine traditional European values. They can't logically explain how the new quota system, designed to stimulate the interchange of European programming, could or would overcome national resistance to such telecasts. Nor do they seem particularly bothered by the fact that the cultural characteristics of various European countries are highly individualistic and audiences are not necessarily receptive to imports from other European nations.

In fact, the only tv and theatrical entertainment that seems to be able to cross virtually all frontiers without difficulty comes from the U.S. And, of course, the future holds the satellite transmissions which don't appear covered by quota restrictions.

Helping local production

Antoine de Clermont-Tonnerre, who heads up Revcom and Les Films Ariane, takes the view that quotas would help build local production, particularly in the smaller countries. That view is strongly disputed by the Americans and many Europeans.

In contrast, Klaus Hallig, who represents Germany's Beta Group, doesn't see the need for quotas, nor does he feel that they would boost production to a significant degree. Both Germany and Denmark oppose a European quota as impinging on their national sovereignty.

Europeans, along with the Americans, are caught up in and disturbed by the rapid flow of events that are drastically changing the European tv landscape. The old order, under which governments ran the communications industries and decided what was "good" for the public to see (and hear) — regardless of how boring — is changing but hasn't yet disappeared.

The new webs, fighting hard for audiences, take a much more American-style, ratings-oriented approach to the medium. What matters today in Europe is how many viewers a show can attract, not the measure of culture it conveys. To that point-of-view, quota is anathema.

The Europeans, or at least those who favor quotas, are projecting themselves as guardians of culture, through the realities — perhaps inevitably — are commercial.

The Europeans are quite happy to see the American companies come in on coproduction, which supplies them with the much-needed cash to undertake major projects. Ulterior motive behind sudden American interest in coproduction is largely that it is a convenient way of getting around quota restrictions.

There are Europeans, like Dercourt and Rouilly, who aren't eager for massive coproduction with U.S. interests. They fear the oft-voiced American insistence on the switch to American program values, such as faster pacing, is likely to have a devastatingly negative impact on the distinctive quality of European product.

Valenti Working on Plan to Counter EC's Quotas on TV Programming

VARIETY 9/16/89

Deauville Motion Picture Assn. of America prez Jack Valenti says he is formulating a counterproposal to the European Community's "Television Without Frontiers" directive that calls for quotas on non-Euro tv programs.

"I don't always want to say 'no, no, no.' I want to say 'yes, yes, yes' to a future that would be exciting and successful."

The Hollywood honcho made the remarks to VARIETY during his annual stay at the Deauville film festival. He then went off to give the same message to France's cultural and communications minister, Jack Lang, in Paris. The two were to meet at presstime.

"Quotas are like a virus," Valenti noted. "Once you get it you don't know where it will lead."

Valenti has been fighting the "Television Without Frontiers" directive, which calls for European channels to dedicate a majority of airtime to European programs when practicable.

Now Valenti is countering with a proposal that could lead to a film school - and multi-continental clearinghouse for project development and worldwide distribution of films and tv programs.

"The real issue is, what is the object of quotas? It is to create a robust, healthy production community. You must build a cadre of talented young people, much in the same way that Canada is doing, thanks to Norman Jewison," Valenti said.

Schools, clearinghouse

To do this, "Europe should fund two or three film schools. First the schools, then a fiscal apparatus should be put in place." Eventually this will lead to the creation of a "central clearinghouse for those who are looking for projects and for those who are looking to get into projects."

Valenti will travel to Brussels in mid-October to outline the MPAA's position and "how I view the future."

Meanwhile, in Paris, Valenti is expected to discuss with Lang the possibility of setting up a bilateral Franco-American agreement on granting authorization for filming in the EC countries.

Also on the agenda are quotas. Valenti wants none, but Lang wants stronger quotas than those in the EC directive.

Lang Sounds Deathknell for Euro TV Quotas; France Mulls Option

VARIETY 7/5/89

By BRUCE ALDERMAN

Paris The controversial issue of quotas for European tv programs is now dead, according to France's Minister of Culture Jack Lang.

The minister told VARIETY that the European Community's directive, Television Without Frontiers, is "buried."

"Lang holds the key to the future of the directive, says an EC source. "If Lang persuades his government to continue its opposition to the directive, it is indeed dead." Apparently, the French themselves are divided on the question.

Lang said his government is trying to come up with a new proposal, but quotas will not be in it.

"Instead of quotas we will be emphasizing incentives to increase European production."

While neither Lang nor aides gave details, one idea kicking around Paris is a proposal that would make television stations devote a certain percentage of their turnover to production. Such a measure is already in effect for pay-tv channel Canal Plus.

Any such proposal would have to go through a long and tortuous march to approval on the European level.

The EC directive had caused controversy on both sides of the Atlantic because of the article stating a majority of programs on European tv should be European in origin when practicable.

Americans protested that the measure discriminated against them while European countries were divided between those that favored stricter quotas and those that wanted none at all.

"You know I've always said that the quotas in the directive were not tough enough," Lang asserted when speaking of the directive's death.

The directive was almost pulled off and quotas almost became EC reality. However, France was one of three countries that switched its support June 14 and refused to approve the directive. Greece and Holland were the other two.

Opposition to the directive grew in France shortly before the EC Parliament met in May. Leading members of the European film industry pleaded with Parliament to either strengthen the quotas or abort the directive altogether.

Coalition collapses

The Parliament, which has very little real power, did amend the text. When the EC ministers held their meeting in June, the fragile coalition supporting the directive fell apart. France, Greece and Holland switched their positions. West Germany, Belgium and Denmark maintained their opposition.

Underlining the difficult task of forging a European consensus is that each country had different reasons for opposing the directive.

The next and perhaps decisive step will take place in mid-July, when the Council of Ministers meets to vote on the directive. According to EC rules, a final decision must come by the end of August, three months after the Parliament held its vote, or the directive is automatically aborted.

"Anyone who thinks the countries can come up with a new formula on quotas before that is dreaming," said the same source.

In a related development, French President François Mitterrand was quoted on Italian television as saying he will vigorously use France's turn in the revolving 6-month presidency of the EC in order to push through measures that will fight both Japanese and U.S. domination in television — the former in hardware and the latter in programs.

France assumed the presidency this month.

BBC/Lionheart Business is Roaring With U.S. Coproductions; Emphasis is on Cable

VARIETY 6/28/89

By JOHN DEMPSEY

New York BBC/Lionheart has engineered a new batch of coproduction deals in the U.S. emphasizing the company's major shift toward cable tv.

BBC/Lionheart's most elaborate series — the 12-hour nature documentary "Trials Of Life: Survival In The Animal Kingdom" — has landed a deal with superstation WTBS for cablecasting in the fall of 1991, says Sarah Frank, senior v.p. of coproductions for BBC/Lionheart TV. The writer/host is David Attenborough ("Life On Earth," "The Living Planet"); executive producer is Peter Jones.

Frank says WTBS also has just bought another Attenborough-hosted natural-history docu, "Lost Worlds," about the fossils of man. Mike Salisbury is the producer. WTBS probably will run "Lost Worlds" as two 90-minute specials late in 1989 or early in 1990. (The BBC will schedule it this fall as four 40-minute programs.)

HBO, BFI

HBO has linked up with the BBC and the British Film Institute to produce a 2-hour movie called "Fellow Traveler," about an American screenwriter in the 1940s who's blacklisted for his political views, moves to England and continues to write, under a psuedonym, for the British film industry. Directed by Philip Saville, it stars Ron Silver, Daniel J. Travanti, Hart Bochner and Imogen Stubbs. Frank says the BBC will arrange for a theatrical release in England before pic airs on the BBC. (The BBC has produced two previous movies with HBO, "The Impossible Spy" and "The Yuri Nosenko Story.")

In its first deal with the Lifetime basic-cable network, the BBC is producing a 60-minute documentary that focuses on an unwed pregnant teenager who has decided to have the baby and then choose the couple that will adopt it. Lifetime will sign an American actor to host the program, tentatively titled "Who'll Win Jeanette."

For the Discovery basic-cable channel, the BBC will produce "Safari Live: Africa Watch," consisting of various live feeds over the weekend of Sept. 22-24 from Kenya. Show is "scheduled to coincide with the times of the day when animal activity is at its peak," according to a statement from Discovery. The satellite feeds also will include prerecorded material. Robin Hellier is executive producer.

'Fellow Traveler' a film about a blacklisted writer, with Ron Silver and Daniel J. Travanti

Early next year, Discovery will cablecast "Nature & The State Of Europe," a series of 12 BBC-produced half-hours on the environment.

The one non-cable deal among the BBC's recent coproductions is the joint venture with non-commercial WNET-TV New York for an 8-hour exploration of the natural history of the American continent, under the umbrella title "Land Of The Eagles." Peter Crawford is the executive producer.

Fiscal Year Ending June 30, 1988

$ THOUSANDS

	THEATRICAL	TV/CABLE	VIDEO	1988 TOTAL
EUROPE				
United Kingdom	$ 36,473	$ 20,058	$ 62,285	$118,816
France	31,093	38,357	14,380	83,830
Germany/Austria	39,118	17,671	37,120	93,909
Italy	41,221	38,271	19,159	98,651
Scandinavia	25,662	9,009	38,264	72,935
Spain	19,766	6,167	13,718	39,651
Other Countries	23,803	13,797	15,848	53,448
Europe Total	$217,136	$143,330	$200,774	$561,240
LATIN AMERICA				
Argentina	$ 3,428	$ 2,658	$ 3,647	$ 9,733
Brazil	2,181	1,200	2,444	5,825
Mexico	5,477	1,455	1,683	8,615
Venezuela	1,557	1,918	1,202	4,677
Other Countries	5,813	2,204	2,021	10,038
Latin America Total	$ 18,456	$ 9,435	$ 10,997	$ 38,888
FAR EAST				
Hong Kong	$ 3,833	$ 4,485	$ 2,668	$ 10,986
Japan	50,597	22,996	52,774	126,367
Korea	10,059	3,706	8,537	22,302
Philippines	4,712	2,130	668	7,510
Taiwan	9,428	3,729	7,128	20,285
Other Countries	17,840	4,209	1,704	23,753
Far East Total	$ 96,469	$ 41,255	$ 73,479	$211,203
OTHER				
Australia/New Zealand	$ 41,019	$ 23,761	$ 23,175	$ 87,955
South Africa	7,178	3,595	4,012	14,785
Soviet Union and Eastern Europe	1,384	456	308	2,148
Other Countries	6,729	2,896	3,156	12,781
Other Total	$ 56,310	$ 30,708	$ 30,651	$117,669
OTHER 1988 SALES	$388,371	$224,728	$315,901	$929,000

PERCENT OF TOTAL

	THEATRICAL	TV/CABLE	VIDEO	1988 TOTAL
EUROPE				
United Kingdom	3.93%	2.16%	6.70%	12.79%
France	3.34%	4.13%	1.55%	9.02%
Germany/Austria	4.21%	1.90%	4.00%	10.11%
Italy	4.44%	4.12%	2.06%	10.62%
Scandinavia	2.76%	.97%	4.12%	7.85%
Spain	2.13%	.66%	1.48%	4.27%
Other Countries	2.56%	1.49%	1.70%	5.75%
Europe Total	23.37%	15.43%	21.61%	60.41%
LATIN AMERICA				
Argentina	0.37%	0.29%	0.39%	1.05%
Brazil	0.23%	0.13%	0.26%	0.62%
Mexico	0.59%	0.16%	0.18%	0.93%
Venezuela	0.17%	0.20%	0.13%	0.50%
Other Countries	0.63%	0.24%	0.22%	1.09%
Latin America Total	1.99%	1.02%	1.18%	4.19%
FAR EAST				
Hong Kong	0.41%	0.48%	0.29%	1.18%
Japan	5.45%	2.48%	5.68%	13.61%
Korea	1.08%	0.40%	0.92%	2.40%
Philippines	0.51%	0.23%	0.07%	0.81%
Taiwan	1.01%	0.40%	0.77%	2.18%
Other Countries	1.92%	0.45%	0.18%	2.55%
Far East Total	10.38%	4.44%	7.91%	22.73%
OTHER				
Australia/New Zealand	4.42%	2.56%	2.49%	9.47%
South Africa	0.77%	0.39%	0.43%	1.59%
Soviet Union	0.15%	0.05%	0.03%	0.23%
Other Countries	0.72%	0.31%	0.35%	1.38%
Other Total	6.06%	3.31%	3.30%	12.67%
TOTAL SALES	41.80%	24.20%	34.00%	100.00%

Global Programming Prices

VARIETY has gathered the prices in the following chart from a number of sources within the tv-distribution community. Prices are broken down within each country in seven different categories.

VARIETY omitted a number of countries from the chart either because distributors said they do almost no business there or because the price range in various categories is so great as to be virtually unuseable.

All figures U.S. $	TV MOVIES	HOME VIDEO	SPECIALS	HALF HOUR	ONE HOUR	THEATRICAL MOVIES	HALF-HOUR CARTOONS
CENTRAL AMERICA/LATIN AMERICA/CARIBBEAN							
Argentina Bolivia Paraguay Uruguay	3,800- 6,000	2,900- 3,100	1,900- 2,100	750- 850	1,400- 1,600	3,000- 6,000	1,000- 1,400
Barbados Trinidad Jamaica	700- 900		450- 550	200- 300	400- 600		100- 125 125- 150 150- 200
Brazil	61,000- 16,500	5,500- 6,500	18,000- 22,000	2,000- 3,000	4,500- 6,500	15,000- 30,000	2,000- 3,000
Chile	3,500- 4,500	1,750- 2,250	900- 1,100	375- 425	750- 850		600
Colombia	2,500- 3,500	1,700- 2,200	3,000- 4,000	650- 750	1,000- 1,200	2,000- 5,000	1,000
Guatemala Nicaragua Honduras El Salvador Panama	1,700- 2,200		900- 1,100	475- 525	950- 1,100		650
Mexico	7,000- 8,000	3,000- 7,000	9,000- 11,000	1,000- 1,200	1,900- 2,100	10,000- 50,000	1,500
Peru	1,000- 1,400		900- 1,100	350- 450	700- 900		400- 600
Puerto Rico	5,500- 6,500		2,500- 3,500	750- 850	1,100- 1,300	5,000- 30,000	1,500
Venezuela	3,000- 5,000	1,900- 2,100	3,000- 5,000	900- 1,100	1,700- 1,900	2,500- 5,000	1,200- 1,500
AUSTRALASIA/FAR EAST							
Australia	90,000-120,000	15,000- 60,000	20,000- 60,000	12,000- 18,000	24,000- 36,000	75,000-500,000	2,000- 6,000
Hong Kong	1,700- 2,200	2,500- 3,500	1,000- 2,000	375- 750	750- 1,500	6,500- 10,000	600- 900
Japan	25,000- 75,000	50,000-200,000	35,000- 50,000	4,000- 6,000	14,000- 16,000	60,000-200,000	5,000
Korea	5,500- 8,500	4,000- 12,000	3,000- 4,000	475- 700	950- 1,500	4,000- 15,000	1,000
Malaysia	1,500- 2,500	2,000- 3,500	1,000- 2,000	300- 500	600- 900	2,500- 9,000	
New Zealand	2,500- 5,000	5,000- 10,000	1,400- 3,000	750- 1,500	1,500- 3,000	3,000- 10,000	1,500
Philippines	2,200- 4,500	500- 1,000	2,700- 3,200	900- 1,100	1,100- 2,800	4,000- 7,000	400- 750
Singapore	1,500- 2,500	2,000- 3,000	1,000- 2,000	350- 400	700- 800	1,000- 1,700	375
Taiwan	2,000- 2,500	2,500- 10,000	1,100- 1,300	425- 475	850- 950	4,000- 20,000	600- 800
Thailand	1,100- 1,300	1,000- 3,000	900- 1,100	475- 525	900- 1,100		300- 500
EUROPE							
Austria	6,000- 8,000	5,000- 7,000	4,000- 6,000	600- 700	1,000- 3,000	6,000- 8,000	
Belgium	4,500- 6,500	5,000- 18,000	1,700- 2,200	3,500- 4,500	7,000- 9,000	5,000- 21,000	
Czechoslovakia	1,100- 1,300	1,000	500- 700	2,800- 3,300	550- 650		
Denmark	4,000- 7,000	10,000- 40,000	3,000- 5,000	1,000- 1,500	1,700- 2,200	3,500- 4,500	
Finland	2,700- 3,200	3,000- 15,000	2,200- 2,700	800- 1,000	1,200- 2,200	3,500- 4,000	
France	30,000- 50,000	15,000- 50,000	30,000- 50,000	10,000- 20,000	25,000- 50,000	30,000-150,000	
Germany	35,000- 60,000	25,000-200,000	20,000- 50,000	10,000- 20,000	10,000- 30,000	20,000-150,000	
Great Britain	40,000-100,000	25,000-100,000	30,000- 50,000	8,000- 16,000	30,000- 40,000	50,000-500,000 +	
Greece	2,000- 3,000	6,000- 40,000	1,200- 1,700	500- 600	900- 1,100	3,000- 3,700	
Hungary	1,300- 1,500		900- 1,100	500- 550	1,000- 1,100		
Ireland	1,700- 2,200		900- 1,100	475- 525	950- 1,000	3,000- 6,000	
Italy	15,000- 50,000	15,000- 75,000	20,000- 50,000	4,500- 10,000	10,000- 30,000	20,000-750,000	
Netherlands	9,000- 11,000	5,000- 35,000	7,000- 10,000	2,200- 3,700	5,000- 7,000	5,000- 15,000	
Norway	3,700- 4,200	20,000-100,000	1,900- 2,100	900- 1,100	1,900- 2,100	3,500- 7,000	
Poland	1,700- 1,900		900- 1,100	400- 500	800- 1,000		
Portugal	1,900- 2,100	1,500- 4,000	1,400- 1,600	700- 900	1,500- 1,700	1,200- 2,000	
Spain	9,000- 15,000	6,000- 50,000	4,500- 5,500	750- 2,500	2,500- 5,000	5,000- 25,000	
Sweden	8,000- 8,400	25,000-125,000	3,700- 4,200	1,700- 2,200	3,700- 6,000	10,000- 40,000	
Turkey	2,700- 3,200	1,200- 3,000	1,400- 1,600	800- 900	1,600- 1,800	3,000	
Yugoslavia	1,500- 3,000	1,500	1,500	500- 750	1,000- 1,500	2,500	
CANADA							
English French	145,000-160,000 15,000- 30,000	10% of U.S. 5,000- 20,000	50,000- 70,000 7,500- 15,000	15,000- 35,000 3,000- 7,500	45,000- 60,000 7,500- 15,000	20,000- 60,000 15,000- 35,000	
OTHER TERRITORIES							
Botswana	1,100- 1,300		1,100- 1,300	250- 300	500- 600		
Iceland	1,200- 1,300	500- 5,000	900- 1,100	300- 400	500- 700	500- 1,000	
Israel	2,000- 3,000	2,000- 7,000	1,100- 1,300	550- 650	950- 1,000	1,200- 4,000	
Middle East	2,700- 3,200	3,000- 20,000	1,500- 1,700	650- 750	1,300- 1,500		
South Africa	7,000- 9,000		4,000- 5,000	1,100- 2,000	2,200- 5,000	5,000- 10,000	

New York The following list shows the major film, television, video and related festivals and markets; when held; their address and, where available, telephone, telex and fax numbers; directors; categories or genres, and whether competitive or not, if applicable.

AUSTRALIA

Australian Intl. Film Festival
September
11/30 Drummoyne Avenue, Drummoyne, N.S.W. 2047, Australia
Tel 81-4326
Director: Barbara Fuller
Film/Competitive
VARIETY office: Sydney

Melbourne Film Festival
June 8-23, 1990
G.P.O. Box 2760 EE, Melbourne, Victoria 3001, Australia
Tel 663-1395
Fax 662-1218
Telex 152613 FIFEST
Director: Tait Brady
Film/Non-competitive
VARIETY office: Sydney

Pacific Intl. Media Market
September
184-6 Glenferrie Road, Malvern, Victoria 3144, Australia
Telex 37426
Directors: S. Wanger, P. de Montignie
VARIETY office: Sydney

Sydney Film Festival
June 8-22, 1990
P.O. Box 25, Glebe, N.S.W. 2037, Australia
Tel 692-8793
Telex AA75111 SYDFEST
Director: Paul Byrnes
Film/Non-competitive
VARIETY office: Sydney

Television 2000 (Australian Broadcasting Tribunal)
November
P.O. Box 215, North Sydney, NSW 2060, Australia
Tel 929-6246
Television/Competitive
VARIETY office: Sydney

AUSTRIA

Austrian Film Days
October 16-21, 1990
Columbusgasse 2, A-1100, Vienna, Austria
Tel 604-0126
Fax 602-0795
Telex 75311077 RSP A
Director: Reinhard Puy Pyrker
Film, tv & video
VARIETY office: London

Viennale-Vienna Filmfest Week
March
Uraniastrasse 1, 1010 Vienna, Austria or Würzburggasse 30, 1136 Vienna, Austria
Tel 8291-4515
Fax 8291-2200
Telex 113985
Directors: Helmut Dimko & Veronika Haschka
Film/Non-competitive
VARIETY office: London

BELGIUM

Brussels Intl. Film Festival
January 17-27, 1990
Place Madou 8, Bte. 5, Brussels, Belgium or Palais des Congrès,32, Avenue de l'Astronomie, 1030 Brussels, Belgium
Tel 218-1267
Telex 61460 FESBRU
Directors: Dimitri Balachoff, Theo Kelchtermans
Film/Competitive
VARIETY office: Paris

Festival Cinématographique de Wallonie
October
Rue Nicolas Bosret 16, 5000 Namur, Belgium
Director: A. Ceuterick
Film/Competitive
VARIETY office: Paris

Intl. Festival of Fantasy & Science Fiction Films
March
144 Ave. de la Reine, B-1210, Brussels, Belgium
Tel 61344 ext. 113
Director: George Delmote
Film/Competitive
VARIETY office: Paris

Intl. Festival of Flanders-Ghent
October
Kortrijksesteenweg 1104, B-9820, Ghent, Belgium
Tel 218946
Telex 12750
Fax 219074
Director: J. Dubrulle
Film/Competitive
VARIETY office: Paris

Rencontres Internationales du Jeune Cinema
70 rue Faider 1050 Brussels, Belgium
Film
VARIETY office: Paris

BRAZIL

Festival Internacional de Cinema de Animacao
Rue 62, No. 251-1, Apartment 43 P-4500, Espinho, Brazil
Animation
VARIETY office: Madrid

Festrio-Intl. Festival of Film, TV & Video
November
Rua Paissandu 362 22210 Laranjeiras, Rio de Janeiro, Brazil
Tel 21-2856642/7968
Telex 22084 ETUR BR
Directors: N. Sroulevich A. Marcio
Film, tv, Video/Non-competitive
VARIETY office: Madrid

Mostra Internacioinais de Cinema & Video
September
NCV, Rua Eng. Ewbank Camera, 78 Bela Vista CEP 90.420, Porto Alegre, RS, Brazil
Films, Video
VARIETY office: Madrid

Mostra Internacional de Cinema
October
Al Lorena 937 cj., 302-1424 São Paulo, Brazil
Tel 11-25043
Film/Non-competitive
VARIETY office: Madrid

BULGARIA

Intl. Festival of Comedy Films
May
House of Humour and Satire, P.O. Box 104, 5300 Gabrovo, Bulgaria
Telex 67413
Director: Stefan Furtounov
Film/Competitive
VARIETY office: London

Golden Chest Intl. Television Festival
October
Committee for Television & Radio, Intl. Relations Dept., 4 Dragan Tzankov Blvd., 1040 Sofia, P.R.
Tel 652871/661149
Telex 22581 TV BG
Director: Todor Markov
TV/Market
VARIETY office: London

Sofia Intl., Festival for Children & Young People
July (every 3 years)
1 Bulgaria Square, Sofia 1414, Bulgaria
Telex 22059
Director: Olin Filipov
Film/Non-competitive
VARIETY office: London

Varna Intl. Film Festival
October 1990 (Biannual)
1 Bulgaria Square, Sofia 1414, Bulgaria
Telex 22059
Animation/Competitive
VARIETY office: London

Varna Intl. Festival of Red Cross & Health Films
May (Alt. years)
Central Committee of the Bulgarian, Red Cross, 1 Biruzov Blvd., Sofia 1527, Bulgaria
Telex 23248
Director: Alexander Marinov
Film
VARIETY office: London

BURKINA FASO

Panafrican Film Festival (Fespaco)
February (Alt. years)
B.P. 2505 Ouagadougou, Burkina Faso
Telex 5255
Director: Phillipe Sawadogo
Film
VARIETY office: Rome

CANADA

Annual Conference Of the University Film & Video Assn.
June
Dept. of Film/Video, York U., 4700 Keele Street North York, Ontario M3J 1P3, Canada
VARIETY offices: New York, Los Angeles

Banff Festival of Mountain Films
November
The Banff Centre, Box 1020 Banff, Alta., T0L 0C0, Canada
Tel (403) 762-6451
Fax (403) 762-6422
Director: Bernadette McDonald
Film
VARIETY offices: New York, Los Angeles

Banff Television Festival
June
The Banff Centre, Box 1020 Banff, Alberta T0L 0C0, Canada
Tel (403) 762-5357
Director: Dr. Jerry Ezekiel
Television
VARIETY offices: New York/London

Festival du Cinema Internationale en Abitibi-Temscamingue
October 27 -November 1, 1990
215 Avenue Mercier, Rouyn-Noranda, Quebec J9X 5W8, Canada
Tel (819) 762-6212
Fax (819) 762-6762
Film/Competitive
VARIETY offices: New York, Los Angeles

Festival of Festivals
September 6-15, 1990
69 Yorkville Avenue, Suite 205, Toronto, Ontario M5R 1B8, Canada
Tel (416) 967-7371
Telex 06-219724
Fax (416) 967-9477
Director: Helge Stephenson
Film
VARIETY offices: New York, Los Angeles

Intl. Festival of Films on Art
March
445 rue St. Francois-Xavier, Suite 26 Montreal, Quebec H2Y 2T1, Canada
Director: Rene Rozon
Film
VARIETY offices: New York, Los Angeles

Montreal Intl. Festival of New Cinema & Video
October
3724 Boulevard St. Laurent, Montreal, Quebec H2X 2V8, Canada
Tel (514) 843-4725
Telex 5560074
Director: Claude Chamberlan
Film, Video
VARIETY offices: New York, Los Angeles

Montreal World Film Festival
August 23-September 3, 1990
1455 rue de Maisonneuve Ouest, Montreal, Quebec H3G 1M8, Canada
Tel (514) 848-3883
Telex 05-25472
Fax (514) 848-3886
Director: Serge Losique
Film/Competitive/Market
VARIETY offices: New York, Los Angeles

Ottawa Intl. Animation Festival
October
Canadian Film Institute, 150 Rideau Street, Ottawa, Ontario K1N 5X6, Canada
Telex 636700474
Fax (613) 232-6315
Directors: Frank Taylor & Tom Knott
Animation
VARIETY offices: New York, London

Vancouver Intl. Film Festival
September
303-788 Beatty Street, Vancouver, B.C. V6B 2M1, Canada
Tel (604) 685-0260
Telex 045-08354
Fax (604) 688-8221
Director: Alan Franey
Film
VARIETY offices: New York, Los Angeles

Yorkton Short Film & Video Festival
May
Yorkton, Saskatchewan, Canada
Director: Ian Reid
VARIETY office: New York, Los Angeles

CHINA

China Intl. Sports Film Festival
May
Chinese Olympic Committee, 9 Tiyuguan Road Beijing, China
Telex 22323
VARIETY office: Chicago

Shanghai Television Festival
November 10-15, 1990 (Alt. years)
651 Nanjing West Road, Shanghai Television Station Shanghai, China
Tel 2552000
Telex 33367SHSTV CN
Director: Gong Xueping
Television/Competitive
VARIETY office: Chicago

COLOMBIA

Cartagena Intl. Film Festival
April
Apartado Aereo, 1834 Cartagena, Colombia
Telex 37642
Director: Victor Nieto
Film
VARIETY office: Madrid

CUBA

Havana Intl. Film Festival
December
ICAIC, Calle 23, No. 1155, Plaza de la Revolucion, Havana 4, Cuba
Tel 400-4711
Film
VARIETY office: Madrid

Intl. Festival Of The New Latin American Cinema
December
Calle 23, No. 1155 Vedado Havana, Cuba
Telex 511419
Director: Pastor Vega Torres
VARIETY office: Madrid

CZECHOSLOVAKIA

Intl. Fest. of Films & TV Programs on Environment
May
Ekofilm Secretariat, Konviktska 5 113 57 Prague 1, Czechoslovakia
Telex 122214
Director: Libuse Novotna
Film, tv
VARIETY office: London

Intl. TV Festival (Golden Prague)
June
Ceskoslovenská Televizia, Gorkeho nam 29, 111 50 Prague 1, Czechoslovakia
Telex 121800
Director: Josef Vanek
Television
VARIETY office: London

Karlovy Vary Intl. Film Festival
July 7-19, 1990 (Alt. years)
Jindrisska 34, Prague 1, Czechoslovakia
Tel 2365385-9
Telex 122259
Director: Frantisek Marvan
Film/Competitive
VARIETY office: London

Prix Danube Intl. Festival of TV Programs for Children
September (Alt. years)
Ceskoslovenská Televizia, Miynska dolina 845, Bratislava 45, Czechoslovakia
Tel 327448
Telex 92277
Director: Jan Kocian
Television
VARIETY office: London

DENMARK

Odense Film Festival
July-August (Alt. years)
Vindegade 18 DK-5000, Odense C, Denmark
Tel 131372
Telex 59853
Fax 914316
Directors: Astrid H. Jensen & Jorgen Roos
Films/Competitive
VARIETY office: London

EAST GERMANY

Leipzig Documentary & Short Film Festival
November
Postfach 940, 701 Leipzig, East Germany (only during festival) Chodowiecki Str. 32, 1055 Berlin, East Germany (permanent)
Telex 512455
Director: Ronald Trisch
Documentaries, Animation/Competitive
VARIETY office: London

National Children's Film Festival
February
Chodowiecki Str. 32, 1055 Berlin, East Germany
Telex 512455
Director: Ronald Trisch
Film
VARIETY office: London

National Feature Film Festival of East Germany
May 16-19, 1990
Chodowieck Str. 32, 1055 Berlin, East Germany
Tel 4300617
Director: Ronald Trisch
Film
VARIETY office: London

EGYPT

Cairo Intl. Film Festival
December 17, 1990
Kasr El Nil St. Cairo, Egypt
Tel 3923562/3923962
Telex 21781 CIFFUR
Fax 3938979
Director: Saad Eldin Wahba
Film/Non-competitive/Market
VARIETY office: Rome

FINLAND

Midnight Sun Film Festival
June
Vainämöisenkatu 19A, SF-00100 Helsinki, Finland or K-13 Kanavakatu 12, 00160 Helsinki, Finland
Fax 413541
Director: Erkki Astala
VARIETY office: London

Tampere Short Film Festival
March 7-11, 1990
P.O. Box 305, SF-33101 Tampere 10, Finland
Tel 35681
Telex 22448 TAM SF
Fax 196756
Director: Pertti Paltila
Short Film/Competitive
VARIETY office: London

FRANCE

Amiens Marche Internationale du Film
November
36 rue de Noyon 80000 Amiens, France
Telex 140754
Competitive/Market
VARIETY office: Paris

Annecy Intl. Animated Film Festival
June 22-23, 1990
Jica/ifa, BP 399 74013 Annecy Cedex, France
Telex 309267
Director: Jean-Louis Bortolato
Animation/Competitive/Market
VARIETY office: Paris

Avoriaz Intl. Festival of Fantasy Films
January 13-21, 1990
33 ave. MacMahon 75017 Paris, France
Tel 42-67-71-40
Telex 640736 PROMO DM F
Director: Lionel Chouchan
Film
VARIETY office: Paris

Biarritz Festival du Film Iberique et Latino-American
64200 Biarritz, France
Film
VARIETY office: Paris

Cabourg Intl. Festival of Romantic Films
June
76 rue Jules Guesde, 92300 Levallois, France
Telex 820746
Director: Suzel Pietri
Film
VARIETY office: Paris

Cannes Intl. Film Festival
May 10-21, 1990
71 rue du Faubourg St. Honoré, 75008 Paris, France
Tel 42-66-92-20
Telex 650765 FESTIFI F
Fax 42-66-68-85
Director: Pierre Viot
Film/Competitive/Market
VARIETY office: Paris

Clermont-Ferrand Short Film Festival
February 3-10, 1990
RISC, 26 rue des Jacobins, 63000 Clermont-Ferrand, France
Tel 3916573
Fax 39221193
Films (Short)/Competitive
VARIETY office: Paris

Cognac Intl. Films Festival of the Thriller
April
33 Ave. MacMahon, 75017 Paris, France
Telex 640736
Director: Lionel Chouchan
VARIETY office: Paris

Deauville Festival (American Cinema)
August 31-September 9, 1990
33 Ave. MacMahon, 75017 Paris, France
Telex 640736
Director: Lionel Chouchan
VARIETY office: Paris

European Environmental Film Festival
April
55 rue de Varenne, 75341 Paris Cedex 7
Telex 201220
Films (Environmental)
VARIETY office: Paris

Festival Internationale du Film sur l'Ecologie et l'Environement
April
Domaine de Grammont, Route de Mauguio, 3400 Montpellier, France
Film (ecology/natural resources)
VARIETY office: Paris

Festival de Cinema des Migrations et Immigrations
October (Alt. years)
ASADIN, 9 rue Guenot, 75011 Paris, France
Tel 40-24-04-93
Director: Abel Bennour
VARIETY office: Paris

Festival du Cinema Franco-Britannique
October
Association Travelling, 1 rue du Fourdray, 50100, Cherbourg, France
Film (British)
VARIETY office: Paris

French-American Film Workshop
July
23 rue de la Republique, 84000 Avignon, France
Telex 432877
Director: Jerome H. Rudes
VARIETY office: Paris

Intl. Festival of Women's Films
March 23-April 1, 1990
Maison des Arts, Place Salvador Allende, 94000 Creteil, France
Tel 49-80-38-98
Telex 212352 SITTAIR
Tel 48-62-69-20
Directors: Elisabeth Trehard, Jackie Buet
VARIETY office: Paris

Intl. Audio Visual Program Festival
October
FIPA, 215 Rue de Faubourg St. Honoré, 75008 Paris, France
Tel 45-05-14-03
Telex 630547
Fax 47-55-91-22
Director: Michel Mitrani
VARIETY office: Paris

Intl. Exhibition of Video & TV
September
B.P. 236, 25204 Montbeliard Cedex, France
Tel 81-91-10-25
VARIETY office: Paris

Intl. Road Safety Film Festival
September
B.P. 141, 11004 Carcassonne Cedex, France
Director: J. Roux
VARIETY office: Paris

La Rochelle Intl. Film Festival
June 28-July 8, 1990
28 Boulevard du Temple, 75011 Paris, France
Tel 43-57-61-24
Director: Jean-Loup Passek
Film
VARIETY office: Paris

La Rochelle Sailing Film Festival
October (Alt. years)
c/o Capitainerie du Port des Minimes, B.P. 145-F, 17005 La Rochelle Cedex, France
Telex 790754 LRH-F
Director: Michel Masse
VARIETY office: Paris

Lille Festival of Sports Films
December
21 rue Patou, 59800 Lille, France
Telex 130127
Director: Paul Zouari
VARIETY office: Paris

Lille Intl. Fest. of Short & Documentary Films
March
26-34 rue Washington, 75008 Paris, France
Film (Shorts)/Competitive
VARIETY office: Paris

Mediterranean Cinema Meeting
October
20 rue Azema, 34064 Montpellier, France
Telex 490833
Fax 67-69-29-69
Director: Pierre Pitiot
Film
VARIETY office: Paris

Midem
January 21-25, 1990
179 Avenue Victor Hugo, 75116 Paris, France
Tel 45-05-14-03
Telex 630547 MIDEM F
Fax 47-55-91-22
Director: Xavier Roy
Music/Market
VARIETY office: Paris

MIP/TV
April 20-25, 1990
179 Ave. Victor Hugo, 75116 Paris, France
Tel 45-05-14-03
Telex 630547
Fax 47-55-91-22
Director: Xavier Roy
Television/Market
VARIETY office: Paris

Mipcom
October 11-15, 1990
179 Ave. Victor Hugo, 75116 Paris, France
Tel 45-05-14-03
Telex 630547
Fax 47-55-91-22
Director: Xavier Roy
Television/Market
VARIETY office: Paris

Nantes Festival Des 3 Continents
November 24-December 5, 1990
BP 3306, 44033 Nantes Cedex, France
Telex 700610
Director: Alain Jalladeau
Film/Competitive
VARIETY office: Paris

Rouen Festival du Cinema Nordique
March
91 rue Crevier, 76000 Rouen, France
Tel 35-98-28-46
Telex 771444
Director: Jean-Michel Mongrédien
Film (Nordic)/Competitive
VARIETY office: Paris

Strasbourg Festival du Film
March 15-25, 1990
L'Institut Internationale des Droits de l'Homme, 1 quai Lezay-Marnesia, 67000 Strasbourg, France
Director: Francoise Gros
Film (European)
VARIETY office: Paris

World Sea Festival
September
1 rue de Vaugirard, 75006 Paris, France
Telex 290716F
VARIETY office: Paris

HOLLAND

Audio-Visual Experimental Festival
November
P.O. Box 307, 6800 AH Arnhem, Holland
Tel 511300
Director: R. Bruinen
VARIETY office: Paris

Dutch Film Days
September
1990 Stichting Nederlandse Film-Tagen, Hoogt 4, 3512 GW Utrecht, Holland
Tel 322684
Telex 43776
VARIETY office: Paris

Holland Animation Film Festival
August (Biannual)
Hoogt 4, 3512 GW Utrecht, Holland
Tel 312940
Animation
VARIETY office: Paris

Intl. Documentary Film Festival
December 7-17, 1990
P.O. Box 515, 1200 AM Hilversum, Holland
Tel 36476
Fax 35906
Director: Jan Vrijman/Ally Derks
VARIETY office: Paris

Rotterdam Film Festival
January 25-February 4, 1990 (Alt. years)
P.O. Box 21696, 3001 AR Rotterdam, Holland
Tel 411-8080
Telex 21378 FILMF NL
Fax 413-5132
Director: Marco Müller
Film
VARIETY office: Paris

HONG KONG

Hong Kong Intl. Film Festival
April 6-21, 1990
Level 7, Admin. Building, Hong Kong Cultural Center, 10 Salisbury Road, Tsim Sha Tsui, Hong Kong
Tel 3-7342900
Telex 38484 USCHK HX
Fax 3-665206
Director: Lau Yuk-lin
VARIETY office: Chicago

HUNGARY

Budapest/Hungarian Intl. Sports Film Festival
November
Rosenberg hp.u.l. 1054 Budapest, Hungary
Telex 227553
Director: Katalin Kovacs
VARIETY office: London

ICELAND

Reykjavik Film Festival
May 1991 (Alt. years)
Amtmannsstigur 1, P.O. Box 88 121 Reykjavik, Iceland
Telex 2111 ISKULT IS
Director: Salvor Nordal
VARIETY office: London

INDIA

India Intl. Film Festival (Filmotsav '90)
January 10-20, 1990
Directorate of Film Festivals, Lok Nayak Bhavan, 4th Fl Khan Market, New Delhi, 110003 India
Tel 61-5953/4920
Telex 316762741 FEST IN
Director: Urmila Gupta
Film/Non-competitive/Market
VARIETY office: London

INDONESIA

Asia-Pacific Film Festival
November
c/o Indonesia Motion Picture Assoc., Jl. Raya H R Rasuna Said (Kuningan) 11-10 Jakarta or Indonesia National Film Council, Jalan Merdeka Barat No. 9 Jakarta, Indonesia
Film
VARIETY office: Chicago

IRAN

Tehran Fajr Intl. Film Festival
February
Farhang Cinema, Dr. Shariati Avenue Gholhak, Tehran, Iran
Tel 265086/267082
Telex 214283 FCFIR
Fax 678155
Film/Competitive
VARIETY office: Rome

IRELAND

Cork Film Festival
October
c/o Triskel Arts Centre, Tobin Street, Cork, Ireland or 38 MacCurtain Street Cork, Ireland
Tel 27-1711
Telex 75390 J WOD EI
Fax 27-3704
Director: Michael Hannigan
Film, Animation, Shorts, Docu./Competitive
VARIETY office: London

Dublin Film Festival
October
1 Suffolk Street, Dublin 2, Ireland
Tel 79-2937
Film (Irish focus)
VARIETY office: London

ISRAEL

Jerusalem Film Festival
June
P.O. Box 8561, Jerusalem 91083, Israel
Film
VARIETY office: London

ITALY

Bergamo Film Meeting
July
Via Pascoli 3, 24110 Bergamo, Italy
Telex 333670
Fax 223129
VARIETY office: Rome

Boario Funny Film Festival
September
Piazza Einaudi 2, 25041 Boario Terme, Brescia, Italy
Fax 532280
Director: Franco Cauli
VARIETY office: Rome

Cattolica Mystery Film Festival
June
Via dei Coronari 44, 00186 Rome, Italy
Tel 656-7902
Telex 623092 IMAGO I
Directors: Irene Bignardi, Georgio Gosetti
Film
VARIETY office: Rome

Desenzano Adventure Film Festival
June
Azienda di Promozione Turistica, Piazza Mat teotti 27, Desenzano del Garda, Brescia, Italy
Tel 9141510
Directors: Franco Cauli, Renzo Fegatelli
VARIETY office: Rome

Europa Cinema (Viareggio)
September
Via Giulia 66, 00186 Rome, Italy
Tel 686-7556
Telex 623092 IMAGO I
VARIETY office: Rome

Festival Internazionale Cinema Giovani
November 9-17, 1990
Piazza San Carlo 161, 10123 Torino, Italy
Tel 547171
Telex 216803
Fax 519796
Director: Gianni Rondolino
Film/Competitive
VARIETY office: Rome

Florence Festival Dei Popoli
November 24 -December 2, 1990
Via Dei Castellani 8, 50122 Florence, Italy or Via Fiume 14, Florence, Italy
Tel 212771
Telex 575615
Fax 213698
Directors: Franco Lucchesi, Mario Simondi
Film (Documentary)
VARIETY office: Rome

Florence Film Festival
May
c/o Assessorato alla Cultura, Via Sant'Egidio 21, 50122 Florence, Italy or Via Martiri del Popolo 27 50122 Florence, Italy
Tel 245869
Director: Fabrizio Fiumi
Film (Indies)
VARIETY office: Rome

Giffoni Fest. of Cinema for Children & Youth
July
Casella Postale 1, 89045 Giffoni Valle Piana, Salerno, Italy
Tel 868544
Telex 721585
Director: Claudio Gubitosi
VARIETY office: Rome

Intl. Competition for Mountain & Exploration Films
May
Centro S. Chiara, Via S. Croce 67, P.O. Box 402, 38100 Trento, Italy
Tel 37832
Director: Emanuele Cassara
Film
VARIETY office: Rome

Intl. Exhibition of Music, Hi-Fi & Video
September
Via Domenichino 11, 20149 Milan, Italy
Tel 313627
Telex 4980330
Director: Roberto Pina Berchet
VARIETY office: Rome

Mifed Multimedia Market
October
Largo Domodossola 1, 20145 Milan, Italy
Tel 499-7267
Telex 331360 EAFM
Fax 499-7274
Director: Patrizia Martellini
Film/Market
VARIETY office: Rome

Montecatini Intl. Biennale of Tourist Films
October (Alt. years)
Via Sistina 27 00187 Rome, Italy
Telex 621611
Director: Antonio Conte
VARIETY office: Rome

Palermo Intl. Sports Film Festival
October
Via Emanuele Notarbartolo 1, 90141 Palermo, Italy
Telex 911006
Fax 625-6256
VARIETY office: Rome

Pesaro Intl. Festival of New Cinema
July 1-10, 1990
Via Yser 8, 00198 Rome, Italy
Telex 6-869524
Fax 624596
Directors: Lino Micciche, Adriano Apra
VARIETY office: Rome

Pordenone
October
Cinemazero, Viale Gregoletti 20, 33170 Pordenone, Italy or La Cineteca del Friuli, via Osoppo 26, 33014 Gemona, Italy
Tel 34048
Telex 450059
Fax 980458
Director: Davide Turconi
Film (Silent)
VARIETY office: Rome

Prix Italia
September
RAI Radiotelevisione Italiana, Via Del Babuino 9, 00187 Rome, Italy
Tel 6862797
Telex RAI PI 614432
Fax 6052797
Director: Piergiorgio Branzi
VARIETY office: Rome

Rimini Intl. Cinema Review
September
Via Gambalunga 27, 47037 Rimini, Italy
Telex 563170
Fax 41-70-44-11
VARIETY office: Rome

Salerno Intl. Film Festival
October
Casella Postale 137, 84100 Salerno, Italy
Telex 720135
Fax 237288
Film
VARIETY office: Rome

San Remo/Bergamo Intl. Exhibition of Author Films
March
Rotonda dei Mille 1, 2400 Bergamo, Italy
Telex 300408
Director: Nino Zucchelli
Film/Competitive
VARIETY office: Rome

SICOF Intl. Exhibition
March
(Alt. years) Via Domenichino 11, 20149 Milan, Italy
Telex 313627
Director: Roberto Pina Berchet
VARIETY office: Rome

SIM-HI-FI '90
September 6-10, 1990
Largo Domodossola 1, 20145 Milan, Italy
Tel 49971
Telex 331360 EAFM
Fax 499-7375
Director: Tullio Galleno
Hi-fi, Video, Consumer Electronics
VARIETY office: Rome

Taormina Intl. Film Festival
July
Via P.S. Mancini 12, 00196 Rome, Italy or Comitato Taormina Arts, Palazzo Corvaja Taormina, Italy
Telex 611581
Director: Gian Luigi Rondi
VARIETY office: Rome

Teleconfronto (Intl. TV Drama Festival)
May
Via Le Piane 11, 53042 Chianciano Terme, Siena, Italy
Television
VARIETY office: Rome

Turin Intl. Festival of Sports Films
May
Coni-Agis, Via di Villa Patrizi 10, 00161 Rome, Italy
Telex 610199
Fax 884-8079
Director: Paolo Ferrari
Film
VARIETY office: Rome

Venice Intl. Film Festival
September
Ca. Giustinian, San Marco, 30124 Venice, Italy or La Biennale Cinema, Ca. Giustinian, 30100 Venice, Italy
Tel 5200311
Telex 410685
Director: Guglielmo Biraghi
Film
VARIETY office: Rome

Verona Intl. Cinema Week
April
Via S. Giacomo alla Pigna 6, 37121 Verona, Italy
Telex 434339
Fax 590624
Director: Pietro Barzisa
VARIETY office: Rome

Viterbo Film Festival of Nature, Man & His Environment
November
Ente Mostra Cinematografica Internazionale Via di Villa Patrizi 10, 00161 Rome, Italy
Telex 884731
Director: Liborio Rao
Film
VARIETY office: Rome

JAPAN

Hiroshima Animation Festival
August 8-13, 1990
11-1 Nakajima-cho Naka-hu, Hiroshima 730, Japan
Animation/Competitive
VARIETY office: Chicago

Japan Academy Awards
March
Dentsu Kosan Bldg., 1-7-13 Tsukiji Chou-ku, Tokyo 104, Japan
VARIETY office: Chicago

Tokyo Intl. Fantastic Film Festival
September
Asano Daisan Bldg., 2-4-19 Ginza Chuo-ku, Tokyo 104, Japan
Telex J34548
Fax 563-6305
Director: Y. Komatsuzawa
VARIETY office: Chicago

Tokyo Intl. Film Festival
September
No. 3, Asano Daisan Bldg., 2-4-19 Ginza Chuo-ku, Tokyo 104, Japan
Telex J34548
Fax 536310
Director: Koichi Murayama
Film/Competitive
VARIETY office: Chicago

Tokyo Intl. Market for ETV Programs
November
NHK Enterprises, 2-2-1 Jinnan Shibuya-ku, Tokyo 150-01, Japan
Telex J34555
Fax 481-1620
VARIETY office: Chicago

Videx Japan
June
Fairs Intl., Ahks Atrium 201, 11-7 Rokubancho
Chiyoda-ku, Tokyo 102, Japan
Telex J32724
Fax 222-4130
VARIETY office: Chicago

MALTA

Golden Knight Intl. Amateur Film & Video Festival
November
The Malta Amateur Cine Circle, P.O. Box 450
Valletta, Malta
Director: John G. Dacoutrosi
VARIETY office: Rome

MONACO

Monte Carlo Intl. Television Festival & Market
February 11-16, 1990
(Fest. Feb. 6-16) CCAM Boulevard Louis II 98000 Monte Carlo, Monaco or Office du Tourisme, 2a blvd Moulins, 98000 Monte Carlo, Monaco
Tel 304944
Telex 469156
Fax 250600
Director: Andre Asseo
Television/Market
VARIETY office: Paris

NEW ZEALAND

Wellington/Auckland Film Festival
July
Federation of Film Societies, P.O. Box 9544, Courtenay Place, Wellington 1, New Zealand
Tel 850-162
Telex 303386
Fax 801-7304
Director: Bill Gosden
Film
VARIETY office: Sydney

NORWAY

Nordic Film Festival
March 28-April 1, 1990
P.O. Box 356, 4601 Kristiansand, Norway
Tel 21629
Fax 20390
Film
VARIETY office: London

Norwegian Film Festival
August 18-25, 1990
P.O. Box 145, 5501 Haugesund, Norway
Tel 28422
Director: Gunnar Johan Lovvik
Film Competitive
VARIETY office: London

Norwegian Short Film Festival
June
Storengveien 8 b, N-1342 Jar, Norway
Film (shorts)/Competitive
VARIETY office: London

POLAND

Cracow Intl. Festival of Short Films
June
P.O. Box 127, 00-950 Warsaw pl. Zwyciestwa 9, Poland
Telex 813640
Director: Olgierd Sziapzynski
Film
VARIETY office: London

Intl. Scientific Film Festival
November (Alt. years)
Osrodek Postepu Technicanego, ul.M.Buczka 1b skr. poczt 454, 40-955 Katowice 2, Poland
Telex 0312458
Director: Jozef Zyla
VARIETY office: London

Polish Film Festival
September
Piwna 22, P.O. Box Nr. 192, 80-831 Gdansk, Poland
Telex 0512153
Film
VARIETY office: London

PORTUGAL

Cinanima (Intl. Animation Film Festival)
November
Apartado 43, 4501 Espinho Codex, Portugal
Tel 724611/721621
Director: Antonio Ferreira Gaio
Animation
VARIETY office: Madrid

Figueira da Foz
September
Rua Luis de Cameos 106, 2600 Vila Franca de Zira, Portugal
Director: José Vieira Marques
Film/Competitive
VARIETY office: Madrid

Oporto Film Festival (Fantasporto)
February 1-13, 1990
Rua Diogo Brandão 87, 4000 Oporto, Portugal
Tel 320759
Telex 223679
Fax 383679
Director: Mario Dorminsky
VARIETY office: Madrid

Troia Intl. Film Festival
June 8-17, 1990
2902 Setúbal Codex, Troia, Portugal
Tel 44121/44124
Telex 18138 TROIAM P
Fax 44162
Director: Salvato Telles de Menezes
Film/tv/video market & festival
VARIETY office: Madrid

SINGAPORE

Singapore Intl. Film Festival
January (Biennial)
11 Keppel Hill, Singapore 0409
Telex 38283
Fax 2722069
Film/Competitive
VARIETY office: Chicago

SPAIN

Atlantic Film Festival
September
Gran Via 43, 28013 Madrid, Spain
Fax 542-6974
Director: José Maria Marchante
VARIETY office: Madrid

Barcelona Film Festival
Passeig de Garcia 47 3er. 2a, 08007 Barcelona, Spain
Tel 215-2424
Telex 99373
VARIETY office: Madrid

Bilbao Intl. Festival of Documentary & Short Films
November
Colon de Larreategui, P.O. Box 579, 37-4 Dcha., 48009 Bilbao, Spain
Tel 424-8698/5507
Telex 31013
Fax 4423045
Director: Luis Iturri
Film
VARIETY office: Madrid

Festival of Cinema
July
Conde de Lemos 4, lo 28013 Madrid, Spain
VARIETY office: Madrid

Gijon Intl. Film Fest. For Children & Young People
June
Emilio Villa 4, 1 Gijon, Spain or Cerinterfilm, Paseo de Begona, 24 Gijon, Asturias, Spain
Telex 87443/87301
Fax 351724
Director: Juan José Plans
Film
VARIETY office: Madrid

Iberoamerican Film Festival
November
Hotel Tartassos, 21003 Huelva, Spain
Tel 55-245611
Telex 75625
Director: José Luis Ruiz Diaz
VARIETY office: Madrid

Ibervideo
February
Av. Portugal s/n, Apt. 11.011, 28080 Madrid, Spain
Telex 44025-41674
Director: Antonio Peña
VARIETY office: Madrid

Intl. Contest of Agrarian Cinema
April (Alt. years)
Palacio Ferial, P.O. Box 108, 50080 Zaragoza, Spain
Telex 58185
Fax 330649
Director: Eduardo Cativiela Lacasa
VARIETY office: Madrid

Intl. Festival of Amateur Films
October (Alt. years)
P.O. Box 378, Igualda, Barcelona, Spain
Telex 52038
Fax 804-4362
VARIETY office: Madrid

Madrid (Imagfic)
March 30-April 7, 1990
Gran Via 62-8, Madrid 28013 Spain
Tel 2413721/2415545
Telex 42710
Fax 542-5495
Director: Rita Sonlleva
Film
VARIETY office: Madrid

San Sebastian Intl. Film Festival
September
Apartado Correos 397, P.O. Box 397, 20080 San Sebastian, Spain
Tel 429625
Telex 38145 FCSS-E
Fax 285979
Director: Pilar Olascoaga
Film/Competitive
VARIETY office: Madrid

Semanautor Film Festival
April
P.O. Box 24, 29080 Malaga, Spain
Telex 77077
Director: Julio Diamante
VARIETY office: Madrid

Sitges Intl. Festival of Fantastic & Horror Cinema
October
Rambla de Cataluña 81, P.O. Box 93, Barcelona, Spain
Telex 51388
Director: J.L. Goas
Film/Market
VARIETY office: Madrid

Valencia Mediterranean Cinema Exhibition
October
Mostra de Valencia, P. del Arzobispo, 2 acc, 46003 Valencia, Spain
Telex 63427 MVME E
Director: José Luis Forteza
Film/Competitive/Market
VARIETY office: Madrid

Valladolid Intl. Film Festival
October 19-27, 1990
Teatro Calderón, c/o Angustias 1, 1 Apartado Correos P.O. Box 646, 47080 Valladolid, Spain
Tel 30-5700/77
Telex 26304
Fax 30-9835
Director: Fernando Lara
Film/Competitive
VARIETY office: Madrid

SWEDEN

Film Festival of Malmö
September 7-15, 1990
Ostra Ronneholmsvagen 4,21147 Malmö, Sweden
Tel 11-80-16
Telex 32158 PUBTM-S
Film
VARIETY office: London

Gothenburg Film Festival
January 26-February 4, 1990
Box 7079, 40232 Gothenburg, Sweden
Tel 410546/47
Telex 28674 FIFEST S
Fax 31-410063/2
Director: Gunnar Carlsson
Film
VARIETY office: London

Kaleidoscope
April
P.O. Box 2260, S-10316 Stockholm, Sweden
Film
VARIETY office: London

Umea Film Festival
September
Box 42, S-90102 Umea Sweden
Tel 140150
Telex 54084
Fax 132791
Film
VARIETY office: London

Uppsala Film Festival
October 19-28, 1990
P.O. Box 1746, S-75147 Uppsala, Sweden
Tel 162270
Telex 76020
Fax 101510
Film/Competitive
VARIETY office: London

SWITZERLAND

Electronic Cinema Festival
June
P.O. Box 97, CH-1820 Montreux, Switzerland
Telex 453283
Fax 963-7895
Director: R. Jaussi
VARIETY office: Paris

Festival Intl. du Film Nature
November
WWF-Geneve CP28, CH-1212 Grand-Lancy, Switzerland
Film (16m & Super 8)
VARIETY office: Paris

Festival of Third World Cinema
January
Rue De L'Industrie 8, 1700 Fribourg, Switzerland
Telex 942005
VARIETY office: Paris

Locarno Intl. Film Festival
August 2-12, 1990
Via della Posta 6, CH-6600 Locarno, Switzerland or Casella Postale, CH-6600 Locarno, Switzerland
Tel 310232
Telex 846565
Fax 317465
Directors: David Streiff, Gian Carlo Bertelli
Film/Competitive
VARIETY office: Paris

Montreux Golden Rose
May 9-15, 1990
Swiss Broadcasting Corp., Giacomettistrasse 3, CH-3000 Berne 15, Switzerland
Tel 439111
Telex 911534/33161
Fax 31-439474
Director: Jean-Luc Balmer
VARIETY office: Paris

Nyon Festival Intl. Du Film Documentaire
October
P.O. Box 98 CH-1260 Nyon, Switzerland
Telex 28163
Director: Erika de Hadeln
Film (Docu)/Competitive
VARIETY office: Paris

Stars de Demain
June
P.O. Box 418, CH-1211 Geneva, Switzerland
Tel 215466
Fax 219862
Director: Beki Probst
Film/Competitive
VARIETY office: Paris

Vevey Intl. Comedy Film Festival
August
Tel 921-4825
Telex 451143
Fax 731-0429
Director: Jean-Pierre Grey
Film (Comedy)
VARIETY office: Paris

TAIWAN

Taipei Intl. Film Exhibition
October
Film Library, 4/F, 7 Ch'ingato East Road, Taipei, Taiwan
Film
VARIETY office: Chicago

TURKEY

Istanbul Film Festival
March 31-April 15, 1990
Yildiz Kultur Ve N Sanat Meerkezi, 80700 Yidiz-Besiktas, Istanbul, Turkey
Tel 160-3533
Telex 26687 IKSU TR
Fax 161-8823
Director: Hulya Ucansu
Film/Competitive
VARIETY office: Rome

U.K.

Birmingham Film & Television Festival
October
Dept. of Recreation & Community Serivces, Auchinleck House, Five Ways, Edgbaston, Birmingham B15 1DS, England
Tel 440-2543
Fax 440-4372
Film & television
VARIETY office: London

Bristol Animation Festival
October
41B Hornsey Lane Gardens, London N6 5NY, England
Animation
VARIETY office: London

Bristol Film & Television Festival
September
61 Queen Charlotte Street, Bristol BS1 5TX, England
Tel 264339
Director: Richard Pugh
Film & Television
VARIETY office: London

British Industrial and Sponsored Film Festival
June
BISFA, 26 D'Arblay Street, London W1V 3FH, England
Film
VARIETY office: London

Cable & Satellite
March
Montbuild Ltd., 11 Manchester Square, London W1M 5AB, England
Telex 24591
Fax 486-8773
Director: Alison Black
VARIETY office: London

Cambridge Film Festival
July
P.O. Box 17, Cambridge CB2 3PF, England or 8 Market Passage, Cambridge CB2 3PF, England
Tel 462666
Telex 81574
Fax 46255
Director: Tony Jones
Film/Non-competitive
VARIETY office: London

IBC 90 (Intl. Broadcasting Convention)
September (Alt. years)
The Institute of Electrical Engineers, Savoy Place, London WC2R OBL, England
Telex 261176
Compeititive
VARIETY office: London

Intl. Edinburgh Film Festival
August
The Film House, 88 Lothian Road, Edinburgh EH3 9BZ, Scotland
Tel 228-4051
Telex 72166
Fax 229-5501
Director: David Robinson
Film
VARIETY office: London

Piccadilly Film and Video Festival
June
177 Piccadilly, London W1V 9LF, England
Film
VARIETY office: London

The London Film Festival
November
National Film Theatre, South Bank, London SE1, England
Tel 928-3535
Telex 929220
Fax 633-9323
Director: Sheila Whitaker
Film
VARIETY office: London

Tyneside Film Festival
October
10-12 Pilgrim Street, Newcastle upon Tyne NE1 6QK, England
Tel 232-8289
Fax 221-0535
Director: Peter Packer
Film
VARIETY office: London

U.S.

American Film & Video Festival
June
920 Barnsdale Road, LaGrange Park, Ill., 60625, U.S.
Docu and Animated films
VARIETY offices: New York, Los Angeles

American Film & Video Festival
May
920 Barnsdale Road, Suite 152, LaGrange Park, Ill., 60625 U.S.
Director: Kathryn Lamont
Film, Video
VARIETY offices: New York, Los Angeles

American Film Institute Film Festival
April
2021 N. Western Avenue, Los Angeles, Calif., 90027, U.S.
Telex 9013339625
Director: Ken Wlaschin
Film
VARIETY offices: New York, Los Angeles

American Film Institute Video Festival
October
2021 N. Western Avenue, Los Angeles, Calif., 90027, U.S.
Tel (213) 856-7787
Telex 9103339625
Fax (213) 467-4578
Director: Kenneth Kirby
Video
VARIETY offices: New York, Los Angeles

American Film Market
February 22-March 2, 1990
12424 Wilshire Boulevard, Suite 3600, Los Angeles, Calif., 90025, U.S.
Tel (213) 275-3400
Telex 194676
Fax (213) 447-1666
Director: Tim Kittleson
Film/Market
VARIETY offices: New York, Los Angeles

American Independent Feature Film Market
October
21 West 86th Street New York, N.Y., 10024, U.S.
Tel (212) 496-0909
Telex 238790NYK
Fax (800) 248-9630
Director: Karen Arikian
Film (Indie)
VARIETY offices: New York, Los Angeles

Annual Convention of the National Assn. of Broadcasters
April
1771 N Street N.W., Washington, D.C., 20036, U.S.
Telex: 350085
Director: Henry J. Roeder
VARIETY offices: New York, Los Angeles

Baltimore Intl. Film Festival
March
c/o Museum of Art, Baltimore, Md., 21218, U.S.
Director: Victoria Westcover
VARIETY offices: New York, Los Angeles

BPME Gold Medallion Awards
June
6255 Sunset Boulevard, Suite 624, Los Angeles, Calif., 90028, U.S.
Tel (213) 469-9559
Director: Jay Curtis
VARIETY offices: New York, Los Angeles

Chicago Intl. Film Festival
October
415 N. Dearborn Street, Chicago, Ill., 60610, U.S.
Telex 936086
Fax (312) 644-0784
Director: Michael J. Kutza
Film/Competitive
VARIETY offices: New York, Los Angeles

Cine San Juan
October
Apartado 4543, San Juan, P.R., 00905, U.S.
Film, Video
VARIETY offices: New York, Los Angeles

Cinetex
September
16055 Ventura Boulevard, Suite 432, Encino, Calif., 91436, U.S.
Tel (818) 907-9021
Director: Sheldon Adelson
Film, television/Market
VARIETY offices: New York, Los Angeles

Cleveland Intl. Film Festival
April
6200 SOM Center Road, C20, Cleveland, Ohio, 44139, U.S.
Telex 980131
Fax (216) 349-0210
Director: Jonathan R. Forman
Film
VARIETY offices: New York, Los Angeles

Denver Intl. Film Festival
October 11-18, 1990
P.O. Box 17508, Denver, Colo., 80217, U.S. or 999 18th Street, Suite 247, Denver, Colo., 80202, U.S.
Tel (303) 298-8223
Telex 710-1111-406
Fax (303) 297-8326
Director: Ron Henderson
Film
VARIETY offices: New York, Los Angeles

Emmy Awards
September
Academy of
Television Arts & Sciences, 3500 W. Olive Ave., Burbank, Calif., 91505, U.S.
Director: Doug Duitsman
Television
VARIETY offices: New York, Los Angeles

Filmfest DC
April
P.O. Box 21396, Washington D.C., 20009, U.S.
Tel (202) 727-2396
Telex 440732
Fax (202) 347-7342
Director: Marcia Talbowitz
Film/Non-competitive
VARIETY offices: New York, Los Angeles

Hawaii Intl. Film Festival
November
1777 East-West Road Honolulu, Hawaii, 96848, U.S.
Tel (808) 944-7666
Telex 989171
Fax (808) 944-7970
Director: Jeannette Paulson
Film
VARIETY offices: New York, Los Angeles

HEMIS
Film
January (Alt. years)
Intl. Fine Arts Center of the Southwest, 1 Camino Santa Maria, San Antonio, Texas, 78284, U.S.
Director: Louis Reile
VARIETY offices: New York, Los Angeles

Houston Intl. Film Festival
April 20-29, 1990
P.O. Box 56566, Houston, Texas, 77256, U.S.
Tel (713) 965-9955
Telex 317876
Fax (713) 965-9960
Director: J. Hunter Todd
Film/Competitive
VARIETY offices: New York, Los Angeles

Humboldt Film Festival
May
Humboldt State University Arcata, Calif., 95521,, U.S.
Film (Student/Indie -60mins & 16m)
VARIETY offices: New York, Los Angeles

Intl. Film & TV Festival of New York
November
5 West 37th Street, New York, N.Y., 10018, U.S.
Tel (914) 238-4481
Fax (914) 238-5040
Director: Sandy Mandelberger
Film, Video, TV/Competitive
VARIETY offices: New York, Los Angeles

Intl. Tournee of Animation
(touring)
2222 S. Barrington Avenue, Los Angeles, Calif., 90064, U.S. Tel (213) 473-6701
Telex 247770
Fax (213) 444-9850
Director: Terry Thoren
Animation
VARIETY offices: New York, Los Angeles

Los Angeles European Community Film Festival
June
P.O. Box 27999, 2021 North Western Avenue, Los Angeles, Calif., 90027, U.S.
Tel (213) 856-7707
Fax (213) 462-4049
Director: Ken Wlaschin
Film
VARIETY offices: New York, Los Angeles

Los Angeles Intl. Animation Celebration
August 23-30, 1990
Expanded Entertainment, 2222 S. Barrington Avenue, Los Angeles, Calif., 90064, U.S. 145 S. Fair
Fax Avenue, Suite 303 Los Angeles, Calif., 90036, U.S.
Tel (213) 473-6701
Telex 247770
Fax (213) 444-9850
Director: Terry Thoren
Animation/Competitive
VARIETY offices: New York, Los Angeles

Marin County Film Festival
June
Marin County Fair & Exposition Fairgrounds, San Rafael, Calif., 94903, U.S.
Director: Ronald Levaco
VARIETY offices: New York, Los Angeles

Miami Film Festival
February 2-11, 1990
Film Society of Miami, 7600 Red Road, Suite 307 Miami, Fla., 33143, U.S. or 444 Brickell Avenue, Suite 229 Miami, Fla., 33131, U.S.
Tel (305) 377-3456
Telex 264047
Fax (305) 577-9768
Director: Nat Chediak
Film/Non-competitive
VARIETY offices: New York, Los Angeles

Mill Valley Film Festival
October 4-11, 1990
80 Lomita Drive, #20, Mill Valley, Calif., 94941, U.S.
Tel (415) 383-5256
Telex 282696
Fax (415) 383-8606
Film/Non-competitive
VARIETY offices: New York, Los Angeles

Mobius Intl. Broadcasting Awards
February 2, 1990
841 N. Addison Avenue, Elmhurst, Ill., 60126-1291, U.S.
Tel (708) 834-7773
Fax (708) 834-5565
Director: Patricia Meyer
TV and Radio Commercials
VARIETY offices: New York, Los Angeles

National Assoc. Of Theatre Owners (Convention)
October
1560 Broadway, Suite 714 New York, N.Y., 10036, U.S.
Director: Joseph G. Alterman
VARIETY offices: New York, Los Angeles

NATPE (Natl. Assoc. of TV Prog. Exec)
January 16-19, 1990
342 Madison Avenue, New York, N.Y., 10173,, U.S.
Telex 176674
Director: Philip A. Corvo
Television/Market
VARIETY offices: New York, Los Angeles

New York Film Festival
September
Film Society of Lincoln Center, 140 West 65th St., New York, N.Y., 10023, U.S.
Director: Richard Peña
Film
VARIETY offices: New York, Los Angeles

New York Intl. Home Video Market
April
Knowledge Industry Publications Inc. 701 Westchester Avenue, White Plains, N.Y., 10604, U.S.
Homevideo/Market
VARIETY offices: New York, Los Angeles

New York World Television Festival
January
1 East 53rd Street, New York, N.Y., 10022, U.S.
Telex 425560 Jeffrey Fuerst
Television
VARIETY offices: New York, Los Angeles

Palm Springs Intl. Film Festival
January 10-14, 1990
401 South Pavilion Way, P.O. Box 1786 Palm Springs, Calif., 92263, U.S.
Tel (619) 322-8389
Fax (619) 320-9834
Director: Jeannette Paulson
Film (Comedy)
VARIETY offices: New York, Los Angeles

Philadelphia Intl. Film Festival (Philafilm)
July
121 N. Broad Street, Suite 618 Philadelphia, Pa., 19101, U.S.
Tel (215) 977-2831
Director: Lawrence L. Smallwood Jr.
VARIETY offices: New York, Los Angeles

Rivertown (Mpls./St. Paul) Intl. Film Festival
April
University Film Society, Minnesota Film Center, 425 Ontario Street SE, Minneapolis, Minn., 55414, U.S.
Tel (612) 627-4431
Director: Al Milgrom
Film
VARIETY offices: New York, Los Angeles

San Francisco Intl. Film Festival
April 30-May13, 1990
San Francisco Film Society, 1560 Fillmore Street, San Francisco, Calif., 94115, U.S.
Tel (415) 567-4641
Telex 65028 MCI UN
Fax (415) 921-5032
Director: Peter Scarlet
Film, Television, Docu, Short/Competitive
VARIETY offices: New York, Los Angeles

Santa Barbara Intl. Film Festival
March 2-11, 1990
1216 State Street, Suite 200 Santa Barbara, Calif., 93101, U.S.
Directors: Phyllis de Picciotto, Janet Doran-Veevers
VARIETY offices: New York, Los Angeles

Seattle Intl. Film Festival
May
801 East Pine Street Seattle, Wash., 98122, U.S.
Telex 329473
Director: Darryl Macdonald
Film & Animation
VARIETY offices: New York, Los Angeles

Showbiz Expo
June 1-3, 1990
2122 Hillhurst Avenue, Los Angeles, Calif., 90027, U.S.
Tel (213) 668-1811
Fax (213) 668-1033
VARIETY offices: New York, Los Angeles

SMPTE Technical Conf. & Equip. Exhibit
October
595 W. Hartsdale Avenue, White Plains, N.Y., 10607, U.S.
Telex 4995348
Director: Lynnette Robinson
VARIETY offices: New York, Los Angeles

Telluride Film Festival (Colorado)
August
The National Film Preserve, P.O. Box B1156 Hanover, N.H., 03755, U.S.
Tel (603) 643-1255
Directors: Bill & Stella Pence and Tom Luddy
Film
VARIETY offices: New York, Los Angeles

The Intl. Emmy Awards
November
NATAS, 142 W. 57th Street, 16th Fl., New York, N.Y., 10019, U.S.
Tel (212) 489-6969
Telex 175766
Fax (212) 489-6557
Director: Richard Carlton
VARIETY offices: New York, Los Angeles

U.S. Industrial Film & Video Festival
June 8, 1990
841 N. Addison Avenue, Elmhurst, Ill., 60126-1291, U.S.
Tel (708) 834-7773
Fax (708) 834-5565
Director: Patricia Meyer
Sponsored, industrial film and video
VARIETY offices: New York, Los Angeles

United States Film Festival
January
Sundance Institute, 19 Exchange Place, Salt Lake City, Utah, 84111, U.S.
Director: Debbie Snider
Film
VARIETY offices: New York, Los Angeles

USA Film Festival
March
P.O. Box 3105 Dallas, Texas, 75275, U.S.
Film
VARIETY offices: New York, Los Angeles

Wine Country Film Festival
July
P.O. Box 303 Glen Ellen, Calif., 95442, U.S.
Film, Shorts, Docu (indie only)
VARIETY offices: New York, Los Angeles

Women in Film Festival
October
6464 Sunset Boulevard, Suite 660, Los Angeles, Calif., 90028, U.S.
Tel (213) 463-0931
Director: Linda Wadley
Film,
Television
VARIETY offices: New York, Los Angeles

USSR

Intl. Fest. of Documentary & Science Film
January
Sovinfest, 10 Khokhlovsky Pereulok, 109028 Moscow, USSR.
Tel 297-7645
Telex 411263 FEST SU
Director: Yuri Khodjaev
VARIETY office: London

Intl. Film Fest. of Asian, African & Latin American Countries
May 1990 (Alt. years)
10 Khoklovsky Pereulok, 109028 Moscow
Tel 297-7645
Telex 411263
Director: A. A. Turaev
VARIETY office: London

Intl. Golden Fleece
Television Film Festival
c/o Soviet
Television, 12 Korolyov St. 127000 Moscow, USSR.
Telex 212183/411140
Directors: David Shalikashivili, Valentin Lazutkin
VARIETY office: London

Intl. Teleforum of Intervision
September
12 Korolyov St. 127000 Moscow, USSR.
Telex 411140
Director: Evgeney Oksyukevich
VARIETY office: London

Moscow Intl. Film Fest. of Children & Youth Films
March
Sovinfest, 10 Khokhlowsky Pereulok, 109028 Moscow, USSR.
Tel 297-7645
Telex 411263
Director: Yuri Khodjaev
VARIETY office: London

Moscow Intl. Film Festival
July
1991 (Alt. years)
Sovinfest, 10 Khokhlovsky Pereulok, 109028 Moscow, USSR.
Tel 297-7645
Telex 411263 FEST SU
Director: Yuri Khodjaev
Film
VARIETY office: London

Tashkent Intl. Film Festival
May
1990 10 Khokhlovsky Pereulok 109028 Moscow, USSR.
Tel 297-7645
Telex 411263 FEST SU
VARIETY office: London

WEST GERMANY

Berlin Intl. Agricultural Film & TV Competition
January 29-February 1, 1990
(Alt. years) AMK Berlin, Postfach 191740 D-1000, West Berlin 19, West Germany
Telex 182908
Fax 38-2325
Director: Karin von Bargen
Film, Television/Competitive
VARIETY office: London

Berlin Intl. Film Festival
February 9-20, 1990
Budapester Str. 50, D-1000 West Berlin 30, West Germany Tel 254890
Telex 185255
Fax 25-48-92-49
Director: Moritz de Hadeln
Film/Competitive
VARIETY office: London

Cologne Photokina (World's Fair of Image)
October 3-9, 1990 (Alt. years)
Messe-und Ausstellungs-Ges.m.b.H Cologne, P.O. Box 21076, D-5000 Cologne 21, West Germany
Telex 8873426
Director: Hans Wilke
VARIETY office: London

European Film Awards
Permanent Secretariat, Lietzenburger Str. 44, D-1000 Berlin 30, West Germany
Tel 261-1888/89
Telex 186888 EFP D
Fax 243545
Directors: Marion Doring, Aina Bellis
Film
VARIETY office: London

Internationaler Medienmarkt München
November 14-17, 1990
Internationaler Münchner Filmwochen, Turkenstrasse 93, D-8000 Munich, West Germany
Tel 381904-0
Telex 5214674 IMF D
Fax 38190426
Market

VARIETY office: London

Internationaler Filmwochenende Wurzburg
January
Filminitiative Wurzburg, Gosbersteige 2, D-8700 Wurzburg, West Germany
Telex 680070
Film/Non-competitive
VARIETY office: London

Internationaler Festival Der Filmhochschulen München
November 4-11, 1990
Internationaler Münchner Filmwochen, Turkenstrasse 93, D-8000 Munich, West Germany
Tel 381904-0
Telex 5214674 IMF D
Fax 38190426
Film (Students)/Competitive
VARIETY office: London

Hof Intl. Filmdays
October (Alt. years)
Lothstr. 28, D-8000 Munich 2, West Germany or Postfach 1146, D-8670, Hof, West Germany
Tel 1297422
Telex 5215637
Director: Heinz Badewitz
Film
VARIETY office: London

Lubeck Nordische Filmtage
November
Postfach 1889, D-2400, Lubeck, West Germany
Tel 1224105
Fax 1221331
Film (Scandi)
VARIETY office: London

Mannheim Intl. Film Week
October 8-13, 1990
Collini-Center-Galerie, 6800 Mannheim 1, West Germany 463423
Director: Fee Vaillant, Lutz Beutling
Film/Competitive
VARIETY office: London

München Filmfest & Film Exchange
June 23- July 1, 1990
Internationaler Münchner Filmwochen, Turkenstrasse 93, D-8000 München 40, West Germany
Tel 381904-0
Telex 5214674 IMF D
Fax 38190426
Director: Eberhard Hauff
Film
VARIETY office: London

Oberhausen Intl. Short Film Festival
April
Grillostrasse 24, D-4200 Oberhausen, West Germany or Westdeutsche Kurzfilmtage, Christian-Steger-Strasse 10, D-4200 Oberhausen 1, West Germany
Tel 8252652
Telex 856414
Fax 82528159
Director: Karola Gramann
Film (Shorts)/Competitive
VARIETY office: London

Prix Futura Berlin Competition
March (Alt. years)
Sender Freies Berlin, Masurenallee 8-14, D-1000 Berlin 19, West Germany
Telex 182813
Director: Peter Leonhard Braun
VARIETY office: London

Prix Jeunesse Intl.
May (Alt. years)
Bayerischer Rundfunk, Rundfunkplatz 1, D-8000 Munich 2, West Germany
Telex 521070
Fax 59003053
Director: Dr. Ernst Emrich
VARIETY office: London

YUGOSLAVIA

Belgrade Intl. Film Festival
January 26 -February 3, 1990
Sava Centar, Milentija Popovica 9, 11070 Belgrade, Yugoslavia
Tel 222-4961
Telex 11811
Fax 222-1156
Directors: Vojislav Kostic, Nevena Djonlic
Film
VARIETY office: London

Pula Festival
July
Festival Jugoslavenskog Igranog Filma, Marka Laginje 5, 52000 Pula, Yugoslavia
Film
VARIETY office: London

Zagreb World Festival of Animated Films
June 4-8, 1990 (Alt. years)
Nova ves 18 41000 Zagreb, Yugoslavia
Tel 271-355
Telex 21790
Fax 275-994
Director: Borivoj Dovnikovic
Animation/Competitive/Market
VARIETY office: London

Homevideo

Homevideo use, both at home and abroad, continued to enjoy its rising popularity in 1989–1990. VCR purchases as well as tape sales and rentals skyrocketed during the year, revealing a definite change in leisure time pursuits.

Technologically, improvements were made in both equipment and tape quality. Also, a new video ratings handbook established more stringent regulations upon homevideo promoters.

World-Wide VCR Sales Rising at Fast Rate

Variety 6/14/89

London The world's total number of video households is expected to reach 187.32-million by the end of this year, repping a global penetration rate of ⅓ of all tv homes. According to London-based media research journal Screen Digest, sales of VCRs this year will reach 38.7-million units (of which 14-million will be sold to homes already owning video equipment), an increase of 8.7% over last year.

Report predicts that 60.2-million households in the U.S. will have VCRs by the end of the year, repping a 67.4% penetration rate of tv homes. Japan is expected to have 24.1-million homes with VCRs, repping a 79% penetration rate, while the U.K. will have 13.3 VCR households, or 69.5% of tv homes.

Indies Seeking Alternative to MPAA Ratings System

6/14/89

By TOM BIERBAUM

Hollywood Some key video independents, upset about Motion Picture Assn. of America regulation on videocassette ratings, are considering options to the ratings, per one indie spokesman.

Tom Burnett, exec v.p. at Virgin Vision, said these options were likely discussed at the June 13 meeting of the Video Software Dealers Assn.'s Manufacturers Advisory Council meeting.

Virgin and most of the other protesting indies have been submitting their releases to the MPAA for rating. However, Burnett said they are considering issuing product unrated; launching an alternative system or lobbying the VSDA to rep the vid industry's viewpoint more aggressively or to get into the ratings system itself.

That group of vid indies has complained that the MPAA is overly restrictive on vid packaging and ad materials (which the association requires be submitted for approval on all rated films), is applying tougher standards to vid-only releases than to theatrical releases and is requiring that the theatrical rating be rescinded for failure to comply with the regulations in the vid market.

The 2d meeting

Some vid execs were particularly steamed about the MPAA's decision to republish the regs, with only minor modifications, without holding a second meeting to discuss the issues with the indies.

The MPAA met with the independents in January. According to some in attendance, the MPAA said it would meet next with retailers, then with the indies again (along with reps from the studio-aligned program suppliers).

The retailer meeting was held in late April, and the slightly revised regs went to press shortly afterward.

A number of indie vid execs say flatly that they were promised a second meting.

Alvin Reuben, Vestron Video's sales-marketing-distribution senior veep, and Paul Culberg, New World Video prez, have sent letters to the MPAA saying they feel

Despite the Proliferation of VCRs, TV Viewing on Increase in Japan

Variety 6/28/89

By JAMES BAILEY

Tokyo With video players in nearly two-thirds of Japanese households, cassette rentals a $2.1-billion-a-year business and homevideo games selling like rice cakes, it might be assumed that the average Japanese viewer doesn't have much time for commercial tv programming.

Wrong.

According to a White Paper released by Video Research, the Tokyo-based ratings firm, the average household in the Kanto areas (Tokyo-Yokohama area) actually spent two more minutes in front of the tube last year than the year previous: seven hours and 56 minutes between 6 a.m. and midnight.

In the average household in the Kansai area (Osaka-Kyoto area), the increase was 12 minutes, to seven hours and 33 minutes. In the Nagoya area, the average household viewing rate went up two minutes, to seven hours and 41 minutes.

Some general trends:

- With more women in the work force, the percentage of total viewers tuning in to traditionally female-oriented noontime programs dropped from 20% to 17.3% from 1984-88.
- Since more businesses are operating 24 hours a day, more Japanese can stay up watching late-night programming,

With the exception of 4-12-year-olds and males 35-49, all of the eight groups surveyed by Video Research watched more late night programming last year than in 1984 with males 20-34 increasing from 12% to 17%.

On viewers' in list: documentaries and educational programming. On their out list: music, dramas, thrillers, actioners and made-for-tv movies. Most popular type of programming was informational, with viewers watching an average of 57 minutes of it a day.

Trying to give the folks more of what they want, the networks increased their total average daily informational offerings from 499 minutes in 1987 to 708 minutes last year.

the association had committed itself to holding a second meeting. The letters ask the MPAA to explain why the meeting wasn't held.

Pam Horovitz, exec director of the VSDA, stopped short of saying the meeting had been promised but acknowledged that the vid indies had made clear their desire for a second meeting.

But at the MPAA, v.p. of administration director Bethlyn Hand said her staff doesn't recall agreeing to such a meeting and doesn't see the point in it, since the issues have been thoroughly discussed. Hand reiterated her willingness to meet with indies again.

The VSDA's Horovitz said the retailer group has been encouraging its members to support the MPAA system under the assumption that the more vidcassettes are rated, the less pressure will come from parents to legislate vid content regulations.

She does agree with indies that say differences between the video and theatrical markets may require different applications of MPAA standards. Horovitz expressed the hope that the MPAA will continue to keep vid reps involved in the process of determining how to apply its ratings to vidcassette releases.

Virgin's Burnett is the most vocal critic of the MPAA's handling of vid ratings. "We've tried to give them the benefit of the doubt," he said. "But there doesn't seem to be a spirit of cooperation on their part."

Consumer Passion for Renting Vidtapes Has Abated Little, New Survey Says

Variety 8/16/89

By TOM BIERBAUM

Las Vegas Alternate leisure activities — not competing entertainment technologies — lead some consumers to cut back on vid rentals, according to a new study.

That was one of the key findings of a Nielsen Media Research survey presented here last week at the Video Software Dealers Assn. convention by Michael Shalett of Street Pulse Group.

However, the report also shows that most consumers continue to rent vidcassettes as often as they did earlier this year.

The survey also showed a surprising 66% of VCR owners skipped on-cassette commercials. Most earlier surveys that dealt with on-cassette ads — especially ones specially prepared for video — showed far lower rates of "zapping."

This latest Nielsen info could undermine the value of vid ads in the eyes of advertisers, who had expressed concern over zapping before data began to suggest it was a minor phenomenon.

Nielsen conducted the bulk of the survey through the national telephone poll in early June. Shalett presented the data and challenged the VSDA audience to guess the results of the survey questions through computer-keypad responses.

Big underestimation

The most surprising miscalculation, particularly by the audience members from the retailer segment, came from a question on how many VCR owners had purchased a prerecorded tape in the past year. The majority guessed 20-30%, but the answer was 63%.

"The suppliers may be right that there's a purchase market you may be underestimating," Shalett said.

The question about rental activity revealed that 22% of VCR owners said they were renting fewer tapes in early June than they had 3-6 months earlier.

But of those who had decreased activity, the most common reason was other leisure-time activities, mentioned in 70% of the cases. The next-closest reason was that VCR owners were spending more time recording off broadcast or cable, which was mentioned by 30% of the respondents. Business and education activities are cited in 24% of the cases, cable and pay-tv viewing in 22% of the cases and a preference to own tapes in 20% of the cases.

Shalett stressed that the audience, predominantly composed of retailers, expected pay-tv subscribers to cut back on rentals more than basic-cable subs and non-subs. But in fact pay subs came out with the smallest percentage of those nonrenters (29%). Basic subs accounted for 57% and those without cable made up the other 43%. Those percentages roughly parallel sub rates nationally for those three groups.

Other survey findings:

- Renting at least once a month were 61% of the VCR homes.
- Renting once a week were 16% of those homes.
- Most popular class of sell-through title was "major pictures" (39%), followed by kidvid (33%). No other category came close to those percentages.
- Consumers say they're paying mostly in the $20-29.99 range for their sell-through tapes, but they're paying more often in the $10-19.99 range.
- Despite the studios' depth-of-copy campaigns, consumers overwhelmingly say they choose their vidstores because of location (the most important consideration for 37% of renters), followed not by availability of titles desired (17%) but by the selection featured in the stores' inventory (20%).
- Of the renters who choose their titles in the stores, the factors that determine their selections are most often noticing an interesting title (29%) or "just browsing" (26%). The combination of "store display" (12%), in-store promo (4%), word of mouth (20%) and recommendation by a store worker (4%) suggest that the combined force of these more active factors can make a difference, per Shalett.
- Female renters are considerably more likely than male renters to leave the store with more than one tape (69% vs. 50%).
- While most of the respondents said they don't sit through the on-tape commercials, 62% of them said they do watch the previews included at the start of many tapes.
- Average number of VCRs in a current VCR home is 1.23.
- Getting by far the highest percentage of "high-value" responses out of surveyed consumers is "blank tape" (44%) and rental tapes (41%). Lagging far behind were pay-tv (10%), purchased tapes (9%), theatrical films (3%) and pay-per-view (2%).
- Nielsen found that 65% of the time renters did not find in stock all the titles they desired to rent. Of those, 14% requested reservations and 98% of those making reservations ultimately rented the title.

Nets Still Dominate VCR Recording, Per Nielsen

Variety 7/5/89

Hollywood Nielsen Media Research says data for the 1988 calendar year show network affiliates maintained their dominant share of VCR recording activity, while indie-tv stations showed the most growth and cable-originated services the biggest dropoff.

The network stations once again garnered 67% of recording, while the indie stations hit 16% (up from 13% in '87), pay-cable services accounted for 9% (up from 8%), cable-originated services slipped

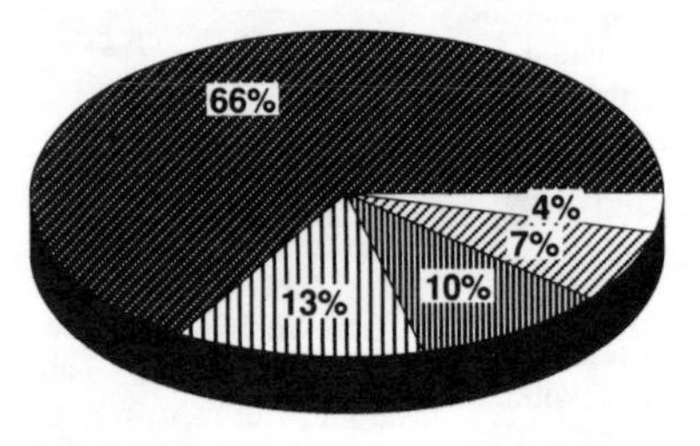

Source: Nielsen Television Index, Feb. 1989

Rank	Program	Network	VCR Households
1	Lonesome Dove Part 2	CBS	2,260,000
2	Lonesome Dove Part 4	CBS	2,079,000
3	ABC Mystery Movie	ABC	994,400
4	CBS Sunday Movie	CBS	904,000
5	Days of Our Lives	NBC	904,000
6	ABC Sunday Night Movie	ABC	732,200
7	L.A. Law	NBC	732,200
8	CBS Tuesday Movie	CBS	732,200
9	NBC Sunday Night Movie	NBC	732,200
10	NBC Monday Night Movie	NBC	632,800
11	All My Children	ABC	632,800
12	The Day After	ABC	632,800

to 5% (down from 8%) and PBS finished with 3% (down from 4%).

Primetime continued to account for 41% of recording, followed by daytime (29%), latenight (7%), the post-1 a.m. overnight hours (6%), early fringe (6%), weekend afternoons and evenings (7%) and other (5%).

Recording with the set turned off accounted for 51% of VCR activity, recording while the set was on the same channel accounted for 29% and recording while the set was on a different channel settled for 20%.

Other stats in Nielsen's annual "Report On Television:" the average U.S. tv household can receive 27.7 channels; 77% of the residents of tv households are now adults over 18, compared with 67% in '70, and the average number of hours of tv usage per day for the '87-'88 season fell 10 minutes, to six hours, 55 minutes.

—Tom Bierbaum

New Vid Ratings Handbook Boasts Same Rules, More Clearly Specified

VARIETY 5/10/89

By TOM BIERBAUM

Hollywood The Motion Picture Assn. Of America is preparing a new handbook that'll outline regulations for the use of MPAA ratings on vidcassettes and the approval procedures for the vid packaging and promo materials of rated films.

The good news for video indies is that the regs will give vid companies more specific guidelines as to use of packaging and promo materials.

But the bad news, at least for a few disgruntled indies and retailers, is that the rules are essentially the same as the ones in existence.

MPAA execs huddled with the vid indies in January and with a rep of vid retailers in April to discuss concerns over the regs. According to MPAA v.p. and ad-administration director Bethlyn Hand, those sessions allowed the association and vid reps to "listen, learn and educate."

Despite some dissatisfaction about the rules, the MPAA feels the regs are appropriate and will minimize the chances of government interference in motion picture and vidcassette content. Twenty-four states are considering bills that would involve the government in the content and promotional matters addressed by the MPAA ratings, Hand noted.

Complaints about the MPAA system, as it relates to video, lately have centered on the requirement that rated pics have packaging that most parents would not object to.

Audio/Video Plus topper Lou Berg, prez of the Video Software Dealers Assn., attended the recent retailers meeting with MPAA reps. He says he came away with some "truth-in-advertising" concerns regarding the MPAA packaging regs.

Berg fears those MPAA restrictions will create a situation in which renters will be misled by mild packaging on "slice-and-dice" films and other graphic pics.

Retailers are the ones who get the heat when packaging is objectionable and when an inoffensive box misleads a renter, Berg said.

He said he'd rather see packaging that reflects the content of more graphic films, but with package placement on higher shelves, where children won't see images parents might object to.

Hand suggests that vid companies need to package and promote their films as theatrical companies always have — with creative materials that communicate the not-for-kids nature of a film without alarming artwork. She cited the promo push for "Pet Sematary" as an example of that kind of campaign.

VCR and Paycable Penetration

VARIETY 9/13/89

VCR and paycable penetration in tv households reached 65.8% and 29.4%, respectively, in July, up from 15.8% and 25.9% five years ago, according to Nielsen Media Research. Estimates are based on responses from diary questions from Nielsen's local-market service.

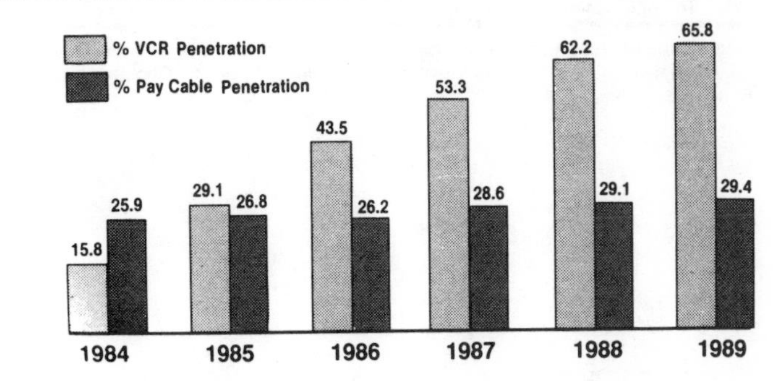

Special-Interest Vid to Blossom in 1990

VARIETY 12/20/89

New York Special-interest video will make a $95-million leap forward in 1990, but vidstores will see their market share dwindle as direct response picks up, according to a study on the nontheatrical market.

The study, released by Cambridge Associates, a Stamford, Conn., industry watchdog, predicts that the special-interest vid market will balloon from $325-million this year to $420-million in 1990. But while retail outlets are likely to see an increase in total sales in the made-for-vid area, video specialty stores will sell only one of every four nontheatrical vids sold next year. Other retail channels, meanwhile, will stay flat or show only a marginal gain next year.

The biggest jump is expected to come from the direct-response arena, which includes mailings, consumer ads with phone numbers for placing orders and cable shopping networks. Cambridge says direct response easily will command the highest special-interest market share with 34% of the volume, up from 31% this year and 26% in 1988.

"Direct response is getting the business by default," says Richard Kelly, prez of Cambridge. "When a special-interest programmer tries to get his video sold at retail he often runs up against distributors and rackjobbers who are just not interested. They often think the video is *too* specialized. For programmers, the answer becomes fairly obvious: Sell the video through methods that will reach the target audience."

Kelly says retail outlets are understandably committed to stocking vids with the widest possible appeal. "It would be crazy for them to tie up shelf space with an esoteric tape that has a very limited audience," he says.

The trend toward relying less on retail outlets to move nontheatrical product is reflected in HBO Video's plan to increasingly try reaching consumers through such channels as direct response. The company said it will focus its retail efforts only on those tapes with recognizable personalities and a broad target audience.

The Cambridge study indicates the supermarkets are the only retail outlets to see gains in market share, from 3% in 1988 to 6%.

Still, Kelly notes, only certain types of video, particularly kidvid and health and fitness titles, seem to do well in supermarkets.

Kelly says the shift in distribution for special interest is not bad news for producers, it simply means they have to be more aggressive about marketing their tape. The fact that the overall theatrical market is climbing from $210-million in retail sales in 1988 to an estimated $420-million next year should be of comfort to anyone involved in the special-interest area, he says. *—Al Stewart*

Top Video Rentals 1989

Variety 1/24/89

Rank	Title/Distrib	Release date (1st Monday)	'89 national boxoffice ($ millions)
1	**A Fish Called Wanda** (CBS/Fox)	Feb. 27	60.0
2	**Die Hard** (CBS/Fox)	Jan. 30	79.6
3	**Rain Man** (MGM/UA)	Sept. 4	168.8
4	**Coming To America** (Paramount)	March 22	128.1
5	**Big** (CBS/Fox)	March 27	113.3
6	**Twins** (MCA)	June 19	106.4
7	**The Accused** (Paramount)	March 15	29.6
8	**Beaches** (Touchstone)	Aug. 28	55.2
9	**Tequila Sunrise** (Warner)	Aug. 7	39.7
10	**Cocktail** (Touchstone)	April 24	77.1
11	**Dirty Rotten Scoundrels** (Orion)	June 26	41.4
12	**Mississippi Burning** (Orion)	July 31	34.5
13	**Bull Durham** (Orion)	Jan. 30	50.3
14	**Tucker: The Man And His Dream** (Paramount)	April 17	19.6
15	**Gorillas In The Mist** (MCA)	April 17	23.0
16	**Young Guns** (Vestron)	Jan. 9	43.4
17	**Red Heat** (IVE)	Jan. 2	35.0
18	**Dangerous Liaisons** (Warner)	July 17	32.2
19	**Working Girl** (CBS/Fox)	Oct. 9	63.7
20	**The Dead Pool** (Warner)	Jan. 30	37.8
21	**Who Framed Roger Rabbit** (Touchstone)	Oct. 16	154.1
22	**My Stepmother Is An Alien** (RCA/Columbia)	June 12	12.9
23	**The Presidio** (Paramount)	Jan. 30	20.0
24	**Betrayed** (MCA/UA)	March 27	25.7
25	**The Accidental Tourist** (Warner)	July 3	30.2
26	**Bloodsport** (Warner)	Dec. 26 '88	11.7
27	**The Naked Gun** (Paramount)	Aug. 21	78.0
28	**Married To The Mob** (Orion)	Feb. 27	21.3
29	**Midnight Run** (MCA)	Feb. 27	37.4
30	**Major League** (Paramount)	Oct. 9	49.4
31	**Child's Play** (MGM/UA)	May 1	32.8
32	**Mystic Pizza** (Virgin)	April 10	12.4
33	**Alien Nation** (CBS/Fox)	June 19	24.9
34	**Three Men And A Baby** (Touchstone)	Nov. 21 '88	167.7
35	**Bill & Ted's Excellent Adventure** (Nelson)	Sept. 4	39.5
36	**Pet Sematary** (Paramount)	Oct. 16	56.6
37	**The Best Of Eddie Murphy** (Paramount)	Aug. 28	N/A
38	**The 'Burbs** (MCA)	Aug. 14	35.2
39	**Crocodile Dundee II** (Paramount)	March 20	109.2
40	**Big Business** (Touchstone)	March 6	39.6
41	**Short Circuit 2** (RCA/Columbia)	Dec. 19 '88	20.7
42	**The Great Outdoors** (MCA)	Jan. 16	38.0
43	**The Unbearable Lightness Of Being** (Orion)	Jan. 2	10.0
44	**Bat 21** (Media)	April 10	3.8
45	**True Believer** (RCA/Columbia)	July 24	8.7
46	**Scrooged** (Paramount)	Nov. 13	60.3
47	**Three Fugitives** (Touchstone)	Nov. 6	41.0
48	**Batman** (Warner)	Nov. 20	241.0
49	**K-9** (MCA)	Nov. 20	39.0
50	**Ghostbusters II** (RCA/Columbia)	Nov. 27	111.0

Note: Some hot-renting videos are ranked low on this chart because they were released late in the year (e.g., "Batman" at No. 48, released Nov. 20; "Ghostbusters II" at No. 50, released Nov. 27).

Necrology 1989

Abrams, Morris R. (film assistant director, producer) September 18
Adams, Daniel (theatrical lighting designer) June 5
Adler, Norman Abner (lawyer, record executive) December 2
Aikens, Charles (choreographer) July 22
Ailey, Alvin (dancer, choreographer) December 1
Albrecht, Leo (circus owner, performer) August 9
Alexander, Arthur (film producer) April 3
Alexander, Richard (actor) August 9
Alfors, Timothy (assistant script coordinator) June 17
Allen, Bob (singer) April 24
Allen, Clifford Lewis 3rd (dancer) October 13
Allen, Reg (set decorator) March 30
Allen, Ross (music conductor) June 13
Allison, Fran (TV personality) June 13
Altavista, Juan Carlos (actor) July 20
Alvarez, Alfredo (Al) (orchestra leader) November 8
Amato, Christine (dress designer) August 9
Amero, Lem (film director, editor) August 5
Anderson, John Clifford (Cliff) (radio and TV executive) March 27
Anderson, Robert E. (advertising executive) August 21
Andrews, Harry (actor) March 7
Andrews, Nancy (actress, singer) July 29
Andrex (actor) July 10
Antibus, Grace Alice Evans (a.k.a. Jacqueline Evans) (actress) June 22
Appet, Lou (animator) April 3
Arian, Marcel R. (theater manager) December 17
Ariel, Harry (playwright, lyricist, actor) July 6
Arnstein, Arnold (music copyist) December 2
Arthur, Lee (actress, sportscaster) June 7
Atkinson, Oriana (writer) July 31
Attle, John C. (actor, singer, dancer) August 29
Audino, John (trumpeter) November 15
Augettand, Lucien (film production designer) December 13
Avanzo, Renzo (public relations director) March 23
Averino, Olga (lyric soprano, voice teacher) January 17

Bac, Andre (cinematographer) May 31
Backus, Jim (actor) July 3
Bahrenberg, Bruce (film publicist) March 4
Bailey, Brian (lighting designer) March 2
Bailey, Harry (comedian) April 20
Bailey, John (film, TV, theater actor) February 18
Bajek, Gilbert E. (musician) August 19
Baker, Donald H. (theater organist) June 26
Baker, Margaret C. (film distributor) April 26
Balchowsky, Eddie (pianist, artist, poet) November 29
Ball, Lucille (actress, comedienne) April 26
Balogh, Erno (pianist, composer, author) June 2
Banzhaf, Peter G. (TV personality) May 29
Banks, Johnny Sr. (singer) November 28
Barbara, Paola (actress) October 1
Barbo, Salvatore V. (Sal) (tenor) November 25
Bari, Lynn (actress) November 20
Barnes, William E. (literary and theatrical agent) March 9
Barr, Richard (theater producer and director) January 8
Barrett, James Lee (film/TV writer) October 15
Barrier, Ernestine de Becker (actress) February 13
Barrios, Jaime (filmmaker) April 18
Barron, Bill (jazz saxophonist) September 21
Barry, Joan (actress) April 10
Barton, Ben (music publisher) December 8
Barzman, Ben (screenwriter, novelist) December 15
Baum, Robert S. (Bobby) (actor) August 23
Bavier, Frances (actress) December 6
Bayless, Eugene (tenor, professor) September 30
Bazlen, Brigid (actress) May 25
Beach, Donn (restaurant/nightclub owner) June 7
Beam, Alvin (dancer) January 15
Becerra, Luis Gomez (a.k.a. El Tal Gomezbeck) (actor-director) September 11
Becker, Marvin E. (documentary filmmaker) June 4
Beckett, Samuel (playwright) December 22
Beckman, John (film/TV production designer) October 26
Bedford, Richard (film editor) October 16
Begelman, Lee Reynolds (TV/theater producer) June 2
Beinecke, Joy Dewey (dance teacher, consultant) November 15
Bellamy Peter (radio and TV reporter/critic) January 6
Ben Amotz, Dan (writer, actor, artist) October 20
Bendayan, Amador (actor, comic, TV host) August 4
Benjamin, Claude (songwriter) May 2

Name	Date
Bennett, Alvin S. (record producer)	March 15
Bennett, Nancy L. (radio and TV personality)	March 12
Benson, Joe (actor, stuntman)	March 27
Beres, Mark (comic)	January 19
Berg, Susie (writer-editor)	May 7
Berk, Howard (unit publicist)	August 29
Berkeley, Lennox (composer)	December 26
Berle, Ruth (publicist)	April 18
Berlin, Irving (composer)	September 22
Berman (a.k.a. Scott), Andrew (public relations executive)	November 8
Berman, Israel M. (filmmaker)	January 15
Bernard, Ruth (showgirl, TV hostess)	January 6
Bernau, Christopher (actor)	June 14
Bernhard, Thomas (playwright)	February 12
Beruh, Joseph (theatrical producer)	October 30
Beverly, Edna Mae (musician)	February 21
Binion, Benny (nightclub owner)	December 25
Blake, Amanda (actress)	August 16
Blanc, Mel (voice artist)	July 10
Blanchette, Paul H. (cable TV executive)	July 18
Blatt, Jerry (composer, lyricist)	January 19
Bleyer, Archie (musical director)	March 20
Blier, Bernard (actor)	March 29
Bloom, George (actor)	May 8
Boatman, Bob (TV director)	August 28
Boatner, Joseph (singer)	May 8
Boden, Niall (radio presenter)	January 11
Bodnar, John (radio program director, disc jockey)	December 6
Bombyk, David (film executive)	January 20
Bond, Greg (actor, singer)	April 10
Bond, James (ornithologist)	February 14
Bond, Raleigh (actor, author)	August 10
Bond, Ralph (film director)	May 29
Bonner, B. Crenshaw (advertising account executive)	January 11
Boone, Philip S. (advertising executive)	August 4
Boston, William Bernard (singer, director, actor)	November 6
Bouise, Jean (actor)	July 6
Bowne, William Alan (playwright, author)	November 24
Boxall, Frederick (music engineer)	June 8
Boyd, Eddie (radio/TV writer)	December 17
Boyle, Robert Ott (actor)	May 16
Bradway, Wendell (bandleader)	February 22
Braman, Walter (bandleader)	December 3
Brand, George (music editor)	April 24
Brand, Harry (publicist)	February 22
Brandl, Milton B. (radio announcer)	March 9
Brannon, Lois Fletcher (music executive)	December 3
Braswell, John (theater director)	February 16
Brauer, Edward (trumpeter)	March 24
Breckner, Robert W. (radio/TV station owner)	February 12
Brennan, Marguerite Basford (pianist, organist)	March 28
Bresnan, Richard (graphics designer)	August 9
Breuer, Harry (percussionist, xylophonist)	June 22
Brewer, Geoffrey (stuntman)	May 18
Brico, Antonia (musician, conductor)	August 3
Brien, Lige (publicist)	April 13
Bright, John (screenwriter)	September 14
Brister, John T. (filmmaker, arts educator)	June 28
Brittain, Donald (documentary filmmaker)	July 21
Brock, Henry (Heinie) (comedian)	August 20
Brockman, Furn Owen (advertising artist)	September 24
Brookner, Howard (director)	April 27
Brooks, Mary Rogers (actress)	December 13
Brotherson, Eric (actor-singer)	October 21
Brotherton, Robert N. (producer, director, animator, editor)	March 29
Brown, Pamela (TV producer)	January 26
Brown, Tally (actress, singer)	May 6
Brown-Wilkinson, Donald Eric (TV executive)	April 17
Bryant, John (actor)	July 13
Buck, David (actor, scripter)	January 27
Buckley, Emerson (opera conductor, music director)	November 18
Buetel, Jack (actor)	June 27
Bullitt, Dorothy Stimson (radio/TV station owner)	June 27
Burks, Rick (singer, actor)	February 19
Burlinson, John J. Jr. (film exhibitor, advertising executive)	January 30
Burns, Kenneth C. (Jethro) (mandolin player, comedian)	February 4
Butrick, Merritt (actor)	March 17
Buxton, Jimmy (jazz trombonist)	June 28
Byrnes, Robert E. (music business agent)	April 20
Cahan, Herbert B. (TV executive)	April 12
Caidin, Stanley (entertainment lawyer)	July 16
Calder, Len (actor, acting teacher)	February 14
Callanan, Brian John (theater assistant general manager)	May 26
Calvi, Pino (song and film soundtrack composer, TV orchestra director)	January 4
Canfield, Homer (radio executive)	December 25
Capo, Bobby (composer)	December 18
Caprioli, Vittorio (actor, director)	October 2
Carli, Alphonsus J. (theater executive)	September 15
Carmine, Michael (actor)	October 14
Carnon, Barry (advertising executive)	March 27
Carpenter, Freddie (choreographer)	January 19
Carroll, Joseph (author, editor, journalist)	March 7
Carroll, Rick (radio program director, consultant)	July 10
Carter, Peter J. Sr. (nightclub and restaurant owner)	March 12
Cary, Falkland (playwright)	April 7
Cassavetes, John (actor, filmmaker)	February 2
Casstevens, William Evan (director, actor, producer, press agent)	May 22

Catania, James Noel (visual effects operator) August 30
Caulfield, Robert F. (cable news service director) July 17
Cavalier, Kiku Kitagawa (film production coordinator) April 17
Cavallaro, Carmen (pianist) October 12
Cayatte, Andre (filmmaker) February 6
Challee, William (stage and film actor) March 18
Chambers, James (French horn player) January 1
Champin, Kenneth F. (animator) February 25
Chandler, Marjorie Grossel (actress) April 5
Chaney, Clarence Ansel (radio station executive) November 2
Chapman, Graham (writer-comedian) October 4
Chapman, Keith (organist) June 29
Che, Chico (singer-songwriter) March 29
Cherington, Tom (talkshow host, news anchor) October 6
Chiari, Mario (art director) April 9
Childs, Peter (actor) November 1
Ching, William (actor) July 1
Christensen, Harold (dancer) February 20
Christie, Audrey Florence (actress) December 20
Cipollina, John (musician) May 29
Cipriani, Enrico (a.k.a. Rick Anthony) (radio host) December 2
Ciro, Steve (dancer, choreographer) June 14
Cisney, Marcella (theater director/administrator) December 8
Civil, Alan (French horn player) March 19
Clark, Dort (actor) March 30
Clark, Jerry (actor, singer, comedian) August 27
Clark, John S. (radio executive) September 19
Clark, Richard A. (radio executive) September 20
Clarke, T.E.B. (screenwriter) February 11
Clough, Matilda (Goldie) Stanton (theater management) April 3
Cobb, Arnett (saxophonist) March 24
Coleman, Myer (movie shipper) November 29
Coley, Thomas (actor) May 23
Colin, Jean (singer, actress) March 7
Colin, Sid (guitarist, lyricist) December 12
Collins, Lucille James (jazz singer, pianist) August 4
Como, William (magazine editor) January 1
Compinsky, Manuel (violinist, educator) January 8
Comstock, Alice (Betsy) (advertising and marketing executive) February 21
Condon, John F.X. (announcer) October 13
Conover, Don (festival producer) April 3
Cook, Ray (musical composer, arranger) March 20
Cook, Reagan (theater set designer) September 4
Cooley-Coehrs, Alma (pianist, organist) October 15
Coonan, Sheila M. (actress) March 28
Cooper, Chester R. (TV producer, programmer) February 8
Corbett, Harry (puppeteer) August 17
Cortez, Mildred (actress, screenwriter) May 16
Cortini, Bruno (film director) October 29
Costa, Bob (radio and TV actor, public relations executive) February 19
Cottler, Irv (drummer) August 8
Coulouris, George (actor) April 25
Coult, William Henry (radio technology pioneer) May 16
Covan, Willie (dancer) May 7
Cowall, Ben (entertainment/sports reporter) June 17
Cox, Joseph Cunningham Jr. (advertising executive) January 7
Crabe, James (cinematographer) May 2
Craft, Roy D. (publisher, publicist) December 25
Crane, Georgia (nightclub owner) October 8
Crawford, Charles D. (film executive) September 29
Credidio, Louis A. (film executive) June 10
Crinkley, Richmond (theater executive) January 29
Cronin, Bernyce (talent agent, acting coach, singer) May 17
Crosby, Lindsay (singer) December 11
Crossman-Hecht, Eva (pianist) October 16
Cruikshank, Lawrence (talent agent, production executive) December 25
Cuervo, Jose Martinez (set designer) March 2
Cummings, Jack (film producer) April 28
Curtis, Harry M. (costume designer) August 6
Cyr, Paul J. (playwright, theater ticket agent) March 12

Da Costa, Morton (play director) January 29
Dali, Salvador (artist, filmmaker) January 23
Dalitz, Morris B. (Moe) (casino owner) August 31
Dallas, Jeffrey (lighting designer) September 25
Dalrymple, Ian (filmwriter, director, producer) April 28
D'Ambrosia, Thomas (a.k.a. Vinnie De Carlo) (actor) July 1
D'Angelo, Salvatore J. (vocalist) March 19
Daniels, Joseph Arthur (entertainment booking agent) May 5
Daniels, Marc (TV director) April 23
Daniels, Peter (pianist, conductor, arranger) February 8
Dansereau, Muriel Tannehill (opera singer, voice teacher) May 13
Dapporto, Carlo (actor, comedian, variety revue performer) October 1
Davenport, William (comedy writer) May 13
David, Carl (producer) April 18
Davis, Bette (actress) October 6
Davis, Bowlin (actor) August 4
Davis, Clifford (TV critic) November 24
Davis, John (ballet dancer, choreographer) September 24
Davis, Lee (actor, singer, dancer) December 2
Davison, William (Wild Bill) (jazz cornet player) November 14
DeAngelus, Alfred (magician, broadcaster, advertising/PR executive) January 4
deAntonio, Emile (filmmaker) December 15
DeBonee, Daniel (stage manager) July 6
DeCarlo, Don (theater agent) December 15

Deeney, Cecil (news editor) January 15
de Grazia, Julio (actor) May 17
DeGroote, Steven (pianist) May 22
del Carril, Hugo (singer, actor, director, producer) August 13
Dell'Isola, Salvatore (theater music director) March 13
de Michele, Mark (property man) May 23
Dennis, Nigel (novelist, playwright, critic) July 19
Dermota, Anton (tenor) June 22
Derwent, Lavinia (broadcaster, writer) November 25
de Santis, Inma (actress) December 21
DeSantis, Joseph V. (Joe) (actor) August 30
Deschevaux-Dumesnil, Suzanne (pianist) July 17
De Vecchi, Mario (film distributor) October 28
DeVol, Grayce (dancer) February 1
de Zarate, Americo Ortiz (filmmaker, assistant director) October 13
Dickens, Homer (publishing executive) February 5
Dickson, Eleanor Shaler (singer, actress) December 22
Diffring, Anton (actor) May 20
Dignam, Mark (actor) September 29
Dittmann, Dean Gus (musical performer, actor) January 28
Doan, Richard K. (radio/TV writer) May 31
Dobish, Sidney (cameraman) October 27
Doktor, Paul (violist, teacher) June 21
Domenech, Raquel (singer, nightclub performer) June 18
Domnick, Ottomar (filmmaker) June 14
Donati, Dolores Jean (mezzosoprano) November 9
Doniol-Valcroze, Jacques (filmmaker) October 5
Donnelly, Robert T. (a.k.a. Bud) (TV program sales executive) November 3
Donovan, Kevin (film producer) May 5
Dorning, Robert (actor) February 20
Dorsey, John J. (TV director) October 31
Douglas, Jack (writer, humorist) January 31
Drake, Galen (radio commentator) June 30
Drake, Richard (a.k.a. Pee Wee) (radio entertainer) October 22
Drane, Gary (actor, director, puppeteer) August 3
Dreicer, Maurice C. (radio personality, program developer) August 23
Drinkwater, Terry (TV news correspondent) May 31
Drybread, Claude S. (a.k.a. Crusty) (bandleader) March 8
Dryhurst, Edward (film producer) March 7
Dudarew-Ossetynski, Leonidas (drama teacher, actor, stage director) April 28
Du Maurier, Daphne (author) April 19
Duncan, Jeff (dancer, choreographer) May 26
Dunn, Benny (comic, nightclub owner) April 16
Dunne, Burt (executive assistant, story editor) August 27
Durante, Vito (dancer, choreographer, teacher, stage manager) May 6
Dvarackas, Peter A. (writer, editor, publicist) July 24
Dyne, Michael (playwright) May 17

Eames, John Matthew (actor) June 13
Earle, Kevin David (theater entrepreneur, producer) November 23
Ebenstein, Arthur A. (film insurance broker) September 22
Eberson, Drew (theater architect) July 8
Echelson, Robert H. (journalist) November 14
Eddy, William Crawford (TV engineer) September 16
Edwards, Chrissie (a.k.a. Chrissie White) (actress) August 18
Egilmez, Ertem (film producer, director) September 21
Eglevsky, Leda Anchutina (ballerina) December 15
Ehre, Ida (actress) February 16
Ehrlich, George (broadcast sports personality) October 9
Ekstrom, Parmenia Migel (p.M.) (biographer, ballet historian) November 14
Eldridge, Roy (trumpeter) February 26
Elliott, Leonard (actor, comedian) December 31
Elterman, Samuel (wardrobe supervisor, dresser) March 13
Endicott, Betty Wolden (TV executive) August 19
Enters, Angna (dancer, mime, painter, writer) February 25
Ertegun, Nesuhi (record producer) July 15
Etess, Mark Grossinger (casino developer) October 11
Evans (a.k.a. Grigsby), John (TV anchorman) January 11
Evans, Lester (musician) October 17
Evans, Maurice (actor, theatrical producer) March 12
Evans, Peter (actor) May 20

Fain, Sammy (songwriter) December 6
Fall, Charles (broadcaster, adman) February 18
Farnum, William Jr. (publicist) August 15
Farrell, Timothy (a.k.a. Timothy Sperl) (actor) May 9
Farros, Harry A. (theater exhibitor) January 26
Fehlman, Lester (Bud) (singer, drummer, dancer, sound technician) August 5
Feibelmann, Helen (personal assistant) December 25
Feurey, Benita S. (consumer reporter) August 2
Fieger, Addy Oppenheimer (composer) September 22
Field, Ron (choreographer, director) February 6
Finley, Evelyn (actress) April 7
Fiorito, Tony (dancer, costumer) October 6
Fishburn, James (stage/TV producer) October 20
Fisz, S. Benjamin (film executive) November 17
Fitzgerald, Pegeen (radio personality) January 30
Flood, Gerald (stage/TV actor) April 12
Flowers, Sally (TV personality) February 13
Forella, Ronn (choreographer, performer) January 18
Forrest, William H. (actor) January 26
Foster, Bill (newscaster) July 27
Fowle, Frank 3rd (actor) January 19
Fowler, David (arts administrator) March 11
Frankel, Max (film property master) October 30
Freedman, Leo (hotel/theater developer) April 8

Freeman, George Anthony (a.k.a. Buddy) (film projectionist)	December 16
Freeman, Norman (film executive)	August 5
French, Norma (coloratura soprano)	June 26
French, Victor (actor)	June 15
Friedberg, Gertrude T. (playwright)	September 17
Frosch, Aaron, R. (entertainment lawyer)	April 29
Frome, Milton (comedy and musical performer)	March 21
Fuldheim, Dorothy (TV commentator)	November 3
Furey, Ed (writer, news director, TV producer)	April 18
Galindo, Marco Aurelio (scriptwriter, director)	September 10
Galindo, Pedro (singer, scriptwriter, actor, producer)	October 8
Gallop, Richard C. (film executive)	November 23
Gambaroff, Sergio (international film agent)	February 2
Gamer, Henry (actor)	February 15
Garces, Mauricio (actor)	February 27
Garde, Betty (actress)	December 25
Gardner, Hy (showbusiness gossip)	June 17
Garisa, Antonio (actor)	August 8
Garnier, Jacques (dance director, choreographer)	June 10
Garrett, Roger (organist)	May 8
Garst, Elizabeth Marie Schwarz (a.k.a. Sunny Dale) (stage performer)	July 10
Gass, Marc J. (TV director)	November 25
Gay, Connie B. (radio station executive, talent agent)	December 3
Geddes, Virgil (playwright)	May 24
Geer, Lenny (stuntman, actor)	January 9
Gentry, Britt Nilsson (actress)	July 27
Geronimi, Clyde (film animator)	April 24
Gerringer, Robert (actor)	November 8
Geruschat, Carl (pianist)	May 19
Gethers, Steven (TV writer/director)	December 4
Giantvalley, Scott (playwright, poet, performance artist)	March 2
Gifford, Alan (actor)	March 20
Gilbert, N. Scott (conductor, musician)	October 12
Gilbert, Peter (cable TV executive)	March 26
Gilfond, Edythe (costume designer)	August 21
Gilkerson, John C. (designer, puppeteer, singer)	December 24
Gilliam, Roscoe (performer, playwright)	October 27
Gimpel, Jakob (pianist)	March 12
Giuffre, Gus (newsman, TV movie host)	June 26
Glassman, Philip I. (boxing manager and promoter)	February 27
Glaudi, Lloyd (a.k.a. Hap) (TV sportscaster)	December 29
Gluskin, Ludwig (drummer)	October 13
Goldman, Milton (talent agent)	October 4
Goodman, Francis (photographer)	September 29
Goodwin, Nancy (TV scriptwriter)	July 23
Goosen, Lawrence (theater producer, manager)	October 9
Gordon, Ashley (theater entrepreneur, producer)	September 16
Gordon, John M. (theater editor, drama critic)	April 4
Gordon, Martin (publicist, personal manager)	June 21
Gordon, Max (nightclub owner)	May 11
Gorman, Lynne (actress)	November 1
Gosa, Jim (radio disc jockey)	December 18
Gottlieb, Dick (TV director)	June 24
Graff, E. Jonny (radio/TV veteran, film executive)	January 31
Grams, Harold (a.k.a. Hod) (announcer, broadcast executive)	September 28
Gray, Perry (set designer)	December 4
Green, Dock (singer)	March 10
Green, John (Johnny) (film composer, conductor, musical director)	May 15
Green, Nancy (theater producer, TV casting director)	December 16
Greenberg, Sanford (accounting executive)	May 27
Greene, Allen (agent)	August 11
Greene, Billie (talent agent)	December 4
Greenham, Vivian (Chris) (film sound editor)	January 21
Greenspun, Herman M. (Hank) (publisher, cable TV owner)	July 22
Greitzer, Sol (violist)	August 31
Grether, Elmer E. (property master)	September 9
Griffin, Bessie (singer)	April 10
Griffin, Thomas S. (string/woodwind player)	May 22
Griffith, Timothy J. (gaffer)	October 11
Grimes, Lloyd (Tiny) (guitarist)	March 4
Gross, Roland (film and TV editor)	February 11
Grossinger, Paul (entertainment buyer, hotel owner)	April 7
Grossman, George (music associate professor emeritus)	January 20
Guard, Pamela (a.k.a. Yana) (singer, personality)	November 21
Guggenheim, Adelaide (film accountant)	April 17
Gullette, William Brandon (film executive)	December 1
Gundelfinger, Alan M. (film technician)	January 26
Gunn, Bill (playwright, screenwriter, director, actor)	April 5
Gustafson, Walter A. (sound effects technician)	May 30
Haberman, Don (film and TV writer)	May 4
Hafid, Ali (drummer)	September 22
Haig, Jack (comic, actor)	July 4
Haines, William Wister (novelist, playwright, screenwriter)	November 18
Haislmaier, Jerry (film director/producer)	June 14
Halicki, H.B. (Toby) (filmmaker)	August 20
Hall, Henry (danceband leader, radio broadcaster)	October 28
Halliwell, Leslie (film reference writer)	January 21

Halpern, Dina (actress) February 17
Halsted, Fred (director, performer) May 9
Hamblen, Stuart (singer, songwriter) March 8
Hamilton, David (dancer, producer, director) April 15
Hammond, Bunch (jazz bassist) December 13
Hammond, Walter (special effects man) July 18
Hansen, Millard (radio news anchor) June 15
Harburg, Edelaine (actress) March 26
Harper, Charles T. (actor, teacher) July 4
Harper, Frances M. (drummer, singer) November 20
Harrell, Robert F. (film distributor, theater manager) April 29
Harren, William J. (choreographer, modern dance artist) October 4
Harrington, Jonathan Malcolm (Johnny) (jazz bassist) September 28
Harris, Paul McCormick (Mac) (theater director) November 22
Harriss, R.P. (newspaper critic) September 26
Hart, Harvey (TV director) November 21
Hartigan, Albert G. (TV producer/executive) July 7
Hartley, Neil C. (music consultant) January 3
Hartstein, Kim Stephen (personal manager) February 21
Hawk, Bob (radio actor/announcer) July 4
Hayes, Grace (film and nitery personality) February 1
Haymer, Johnny (actor) November 18
Haymes, Robert William (composer, actor, audiovisual producer) January 28
Heider, Wally (music studio executive) March 24
Heilweil, David (professor, producer-director) February 15
Hemingway, Frank (radio TV newscaster) June 25
Hemsley, Winston DeWitt (dancer, choreographer) May 9
Hendler, Gary (film executive) June 29
Henrichsen, Borge Roger (composer, pianist, bandleader) July 20
Henry, Bill (TV executive) March 14
Henshaw, Wandalie (actress, director, scholar) April 29
Herbert, Pitt (actor) June 23
Hernandez, Sergio (writer, script reader, actor) June 16
Hesler, G. Christian (actor) June 11
Heywood, Eddie (composer, pianist) January 2
Hibbs, Harry (singer, accordian player) December 21
Hickman, Norman (author, film producer) February 24
Highfill, J. Allen (costume designer) July 5
Hill, Lord Richard (broadcasting executive) August 22
Hirsch, John (artistic director) August 1
Hite, Kathleen (radio and TV writer) February 18
Hodge, Van (violist) April 22
Hoffman, Ben Jr. (musician) December 2
Hoffman, Herman (film director) March 26
Hoffman, Jack (lyricist, music publisher) January 16
Hohman, Charles L. (actor) February 16
Holloway, Jean (screenwriter) October 18
Holt, Jason (actor, singer) March 6
Honeycutt, Ann (author, radio producer) September 26
Hooten, Mickey L. (TV executive) December 24
Horowitz, Vladimir (classical pianist) November 5
Hoppe, Robert (set designer) July 25
Horne, Elliot (jazz writer, press agent) August 29
Horvatich, Rudy (makeup artist) September 2
Hostettler, Andy (actor, dancer, choreographer) December 25
Houle, Daniel (actor) December 2
Hovey, Serge (composer) May 3
Howard, Robin (dance director) June 11
Howard, William (film reproduction executive) April 14
Hubbard, Shirley Anne (entertainment executive) August 25
Hulburd, Nelda Marshall (singer, dancer) January 26
Hunt, Charles W. (entertainment agent) April 23
Hurley, Lu (helicopter traffic reporter) July 23
Hursley, Frank M. (radio and TV writer) February 3
Hyams, Barry (theater producer/publicist) September 1
Hyde, Stephen F. (casino executive) October 11

Image, Jean (animation film producer-director) October 21
Impaglia, Charles (TV programmer) August 20
Ingram, Bill (TV news anchor) December 28
Irvine, Diana R. (singer) May 7
Ivens, Joris (filmmaker) June 28

Jackson, Ray (actor) October 25
Jacobs, Philip (film distributor) April 18
Jacobson, David J. (marketing executive) October 23
Jaffe, Allen (actor) March 18
James, Emry (actor) February 5
Jani, Robert F. (stage producer) August 6
Jarvefelt, Goran (opera director) November 30
Jenkins, Sarah Malvina (actress, singer, book reviewer) December 9
Jenks, Alexandra Grow (opera association executive) May 16
Jennings, Donald (conductor, musical director) October 2
Jennings, Elinor F. (religious broadcasting director) July 1
Johnson, Helen (political film producer) January 20
Johnson, Mason (actor, dancer, production stage manager) September 30
Jones, David (publicity executive) May 6
Jones, Martha (Marti) (TV executive) November 28
Jones, Reed (performer, choreographer) June 19
Jordan, Jim (graphic artist, commercial designer) October 12
Jorgensen, Christine (TV/nightclub personality) May 3
Joseloff, Stanley S. (legal advisor, songwriter, radio producer) October 1
Joseph, Bernice Kamsler (pianist, singer) August 9
Joseph, Pleasant (a.k.a. Cousin Joe) (jazz performer) October 2
Joyce, Ruel Luster (bassist) May 13

Julian, Danvers B. (bandleader, trombonist) April 27
Jurriens, Henny (art director) April 9

Karalis, Peter (nightclub owner) June 5
Karamanos, Chris Nicholas (hotel owner, casting/catering executive) June 8
Karmark, Henning (film producer) January 11
Karr, James (dancer) November 11
Katz, Stephen (director) November 16
Kaufman, Hazel (dancer) December 28
Kaufman, Les (publicist) January 9
Kay, Ellingwood (Bud) (story editor) December 21
Kay, Graeme (publicist, journalist) November 26
Kaye-Martin, Edward (actor, teacher, director) August 13
Kearsley, Barry Andrew (stage manager) October 7
Keller, Alfred S. (cinematographer) September 6
Keller, Sam (theatrical distribution supervisor) May 8
Kelly, Dan (announcer) February 10
Kelly, Rex (guitarist) February 24
Kennedy, David L. (musician, composer, arranger) June 1
Kenner, Warren (actor) March 21
Kent, William Brad (bandleader) July 12
Kerasotes, Nicholas (theater executive) October 9
Kessler, Ken M. Jr. (assistant director) January 31
Khmara, Ilia (folksinger, actor) August 5
Killen, Thomas Albert (theater writer/columnist) August 12
Kimbrough, Emily (author, lecturer, radio commentator) February 11
King, Abby (bandleader, impressario) April 18
King, Frank (film producer) February 12
King, Peter (dramatist, radio producer) February 1
Kingsley, Lee Goode (music publicity associate) January 30
Kirby, Jim (film buyer) June 24
Kirkwood, James (actor, novelist, playwright) April 22
Kirvay, George Spencer (publicist, promotion executive) November 18
Koford, Luella (a.k.a. Blue) (clothes designer, hair stylist) June 7
Kogan, Herman (newspaper editor, drama and book critic) March 8
Kohn, Bernard A. (music executive) December 20
Koltes, Bernard-Marie (dramatist) April 15
Konitsiotis, Klearchos (film producer) October 9
Koopman, George A. (technological entrepreneur, stunt coordinator) July 19
Kopolus, Gus (concession supplier) February 11
Korn, Bennet H. (TV executive) October 31
Kotecki, Dennis (actor, singer) April 20
Kovacs, Katherine Singer (journal editor) May 12
Kramer, Mandel J. (actor) January 29
Kramer, Ronda M. (Ronnie) (film production coordinator) November 22
Kraus, Harold (set electrician) May 30
Krawiec, John (film executive) February 20
Kreel, Kenneth (dancer) December 3
Krim, Seymour (writer, film critic) August 30

Laemmle, Bertha (interior designer) August 8
Laemmle, Max (arthouse owner) January 20
Laine, Edvin (film/stage director) November 18
Lamborn, Ronald A. (radio executive) November 12
Lampkin, Charles (actor) April 17
Langberg, Ebbe (producer, director, actor, theater manager) February 3
Laporte-Coolen, Catherine (TV journalist) January 24
Larkin, William (comedy writer) August 3
Larsen, Seth Beegle (sound engineer) July 27
Larson, G. Bennett (radio and TV newscaster) March 18
LaShelle, Joseph (cinematographer) August 20
Lasker, Jay H. (music executive) June 11
Lasky, Marlene (film/theater historian) August 9
Lasky, Philip G. (TV executive) November 14
Lazarus, Charles (journalist) January 30
LeBorg, Reginald (film director) March 25
Le Cain, Errol (animator, designer) January 3
Lee, Billy (actor) November 17
Lee, Brian (actor, writer) July 25
Lee, Margaret (Dany) (Cannes Film Festival hostess) January 24
Lee, Maryat (playwright, producer) September 18
Leeson, Cecil (saxophonist) April 17
Lembke, Robert E. (TV quizmaster) January 14
Lemmon, John W. (radio station executive) March 29
Lenny, Jack (theatrical manager and booker) March 12
Leonard, Mary Anne (entertainment editor) October 26
Leone, Sergio (filmmaker) April 30
Leone-Moats, Alice (author, foreign correspondent, columnist) May 14
Leonidoff, Leon (music arranger) July 29
Lepiner, Michael (TV/film producer) January 22
Lerner, Sammy (songwriter) December 13
LeVeque, Edward (actor) January 28
Levine, Henry (a.k.a. Hot Lips) (musician) May 6
Levine, Joseph I. (theatrical producer) October 17
Levine, Jules (publicist) October 31
Levine, Nathaniel (film producer, distributor) August 6
Lewis, Ednyfed (Ed) (radio performer, program manager) September 29
Liang, Teresa C. (Terry) (public relations executive) August 17
Liazza, Stella de Mette (opera singer) May 17
Ligero, Miguel (actor) February 1
Lillie, Bea (actress, singer) January 20
Lillie, John A. (film insurance executive) April 29
Linen, James 4th (media executive) July 1
Lion, Margo (actress) February 25
Little, Billy (accordian player) September 23
Lland, Michael (dancer) January 20
Logan, Nedda (actress) April 1

Lonergan, Arthur (production designer, art director) January 23
Long, James (peter) (theatrical producer) August 11
Long, John F. (stage actor) April 2
Lopatin, David S. (publicist) August 4
Lovett, Samuel (Baby) (drummer) January 23
Low, Charles P. (nightclub owner) August 2
Low, Warren (film editor) July 27
Lowery, William G. (Jerry) (convention center executive) July 18
Loxton, David R. (public TV producer) September 20
Luce, Claire (actress, dancer) August 31
Lucente, Dominick Sr. (film theater manager) January 23
Lucero, Enrique (actor) May 9
Luckham, Cyril (actor) February 7
Ludecke, Wenzel (sound dubbing artist) September 5
Luhrman, HenryRogers (theatrical publicist) April 11
Lupi, Roldano (actor) August 13
Lupo, Michelle (film director) June 27
Lyman-Farquhar, Christine Ramsey (actress, professor) June 25
Lynch, Liam (playwright, novelist) April 25
Lynn, George (composer, music director) March 16
Lyons, Richard E. (film producer) March 18

MacColl, Ewan (folksinger, songwriter, playwright) October 22
MacCulloch, Nancy (Bunny) (film editor) May 29
MacDonald, Angus L. (Jimmy) (pianist, composer) July 8
MacDonald, Charles (magic show producer) January 31
MacDonald, Craig (stage and TV producer) February 24
MacDonald, John Bristol (theater director) June 15
Mack, Marion (actress) May 1
Macliss, Paco (costume designer) December 31
Maddox, Calvin (engineer) October 4
Maguire, Kathleen (actress) August 9
Mahoney, Jock (actor) December 14
Maile, Vic (record producer, engineer) July 11
Major, Lloyd Harrison (public affairs broadcaster) March 12
Malamood, Herman (tenor) September 23
Malinda, James Anthony (designer) August 9
Malone, Ted (radio broadcaster) October 20
Manes, Gina (actress) September 6
Mangan, Helen (nightclub owner) August 22
Mangano, Silvana (actress) December 16
Manito, Louis J. Jr. (TV news producer) October 12
Mann, William (music critic) September 5
Manteo, Miguel (a.k.a. Papa Mike) (puppeteer) September 13
March, Alex (director) June 11
Marion, John Patrick Martin (pianist) October 10
Marks, Albert A. Jr. (pageant producer) September 24
Maroto, Eduardo Garcia (film director, production manager) November 27
Marsh, Harry Thomas (a.k.a. Skeets) (jazz drummer) October 5
Marsh, Tiger Joe (actor) May 9
Marshall, Andrew 3d (actor, playwright) August 10
Marshall, Betty Voigt (publicity agent, newspaper editor) January 22
Mart, Robert (TV exective) June 4
Martus, Ida (theatrical program coordinator) March 23
Marzi, Franca (actress) March 6
Massie, Reginald (animator, art director) September 26
Mastropaolo, Charles T. (drummer) February 7
Matsuda, Yusaku (actor) November 6
Matuszak, John (actor) June 17
Maurer, Steven (film sales manager) March 23
Maurice, Dick (entertainment editor) November 9
Maxwell, Bill (film staff coordinator) March 2
Mazzucco, Robert (playwright) November 6
McAnally, Ray (actor) June 15
McBain, Kenneth (director, producer) April 22
McBride, Raymond E. (entertainment journalist) October 15
McCallum, John Jr. (pianist, musical director) January 23
McCann, Rodney E. (news anchor, video producer) February 13
McCarthy, Mary (novelist, critic) October 25
McClary, Bill (actor, singer, dancer, musician) January 6
McCoy, Maria (costume designer, makeup artist) April 11
McGee, Dennis (fiddler) October 3
McGuiness (Mack), Bernard (radio announcer) February 1
McGuire, Jon Brandon (actor) February 28
McHaynes, Johnnie L. (Mack) May 25
McIlwraith, David H. (dancer, producer) April 15
McKinley, Cal (big band guitarist) September 11
McMichael, George E. (a.k.a. Judd) (jazz singer) October 29
McMillan, Kenneth (actor) January 8
McNellis, Maggi (talkshow hostess) May 24
McReynolds, William C. (broadcast executive) August 27
McWeeney, Tom (co-op advertising employee) October 8
Medina, Vicente (impressario) March 25
Meillon, John (actor) August 11
Melnitz, William (theater professor) January 12
Meltzer, Allan (publicist) June 17
Melville, Sam (actor) March 9
Menendez, Jose (entertainment executive) August 20
Menke, Don (broadcasting executive) February 27
Menschell, Bernard (film exhibitor) April 24
Merli, Maurizio (actor) March 10
Meyer, Nancy Frazer (newscaster) May 4
Meyers, Timothy Francis (actor, director, playwright, teacher) March 14
Miah, Hasson, William (drummer) April 26
Michaelis, Ruth (mezzo-soprano) December 3
Michel, Andre (director) June 5

Miehe, Ulf (film director, screenwriter, crime novelist) July 13
Miele, Louis (magazine publisher, editor-in-chief) August 6
Milanov, Zinka (soprano) May 30
Miller, Frank Jr. (jazz composer, pianist, teacher) July 13
Miller, Randall V. (musician) February 18
Millington, Rodney (actor, casting director) September 19
Mills, Bruce E. (pianist, vocalist, composer) April 18
Mills, Herbert (vocalist) April 12
Milner, Arthur Sylvan (broadcaster, writer, artist) June 11
Milotte, Alfred G. (nature-film photographer) April 24
Milotte, Elma (nature-film photographer) April 19
Milton, Billy (vaudeville entertainer, composer, actor) November 23
Milton, Howard (theater accountant) September 21
Minniear, Harold H. (studio teacher) April 16
Mitchell, Bobby (singer) March 17
Mitchell, James Jr. (still photographer) April 11
Molese, Michele (singer) July 5
Momblow, Donald L. (jazz musician) October 7
Monaco, Eitel (Italian film industry pioneer) February 4
Moore, Isabel (novelist, screenwriter, story editor) April 24
Moore, Sam (radio writer) October 13
Morgan, Brian G. (actor, director, playwright, drama teacher) February 5
Morgan, Mary (radio and TV personality) January 12
Morgan, Rex S. (TV host) February 21
Morgan, Stanley (singer) November 21
Morin, Alberto (actor) April 7
Morris, William Jr. (advertising executive) November 3
Morrison, Ernie (a.k.a. Sunshine Sammy) (actor) July 24
Morrison, Herb (radio announcer) January 10
Morton, Gertrude (violinist) January 14
Moss, Arnold (actor) December 15
Mueller, Dorothy (a.k.a. Cookie) (actress, columnist, art critic) November 10
Mullins, Greg (talent agent) December 21
Munro, Ronnie (music arranger) July 3
Munson, Byron (wardrober) July 28
Murphy, Edward A. (French horn player) March 5
Murphy, Rose (a.k.a. Chee Chee) (jazz pianist, singer) November 16
Murphy-Ballard, Susan (newspaper columnist) July 1
Murray, Colin (comedian, singer) April 15
Murray, James (radio personality) March 12
Murray, Lyn (composer, conductor, musical arranger) May 20
Murray, Patty (disk jockey) March 18
Musazzi, Felice (theater actor, director) August 4

Nadeau, J. Nicholas (theatrical designer) August 21
Nasser, James E. (theater exhibitor) June 28
Neidorf, Ross Lee (actor, playwright) March 4
Neighbors, David (singer, actor, director) April 17
Nemenoff, Genia (pianist) September 17
Neuss, Wolfgang (actor) April 5
Newborn, Phineas (jazz pianist) May 26
Newhouse, Mitzi E. (philanthropist) June 29
Newman, Lionel (music composer, conductor) February 3
Newman, Margy (theater founder) July 29
Niemeyer, Harry (publicist) May 11
Nightingale, Earl (radio commentator/actor) March 25
Nilsen, Margit (stage performer) September 13
Nix, Herbert A. Jr. (jazz pianist, organist) July 4
Noel-Noel (actor, songwriter-performer) October 5
Nord, Eric (a.k.a. Big Daddy) (nightclub owner) April 26
Norman (a.k.a. Nutting), Dick (radio talkshow host) January 26
Norman, Jerry (dancer, choreographer) September 13
Novins, Stuart (TV news correspondent) December 7
Noyes, Thomas Ewing (actor, producer) October 28

O'Brien, Sherry (announcer) April 10
O'Davoren, Vesey (actor) May 30
O'Day, Nell (dancer, actress) January 3
O'Dell, Scott (children's book author) October 15
O'Hanlon, George (actor, writer) February 11
O'Leary, Hugh Thomas (radio and TV sales executive) July 25
Olembert, Theodora (filmmaker) December 7
O'Rourke, Frank (writer) April 27
Ogdon, John (piano virtuoso) August 1
Oliverio, John (artist and wardrobe assistant) February 22
Olivier, Laurence (actor) July 11
Olivo, Bob (Ondine) (actor) April 27
Olsen, Kenneth M. (festival/theatrical producer) February 21
Organ, James Thomas (ventriloquist-puppeteer) June 17
Orrin, Dale Linx (dancer, teacher, choreographer) May 5
Osborne, May Allison (actress) March 27
Ostriche, Muriel (actress) May 3
Oswald, Gerd (director) May 22
Otto, Donald W. (radio director) February 3
Ozeray, Madeleine (actress) March 29

Paice, Eric (stage/TV writer) July 6
Painleve, Jean (documentary filmmaker) July 2
Palanchian, John (violist) December 2
Pallanca, William (film importer) October 13
Palmieri, Edmund L. (entertainment law judge) June 15
Panette, Jackie (singer) June 7
Panitz, Murray W. (flutist) April 13
Pankin, Elle (radio talkshow host) September 24
Papkin, Mary Ellen (a.k.a. Princess Yvonne) (mentalist) March 4
Pardave, Alberto Catala (actor) March 10

Parenti, Franco (actor, director) April 28
Parlett, Thomas Ward (theatrical company manager) April 2
Pass, Lenny (actor) August 18
Pastor, Ethel (film exhibitor, booker, distributor) July 9
Patane, Giuseppe (music conductor) May 30
Patrick, Robert (theater executive) June 23
Patterson, R.H. (a.k.a. Pat) (special effects man) October 24
Patton, Macon (TV station owner) April 20
Paulson, Al (TV announcer) November 24
Payne, John (actor) December 6
Peckham, Charles, S. (director, producer) April 1
Pelayo, Luis Manuel (actor) July 26
Peppercorn, Carl (film sales executive) November 5
Perrotti, Helen (opera singer, teacher) October 6
Perrin, Vic (actor) July 4
Pesce, Ruben (theater director) July 3
Peters, Didi Daniels (producer, publicist) February 25
Picker, Arnold M. (film executive) October 8
Pigaut, Roger (actor, director) December 25
Pines, Kevin (personal manager, producer, casting director) January 11
Piper, John William (film executive) October 23
Piro, Killer Joe (dance instructor) February 7
Pittman, Max J. (propmaster) January 14
Platt, David (program buyer) April 28
Podoloff, Nathan (arena executive) September 12
Pompeii, James S. (actor, singer, guitarist, dancer) November 20
Poole, John (lawyer, broadcasting executive) April 14
Post, Carl (press agent, concert pianist) September 20
Post, William Jr. (actor) September 26
Prado, Damaso Perez (bandleader) September 14
Prado, Raul (singer) April 8
Priest, Stephen (music video director, TV producer-director) October 20
Pringle, Aileen (actress) December 16
Prioli, Frank (pianist) August 7
Pritchard, John (music director) December 5
Prokofiev, Lina (singer) January 3
Psacharopoulos, Nikos (theater director) January 12
Puebla, Carlos (singer, songwriter) July 12

Quarles, T. David (TV executive) November 25
Quayle, Anthony (actor, director) October 20
Quertermous, Charlie (actor) April 17
Quigley, Robert (actor-writer, producer) November 27
Quine, Richard (actor, director) June 10

Radner, Gilda (comedienne) May 20
Ragaway, Martin (comedy writer) April 20
Rankin, Dugald (filmmaker) November 12
Raposo, Joe (composer, TV producer) February 5
Rapport, Samuel (stage manager) May 1
Rathsack, Heinz (film archivist) December 12
Rawley, Ernest (theater manager) January 16
Reichner, S. Bickley (Bix) (lyricist, composer, music publisher) April 8
Reid, Vivian (actress) July 20
Regner, Elmer (theater producer) March 8
Resnick, Irving (Ash) (casino exec) January 18
Reuss, Jeffrey A. (cable marketing manager) July 20
Ricciardelli, James P. (theater production manager) December 17
Rice, Alfred (entertainment lawyer) March 29
Rice, Mary Alice (showgirl) October 9
Rice, Michael S. (TV executive) September 6
Rich, Mickey (TV/film writer) July 18
Richards, Bill (band booker) August 3
Richebe, Roger (film producer, director, distributor) July 10
Riese, Thorbjorn (theater exhibitor) April 30
Riolo, Anthony (arts administrator) March 28
Riser, Frank S. (property master, publicist) December 29
Rive, Richard (author, dramatist) June 5
Rivkin, Joe (casting director) February 3
Robbins, Helen E. (film lab technician) January 16
Robbins, Joseph (set construction manager) July 21
Roberts, Christopher (theatrical general manager) April 29
Roberts, Marguerite (screenwriter) February 17
Robi, Paul (recording artist) February 1
Rodensky, Shmuel (actor) July 18
Rodrigues, Lael (film director, producer) February 9
Rogers, Robert F. (TV news writer, producer) September 24
Roman, Antonio (film director) June 16
Rondi, Brunello (film director, scriptwriter) November 7
Roos, Joanna (actress, playwright) May 13
Rosenbloom, Milton M. (copyright lawyer) October 1
Rosengarten, Clae (youth orchestra founder) February 22
Ross, Angelo (film editor) September 23
Ross, Art (radio performer, emcee, program director) June 28
Ross, Scott (harpsicordist) June 13
Rostvold, Bjarne (jazz drummer-percussionist) July 12
Rothschild, Rosemary (writer, publicist) March 15
Rougeot, John F. (best boy, key grip) July 12
Rouquier, Georges (filmmaker) December 19
Rouziere, Jean-Michel (play producer) February 13
Rowicki, Witold (conductor) October 1
Rubell, Steve (nightclub owner) July 25
Rudin, Nat (pianist, organist, accordianist) February 25
Rudy, Eduardo (actor, radio announcer) December 6
Rugoff, Donald (theater exhibitor) April 28
Rush, Art (personal manager) January 1
Ryman, Herbert Dickens (artist) February 10

Sachs, Susan (dancer, actress, teacher) October 20
Salacrou, Armand (playwright) November 23
Salem, Albert (theatrical manager) November 16
Salkow, Irving (talent agent) September 23

Salce, Luciano (film director) December 17
Sambrook, A.B. (radio pioneer) January 1
Sanger, Elliot (radio executive) July 9
Santuccio, Gianni (actor) September 29
Sauers, Patricia (actress-comedienne) March 25
Sauguet, Henri (composer) June 21
Sayer, Philip (actor) September 19
Schakne, Robert (TV news correspondent) August 31
Schaeffer, Rebecca (actress) July 18
Schaffner, Franklin J. (director) July 2
Schechter, Abel Alan (Abe) (TV news director, producer) May 24
Schnitzer, Veola (Vee) (film publicist) May 24
Schorr, William W. (Willie) (director, producer) June 18
Schroth, Carl-Heinz (actor) July 19
Schulman, Rose (acting teacher) July 30
Schuman, Donald M. (advertising executive) March 30
Schwimmer, Walter (radio/TV producer) September 1
Scott, Walter (set decorator) February 2
Sebastian, Georges (conductor) April 12
Seiderman, Maurice (makeup man) July 18
Sektberg, Willard (pianist, organist, vocal coach) July 28
Sell, Stephen (music theater director) May 26
Sellen, Leslie C. (pianist, organist, saxophonist, cellist) August 5
Seller, Thomas (screenwriter) October 28
Senz, Ira (wigmaker, makeup artist) August 15
Serato, Massimo (actor) December 22
Sergio, Lisa (radio commentator) June 22
Sfat, Dina (actress) March 20
Shafer (a.k.a. Arlen), Roxanne (actress) February 22
Shain, Percy (TV critic) October 16
Shapiro, Irvin (film importer, broadcast scriptwriter) January 1
Shaw, Arnold (author, composer, music researcher) September 26
Shaw, Woody (trumpeter) May 10
Shelley, Dave (actor) June 27
Shenar, Paul (actor) October 11
Shepler, Rev. Clayton B. (broadcast executive) September 24
Sherman, Connie (soprano) December 3
Sherman, Hiram (stage actor) April 11
Sherman, Madeleine Fairbanks (stage performer) January 26
Sherwood, Lydia (actress) April 20
Sherwood, Madeline H. (actress) April 4
Shirley, Bill (singer, performer) August 27
Shirley, Jimmy (jazz guitarist) December 3
Shonfield, Phil (film sales executive) March 11
Shyre, Paul (playwright, director) November 19
Silber, Nadave (advertising publicity manager) June 12
Silver Joe (actor) February 27
Silverman, David (film executive) December 31
Silverman, Mark (film producer) February 3
Simenon, Georges (novelist) September 4
Simmons, John (vocalist, music director) March 16
Simon, Cyril (theater director) July 17
Simpson, Bob (jazz pianist) October 18
Sims, Harry (songwriter) February 9
Slaff, George (entertainment lawyer) March 24
Smack, Michael (set electrician) November 1
Smith, Brendan (actor, playwright, producer) October 31
Smith, Dalton (trumpeter) April 26
Smith, Edgar P. (magazine/broadcast executive) October 10
Smith, Jack (filmmaker, actor, playwright) September 18
Smith, Julia (composer, pianist, teacher) April 27
Smith, Lloyd (a.k.a. Fatman) (comedian, musician, singer) March 10
Smith, Mary A. (cable company regional manager) February 26
Smith, S. Mark (radio and TV executive) May 10
Snyder, Clorice V. (singer) June 12
Soucie, Jerry (makeup artist, hair stylist) July 26
Southern, Richard (theater consultant, scenic designer) August 1
Spekter, Edythe Schneider (singer) March 14
Spencer, Lena (coffee house owner) October 23
Speronis, Stephen Louis (TV commentator, professor) June 25
Spinell, Joe (actor) January 13
Spitz, Hank (film producer, production manager) June 6
Spolan, Michael (producer) June 20
Spragg, Graydon H. (newsman) November 8
Squire, William (actor) May 3
Stanley, Edward (public affairs director) May 17
Stark, Johnny (impressario) April 24
Starrett, Jack (actor, director) March 27
Steinmetz, Bill (publicist) August 31
Stephenson, William (British spy) January 21
Stern, Ernest (exhibitor) March 4
Stevens, Albert (a.k.a. Bob Baker) (disk jockey) May 11
Stevens, Lee (advertising executive) February 2
Stevens, Melissa (writer, theatrical producer) February 19
Stevens, Robert (TV writer, director) August 7
Stewart, Ian (bandleader) July 30
Stine, Whitney (film book writer) October 12
Stinson, William R. (music publisher) February 10
Stone, Irving (novelist) August 26
Stout, Bill (TV reporter, commentator) December 1
Strauss, Bill (announcer, news director) February 25
Streeper, Walter E. (theater executive) December 5
Strotchkov, Mark (film executive) December 10
Stuart, Jerri Rogers (dancer, theatrical agent) February 6
Sudbury, Sherman (a.k.a. Montyne) (acrobat, artist/sculptor) March 17
Sullivan, Joy M. (TV account executive) February 23
Sundin, Michael (actor) July 24
Sverdlin (a.k.a. Goodman), Hannah Grad (radio and TV writer) January 19
Swallow, Gwendoline (Wendy) (dancer, choreographer) April 7

Sweet, Kendrick (postproduction supervisor, sound effects editor) June 2
Swigart, Ruth Robison Bailey (actress, stage producer) September 20
Switzer, Phyllis (TV executive) June 28
Sydnor, Earl (actor) July 9

Tafoya, Alfonso (actor, announcer, TV broadcaster) September 22
Talvela, Martti (singer) July 22
Tate, John (a.k.a. John McCullion) (actor) April 15
Tavernier, Rene (writer) December 16
Taylor, Edwin Byrne (film editor) January 13
Taylor, Robert (a.k.a. Pops) (road manager) March 22
Templeton, John F. Sr. (actor) September 10
Terazawa, Janis (TV reporter) May 25
Terris, Norma (actress) November 15
Tetreault, Nelson (TV cameraman) August 19
Tezuka, Osamu (cartoonist) February 9
Thamar, Tilda (actress) April 12
Thayer, Walter Nelson 3rd (cable TV owner) March 4
Thirkell, Lance (diplomat, TV station manager) January 10
Thomas, Ann (actress) April 28
Thomas, Madoline (actress) December 30
Thomas, Maxine (publicist, manager) August 13
Thompson, George Selden (writer) December 5
Thompson, Patricia M. (entertainment executive) November 23
Thompson, Tauno T. (trumpeter) June 23
Thomson, Virgil (critic, composer) September 30
Thor, Dan (actor) September 2
Thorpe-Bates, Peggy (actress) December 26
Tiedemann, John (broadcast sales executive) August 1
Timm, Doug (music composer) July 21
Tipton, Billy (saxophonist/pianist) January 21
Tirelli, Teresa (singer, actress) June 16
Tobin, Margaret Batts (opera board member) August 3
Todd, Lori Mattis (actress, singer, dancer) April 9
Tolchin, Arthur (broadcaster, theater operator, film distributor) January 25
Tovstonogov, Georgi (theatrical producer) May 24
Trailovic, Mira (theater director) August 7
Trapani, Enzo (TV director) November 14
Traylor, William (actor, drama coach) September 23
Treen, Mary (actress) July 20
Tregoe, William L. (actor, stage director) April 17
Trinder, Tommy (comic) July 10
Tsarouhis, Yiannis (artist, designer) July 20
Tubbs, Vincent (publicist) January 15
Tucker, Julius L. (radio and TV minister) March 2
Tucker, Tommy (bandleader) June 11
Tugend, Harry (screenwriter, producer, film executive) September 11
Turner, Kenneth P. (musician) October 7
Tute, Warren (playwright, novelist, TV executive) November 26
Tuttle, Frank Day (producer-director) May 22

Ungeheuer, Gunther (actor) October 14

Vail, Lillian (dancer) April 1
Vale, Freddie (actor, theatrical agent) December 15
Van Cleef, Lee (actor) December 16
van Collem, Simon (film reviewer) June 21
Vandergriff, Robert (production stage manager) October 25
Vanel, Charles (actor) April 15
Van Pinxteren, Hans (cinema owner) October 14
Varden, Norma (actress) January 19
Vargas, Clifford E. (circus owner, producer) September 4
Vargas, Fernando (recording studio executive) February 20
Vargas, Pedro (singer) October 31
Vellani, Antonio (Toni) (filmmaker, director) November 13
Verzola, Louis Charles (music executive) September 15
Vesperini, Emilio (film executive) February 8
Victor, David (TV writer, producer) October 19
Vidalin, Robert (actor, radio announcer) December 3
Villi, Olga (actress) August 14
Vincent, Romo (comedy and vaudeville actor) January 16
Viner, Mervyn J. (film exhibitor) January 18
Vinson, Clyde M. (voice, movement, and acting teacher) February 5
Violi, D. William (dealer in historical recordings) January 26
Virgilio, Nicholas (radio d.j., haiku poet) January 3
Vivo, Jose (actor) July 26
Vogler, Herbert (stage manager) August 23
von Berzeviczy-Pallavicini, Federico (artist, theater-set designer) November 13
von Cziffra, Geza (screenwriter, director, novelist) April 28
von Karajan, Herbert (symphony orchestra conductor) July 16
Voorhees (conductor, musical director) January 10
Vorisek, Richard J. (sound director) November 7

Waggoner, Keith (theater manager) December 23
Wahrer, Steve (drummer) January 21
Waits, Frederick Dawud (Freddie) (jazz drummer) November 18
Walker, Brian (production executive) September 16
Walker, Elizabeth Kay (public TV station manager) January 13
Walker, Richard (actor) August 26
Wall, Sam (publicist) January 8
Wallace, Al (orchestra leader) December 22
Wallace, Lee (film casting director) October 31

Wallace, Mary Lewis (singer, booking agent) August 21
Waller, John (dancer, art dealer) November 4
Walsh, Leland Jr. (ballet dancer) April 14
Walt, Sherman (bassoonist) October 26
Walton, Richard (Dick) (assistant film director) May 17
Wamboldt, Helen Jane (broadcasting executive, singer) April 28
Ward, Jay (animator) October 12
Warnick, John P. (radio news director) November 19
Washington, Bernie (bandleader, disc jockey) March 5
Watkins, June (stage performer) May 26
Watson, Douglas C. (musician) April 25
Watson, Douglass (actor) May 1
Watters, Lu (trumpeter) November 5
Weaver, Carl Earl (actor) July 13
Webber, Robert (actor) May 17
Weitman, Robert M. (exhibition and production executive) January 20
Wells, Billy (actor) December 18
Welter, Tom (radio traffic helicopter pilot) October 31
Wengerd, Tim (dancer) September 12
Werner, Eleanor R. (TV producer-director) May 20
West, Lockwood (actor) March 28
Westbrook, John (singer) June 16
Westbury, Marjorie (actress) December 16
Wetmore, Joan (actress) February 13
White, Lester (comedy writer) January 25
White, Ronald N. (music executive) September 18
Whitehead, Jack (cinematographer) September 10
Whitley, Keith (singer) May 9
Whitmore, George (playwright, journalist) April 19
Whittington, Harry (novelist) June 11
Whyte, Ron (playwright, editor) September 13
Wichser, Robert (a.k.a. Bobby Wick) (comedian) October 17
Wiegert, Rene (music director) June 26
Wilde, Cornel (actor, director) October 15
Wilder, Leslie Fiske Wilder (assistant sound and film editor) July 26
Williams, Barbara (singer, actress) December 2
Williams, Clark (actor, director, writer) February 13
Williams, Connie (broadcast account executive) July 17
Williams, Guy (actor) May 6
Williams, Lavinia (dancer, teacher) July 19
Willis, Elizabeth (pianist, patron of the arts) October 18
Willis, Joshua (a.k.a. Jack) (jazz trumpeter) October 1
Willis, Marion 3rd (Matt) (actor) March 30
Wilson, Malcom John (a.k.a. Big) (TV movie host, radio disk jockey) October 5
Wilson, Trey (actor) January 16
Winckler, Robert (entertainment lawyer) December 28
Winters, Roland (actor) October 22
Woodbury, Al (orchestrator, arranger, conductor, composer) May 26
Woodbury, Joan (actress) February 22
Wooding, Rodney (publicist) July 23
Woods, Edward (actor) October 8
Wooland, Norman (actor) April 3
Worth, Harry (comedian) July 20
Wright, Ben (actor) July 2
Wright, William A. (actor) December 29

Yacine, Kateb (novelist, poet, playwright) October 28
Yallen, Alfred (radio executive) October 2
Yardley, Michael J. (theater manager) August 16
Yasuda, Ginji (casino owner) December 3
Young, Daniel (accordionist) June 24
Young, Karl E. (concert pianist) November 25

Zacchini, Eddie (circus performer) December 7
Zannas, Pavlos (cineaste, critic, translator) December 7
Zavattini, Cesare (screenwriter) October 13
Zay, Jean (costume designer) August 12
Zegart, Arthur (filmmaker) February 2
Zeigman, William (Ziggy) (personal manager) April 15
Zimmerman, Herman (agent) June 8
Zitarrosa, Alfredo (singer, composer) January 17
Zuckerman, Ira (theater director, professor) April 22
Zwar, Charles (composer-lyricist) December 2

Glossary

A.k.a Old Yank abbreviation for *also known as.*

Ad-pub Advertising and publicity. Used as in *he is ad-pub chief.*

Affil Abbreviation of affiliate, a TV station not owned by but programming material from a network or other station.

Al fresco Italian for *in the open air*, usually as an adjective. Also commonly used in non-showbiz circles, especially as in *al fresco dining.*

Anchor Short for *anchorman* or *anchorwoman*, the key, visible host in presenting news and information to the televiewer.

Ankle Verb form, *to ankle*, means to make one's exit, take leave of, quit (as in one's job).

Auds Abbreviation for auditoriums, occasionally audiences.

Aussie Australian.

Author Verb form, to author, means to write or be the author of a script or book, etc.

Ax To fire (from a job), to cancel or drop (from a TV schedule), etc.

B & W Black and white, as opposed to color (usually TV).

B.O. Probably the most frequently used abbreviation in *Variety* means not *body odor* but *box office.* Reference is not to the cubicle where the theater employee sits and sells you your ticket, but to the cash taken from you. Synonymous with *gross.*

Bally Short for *ballyhoo.* A heavily-publicized announcement or gathering.

Baloney American slang meaning *nonsense.* Popularized in *Variety* in the 1920s.

Bash Commonly used in the U.S. as meaning *party* or *cocktail party.*

Beertown Milwaukee.

*The major portion of this glossary has been provided by various *Variety* staff-writers, and can be found in the following issues: February 4, 1984 (pp. 106, 126); May 6, 1987 (pp. 275, 279, 283, 287, 299, 304); October 12, 1988 (pp. 126, 132) and April 19, 1989 ((pp. 169–173)

Belly-laff A deep, loud, uninhibited laugh. Popularized in *Variety* in the early 1930s.

Bigwigs Important show business owners, executives, distributors and/or exhibitors.

Bimbo Originally a term of endearment, then of jocular disrespect, eventually a reference to a promiscuous woman. Used in *Variety* in the 1920s.

Biopic A biographic picture, or rather a film based on someone's biography.

Bird Telecommunications satellite.

Biz Business. As in the most famous of expressions, *showbiz,* which encompasses everything from a rock concert to a highbrow film.

Blockbuster During World War II, a large bomb that could destroy a whole square block of enemy territory. Now, a huge success, like *Star Wars*, *E.T.*, and *Superman*, etc.

Bobby Soxer Teenager. The term became popular in the 1950s, when wearing white bobby socks became fashionable among adolescents.

Boffo Boxoffice hit (not as big as whammo).

Boffola A big hit or belly laugh. Popular in the 1940s.

Bomb In the U.S., a total failure. In the U.K., a rousing success. As in *The film bombed at the box office.*

Boom A huge success. As in *sales boom.*

Bow As used in *Variety,* usually means to *open*, *debut* or *premiere.* As in a film *bowing* in fifty theaters.

Buck The U.S. dollar. As in a film grossing big *bucks.*

Canto *Variety* synonym for a week or similar short period of time.

Chantoosie A girl singer. From the French, *chanteuse.*

Chi Chicago.

Chick An attractive, pert, lively girl. Popular in *Variety* in the mid-1930s.

Chirp A female singer. Introduced in Variety around 1930.

Chopsocky A martial arts film.

Cleffer A songwriter. Also *clef*, to write a song. First used by *Variety* around 1948.

Click Perfect. Used when referring to planned events which have been executed flawlessly. As in a *click premiere*.

Cliffhanger A melodramatic adventure or suspense film.

Coin Money. *Variety* has traditionally avoided the word *money*, preferring usually *coin* as a synonym, as in *coin received for preproduction was $1-million.*

Combo A combination or cooperative. As in a *combo of companies.*

Confab Abbreviation for *confabulation*, meaning a chat, gossip session, conversation. Also a trade convention.

Coprod Short form of *coproduction.*

Corny Sentimental, obvious, old-fashioned, out-of-date. Popular in *Variety* in the mid-1930s. Derived from *cornfed*, meaning music played in a country or rustic style.

Crix Abbreviation for critics.

Cuffo To pay for, or *cover* in expenses. This term was used most often during the 1950s payola scandal where trips to exotic places were often given to disk jockeys, ad executives, etc., in lieu of cash kickbacks.

Deejay D.J. short for *disk jockey*, or on-air talent for a radio station who plays records within a program.

Dereg Deregulation of previous governmental rulings.

Diskeries Record companies, those making *disks*.

Distrib Distributor or distributing. Hence, *distribbery* (a distribution company), *distribbed* (distributed), *distribbing* (distributing), etc.

Docu Documentary film or program.

Down Under Although half the world is in the Southern Hemisphere, *below* the equator, only Australia (a.k.a. "Oz," short for *Aussie*) has seized this appellation.

Ducats Originally a coin used in Renaissance Venice, Holland and Austria. But in *Variety*, a ticket.

Dupe Abbreviation for *duplicate* or *a duplication* thus *illegal duping of a film*. . .

Emcee M.C., a master of ceremonies. Also, as a verb, as in *Johnny Carson emceed the show.*

Exec Standard American abbreviation for "executive." Plural usually *execs*, but occasionally *exex.*

Exhib Exhibitor. One or company showing films. Thus *exhibbing, exhibbed*, etc.

Exploitation Refers to *B* grade pictures. Also, marketing and merchandising of a film.

Fave Favorite.

Feature Filmed entertainment at least 60 minutes in length for theatrical playoff in some territories.

Feevee Pay television, pay-cable, subscription TV. Any program service distributed to home TV sets for a fee.

Femme French for *woman*. Often used in *Variety* as an adjective, thus *femme director* for *female director.*

Fest Festival, usually referring to films, TV, performing arts, etc.

Frame *Variety* synonym for a week or similar short period of time.

Freeloader One who habitually eats, drinks, or vacations at the expense of another or of a company. Popular in *Variety* in the 1940s and 1950s.

G Commonly used in America to designate *grand* or $1,000. This slang expression goes back at least to the 1920s.

GM General manager.

Gaming Gambling.

Gams Female legs. Popular in the early 1940s.

Go-round *Variety* synonym for a week or similar short period of time.

Gotham Nickname for New York City.

HD or HDTV High Definition Television. A new technology that adds at least twice as many scanning lines to the television picture, giving it superior video quality.

HV Abbreviation for homevideo. The medium utilizing videocassettes and videodisks for playing back programming in the home.

H'wood Hollywood.

Happy talk Light-handed news treatment on TV.

Hardtop A regular indoor theater, in contrast to a drive-in.

Hardware Electronic equipment and appliances, such as TV sets, cameras, cassette recorders, etc., as opposed to software.

Headliner The top act or performer on a vaudeville bill or in a nightclub or other revue-type show.

Helmer A director, usually of films, but also refers to television. From a helmsman, the one who steers a ship.

Hick Country-like or farmerish.

High-hat To act superior or patronizingly. To snub. Also a style of performing. Dates from the 1920s.

Homevid Homevideo. The medium using videocassettes and videodisks for playing back programming in the home.

Hoofer Old vaudeville slang for a dancer, one who is on his or her *hooves*, or feet.

Hub Boston.

Huddle The gathering of players on an American football team to decide on the next play. By extension, any conclave or business meeting.

Hype From *hyperbole*, as in overzealous praise, usually employed perjoratively in connection with showbiz product or events.

Icer Ice show, performed on a skating rink.

Indie Independent. Such as in an *indie distrib*, meaning a distributor that is not one of the *majors*.

Info Information.

Ink As a verb, *to ink* means *to sign*, usually a contract. An anachronism from the days before ballpoint pens and computer printouts were used.

Inning A baseball and cricket expression denoting a division of the game during which each team comes to bat once. In *Variety*, a period of time, as in *exclusive product released this inning*... Sometimes *frame*, a bowling expression, is also used.

Italo Italian.

Jet As a verb, *to jet* means *to fly* from one place to another. Replaces the earlier *to superchief*, or take the train from New York to Los Angeles.

Kayo To knock out. Taken from the boxing world, where a victorious fighter *kayos* his opponent. In showbiz, the term is used to indicate the elimination or removal of something/someone.

Keys Key cities. A film business term referring to key city openings for a film week.

Kickoff (sometimes tee-off) American football expression for *starting*. The game begins when the first player *kicks off* the ball. Similarly *tee-off* in golf. Hence in *Variety*, a film festival *kicks off* with a certain picture.

Kideo Children's programming for the homevideo market.

Kidvid Children's video programming, in the older sense of *video*, meaning television.

Kiwi New Zealand or New Zealander.

Lap *Variety* synonym for a week or similar short period of time.

Latino Latin America.

Leerics Sexually suggestive song lyrics.

Legit Legitimate theater, meaning a live stage production, in contrast to cinema or television.

Legs Commercial endurance, particularly at the boxoffice. A film, play, etc., that continues to sell tickets over the long haul is said to have *legs*.

Lens In verb form, to photograph, or sometimes to shoot a film. (From the camera lens.)

Location A film theater *(See situation)*. Also, the place where a film is being shot, as in *western locations*.

Loot Money.

Mainstream Close to the heart of things.

Majors The *major* Hollywood film producer/distributors: MGM/UA, 20th Century Fox, Columbia, Warner Bros., Paramount, Universal, and Disney (Buena Vista).

Megabuck A big-budget film costs lots of money, or *megabucks*; likewise used in other media.

Micro Microwave transmission.

Mini-majors Orion, Tri-Star and Embasssy.

Mitting Applause. *Heavy mitting*, strong applause.

Mogul The head of a major film studio. From the title of the all-powerful emperors of India.

Moppet A small child, usually in the 6–12 age group. The "Muppets" TV show appealed largely to moppets, althought the latter existed millions of years before the former.

Nabes Neighborhood theaters.

Nee French for *born as*; mostly used to denote a woman's maiden name, but in *Variety* we have the Chelsea Theatre (*nee* Odeon).

Neg Abbreviation for *negative*, as in film.

Nets Television networks.

Nip To eliminate or reduce.

Nitery Nightclub.

Nix Yank slang for *to refuse, deny, overrule, ban, forbid, etc.*

Nut Operating expenses to be recovered, often deductible in an exhibitor's contract with distributors.

O & O Expression for *owned and operated,* usually referring to a local TV or radio station owned and operated by a network (in contrast to being an affiliate.)

O-O Old American slang meaning *to give the once over,* i.e., to look over, inspect, ogle.

Oater A Western film. From the oats that horses eat.

Orbiter Telecommunications satellite.

Org Organization, company

Oz Australia.

Ozoner A drive-in theater. From ozone, the gas which constitutes a tiny part of the air we breathe.

PA Press agent. In the U.K., personal assistant.

PR Public relations.

Pact An agreement, or more commonly a contract. Also as verb form *to pact.*

Palooka Any stupid or mediocre person, especially if big and strong. A prize-fight term, based on a once popular comic strip, adopted by *Variety* in the 1920s.

Passion pit A drive-in movie. One of the more memorable terms invented by *Variety.*

Payoff The final result, outcome, or profit of a venture. Originally an underworld expression. Picked up by *Variety* in the 1920s.

Payola Bribery or under-the-table payments. Describes the payments made to disk jockeys to gain increased record airplay.

Pen To write, as in a script.

Philly Philadelphia.

Pics, Pix Motion pictures.

Pickups Films made by one company that have been acquired by a second company.

Pipeline A schedule of screen products in production.

Platter A phonographic record.

Pour In its nominative form, a cocktail party.

Powwow Old American Indian expression for meeting of chiefs to decide crucial matters. By extension, any conclave or business meeting. Alternative is *huddle*, from U.S. football.

Praiser Publicist.

Praiseries Public relations firms.

Preem To premiere, or a premiere. Sometimes the synonym *debut* is used, or more commonly the simple *to open.*

Prep To prepare. Thus, also, *prepping*, *prepped*, etc.

Prez, Prexy President, as of a company, of a festival jury, etc.

Prod. Producer; also less frequently, production.

Product The most general term for any kind of film or films, TV program, etc. Anything that has been produced.

Promo Promotion. As in *promo campaign*, *promote* or *promo reel.*

Pushover Originally a prizefight expression, adopted by *Variety* in the 1920s first as a description of a girl easily seduced, eventually as a description of anyone who is quickly receptive or responsive to persuasion.

Quizzing Questioning.

Reps Representatives. Hence *repping* (representing), *repped* (represented) and even *reppery.*

Retro Retrospective. In a film festival, films made in or *looking back* to earlier times.

Romp *Variety* synonym for a week or similar short period of time.

Round *Variety* synonym for a week or similar short period of time.

Russo Russian.

Sans French for *without.*

Sci-fi Science fiction.

Script Synonym for *screenplay.* But also in verb form, to write a script; thus also *scripter*, *scripted*, *scripting*, etc.

Sesh A session.

Sex appeal The ability of some performers to project desirability across the footlights or through the camera lens. A *Variety* coinage that has passed into the general language.

Sez Cute variant for *says*, which it sounds like.

Sidebar A secondary event, such as a retrospective screening at a film festival, as distinct from the main competing section.

Sitcom Situation comedy, usually TV programming.

Situation Same as *location*, i.e., a theater.

Sked Standard Anglo abbreviation for *schedule.*

Skein A television program series.

Slate A schedule.

Slot An opening, or long narrow groove; but used in *Variety*, it means a position, as someone is appointed to an executive *slot*, a film is given the top *slot* in a festival, or a TV program plays in the 9 p.m. *slot*.

Soap opera Daytime radio and television drama, frequently sponsored by soap and household products advertisers. Hence this *Variety* coinage of the 1930s.

Socko Big hit (not quite as big as boffo).

Software What you play on the *hardware* to be entertained, i.e., programming on video/audio cassettes, or video/audio disks, or home computers.

Solons For those weak on classical history, Solon was a wise Greek (who was always studied in American elementary schools). He was the founder of the Athenian democracy in the sixth century B.C. In *Variety*, a *solon* is a pundit, a man *in the know*, an authority, or at least one who thinks he is. As in *film trade solons*. . . .

Sparkplug To initiate or to ignite.

Specs Among its meanings are *specials*, *spectators*, *spectaculars*, and maybe even *spectacles* (not the kind to improve your vision).

SRO Old Yank abbreviation for *standing room only*, the sign they put outside theaters to tell you there are no seats left.

Stanza *Variety* synonym for a week or similar short period of time.

Starrer Though *Variety* usually frowns on calling anyone a *star*, a *starrer*, meaning *being the star of*, is commonly used. A John Wayne *starrer* is a film in which he is the star.

Stickler A problem or difficulty.

Sticks Rural. Usually referring to a non-urban audience.

Stint *Variety* synonym for a week or similar short period of time.

Subsid Subsidiary.

Syndie In TV, short for *syndication*, or a *syndicator*, meaning an off-network firstrun program distributor, in constrast to the three major American networks.

Tag or **Tap** To appont, as in a job.

Teleblurb A television commercial.

Telepic A feature-length film funded by a network for first exposure on television.

Televish Television.

Terp A dancer, a chorine.

Thesp A thespian, i.e., an actor or actress.

They-went-thatawayer A Western movie. From the common line in the classic posse chase sequences: *They went thataway.*

Thrush A girl singer.

Tizzy A state of confusion or disarray.

Topline To have the top billing in a film or show. Thus *toplining, topliner.* Similar meaning to *topcast.*

Topper The top man in a company or organization, or even in a department.

Toying To consider, as in *toying with an idea*.

Tubthumping To make a lot of noise, to promote. Many, many years ago actors went about the streets thumping on tubs to drum up business.

Turnaround When the orginal sponsor of a project passes on renewal of its option on said project, it goes into *turnaround*, and may be peddled to a new sponsor interested in getting it produced.

Twofers Literally "two for's," meaning two for one; when you get two tickets or shows or films for the price of one.

UK The United Kingdom, meaning England, Scotland, Wales and Northern Ireland.

Unspool To project a film or unravel a tape from its spool. Also *unreel*.

V.P., veep, veepee All mean vice president, the most common title in America for high executive positions, above the director or manager of a department.

VCR Videocassette recorder.

Venue A relatively recent expression that has crept into *Variety* from England, literally meaning a place or neighborhood where a crime is committed or a cause of action arises, but now popularly used as a place of rendezvous, usually a theater, arena, or auditorium.

Vet Veteran.

Vidclips *Clips* (excerpts from a film or TV production) used on or by television for promotional purposes.

Vidisk Videodisk.

Warbler a singer of either sex.

Web A television network.

Whammo A sensation (bigger than boffo).

Whirl *Variety* synonym for a week or similar short period of time.

Whodunit A mystery or detective story.

Wicket A small opening in a door or gate; thus by extension, a boxoffice, or rather the little opening into which you shovel your dollars, pounds, etc.

Wind Abbreviation for *wind up*, or *wrap up*, meaning terminate or complete. Thus a film *wraps* or *winds* when its principal photography is completed.

Wrap Abbreviation for wrap up, meaning terminate or complete. Thus a film *wraps* when its principal photography is completed. *Wind* for *wind up* has the same meaning.

Yarn Old Anglo-Saxon synonym for *story* or *tale.* A sailor would *spin out* his story like the unravelling of a *yarn.*

Yawner A boring show.

Yokel A person who lives in the local vicinity.

VARIETY SOURCEBOOKS

YES, enter my subscription to the **VARIETY SOURCEBOOKS** at a 20% discount, beginning with:

☐ the current 1989-90 volume on BROADCAST-VIDEO at $11.96 — a **20% discount** off the list price of $14.95 (240-99999-9 [full]).*

☐ the 1991 volume on FILM-THEATER (240-99999-9 [register only]).*

**Subscriptions will be billed annually, with shipping and handling charges added.*

No, I do not wish to subscribe at this time, but...

☐ please send me a copy of the current **1989-90 volume only,** on BROADCAST-VIDEO, at the regular list price of $14.95, plus $3.00 for shipping and handling (240-80067-2). **Prepayment by credit card or check must accompany non-subscription orders — please complete the box below.**

Prepaid Order for **VARIETY SOURCE-BOOK I** (240-80067-2) $ 14.95
+ Your State Sales Tax $ ____
+ Shipping & Handling $ 3.00
TOTAL $ ____
☐ My check is enclosed.
☐ MasterCard ☐ VISA ☐ American Express
Card #
Exp. Date
30 DAY RETURN PRIVILEGES VSBI

Please include street address as we cannot deliver to post office box.

Name ____________________
Street ____________________
City/State/Zip ____________________
Signature ____________________

Focal Press ★ 80 Montvale Avenue ★ Stoneham, MA 02180

PLACE
STAMP
HERE

Focal Press
80 Montvale Avenue
Stoneham, MA 02180